'Undergraduate Topics in Computer Science' (UTiCS) delivers high-quality instructional content for undergraduates studying in all areas of computing and information science. From core foundational and theoretical material to final-year topics and applications, UTiCS books take a fresh, concise, and modern approach and are ideal for self-study or for a one- or two-semester course. The texts are all authored by established experts in their fields, reviewed by an international advisory board, and contain numerous examples and problems, many of which include fully worked solutions.

The UTiCS concept relies on high-quality, concise books in softback format, and generally a maximum of 275–300 pages. For undergraduate textbooks that are likely to be longer, more expository, Springer continues to offer the highly regarded Texts in Computer Science series, to which we refer potential authors.

More information about this series at ▶ https://link.springer.com/bookseries/7592

Donald Sannella · Michael Fourman · Haoran Peng ·
Philip Wadler

Introduction to Computation

Haskell, Logic and Automata

 Springer

Donald Sannella
School of Informatics
University of Edinburgh
Edinburgh, UK

Michael Fourman
School of Informatics
University of Edinburgh
Edinburgh, UK

Haoran Peng
School of Informatics
University of Edinburgh
Edinburgh, UK

Philip Wadler
School of Informatics
University of Edinburgh
Edinburgh, UK

ISSN 1863-7310   ISSN 2197-1781   (electronic)
Undergraduate Topics in Computer Science
ISBN 978-3-030-76907-9   ISBN 978-3-030-76908-6   (eBook)
https://doi.org/10.1007/978-3-030-76908-6

This Springer imprint is published by the registered company Springer Nature Switzerland AG
The registered company address is: Gewerbestrasse 11, 6330 Cham, Switzerland

Preface

This is the textbook for the course *Introduction to Computation*, taught in the School of Informatics at the University of Edinburgh.

The course carries 10 ECTS credits, representing 250–300 hours of study, and is taken by all undergraduate Informatics students during their first semester. The course is also popular with non-Informatics students, including students taking degrees in non-science subjects like Psychology. It has no formal prerequisites but competence in Mathematics at a secondary school level is recommended. Around 430 students took the course in 2020/2021. It has been taught in essentially the same form, with occasional adjustments to the detailed content, since 2004/2005.

First-year Informatics students also take a 10-credit course in the second semester that covers object-oriented programming in Java, and a 10-credit Mathematics course in each semester, on linear algebra and calculus. (Plus 20 credits of other courses of their choice.) Together, these courses are designed to provide a foundation that is built upon by subsequent courses at all levels in Informatics. This includes courses in theoretical and practical topics, in software and hardware, in artificial intelligence, data science, robotics and vision, security and privacy, and many other subjects.

ECTS is the **European Credit Transfer and Accumulation System**, which is used in the European Union and other European countries for comparing academic credits. See ▶ https://en.wikipedia.org/wiki/European_Credit_Transfer_and_Accumulation_System.

Topics and Approach

The *Introduction to Computation* course, and this book, covers three topics:

— **Functional programming:** Computing based on calculation using data structures, without states, in Haskell. This provides an introduction to programming and algorithmic thinking, and is used as a basis for introducing key concepts that will appear later in the curriculum, including: software testing; computational complexity and big-O notation; modular design, data representation, and data abstraction; proof of program properties; heuristic search; and combinatorial algorithms.
— **Symbolic logic:** Describing and reasoning about information, where everything is either true or false. The emphasis is mainly on propositional logic and includes: modelling the world; syllogisms; sequent calculus; conjunctive and disjunctive normal forms, and ways of converting propositional formulae to CNF and DNF; the Davis-Putnam-Logemann-Loveland (DPLL) satisfiability-checking algorithm; and novel diagrammatic techniques for counting the number of satisfying valuations of a 2-CNF formula.
— **Finite automata:** Computing based on moving between states in response to input, as a very simple and limited but still useful model of computation. The coverage of this topic is fairly standard, including: deterministic finite automata; non-deterministic finite automata with and without ε-transitions; regular expressions; the fact that all of these have equivalent expressive power; and the Pumping Lemma.

One aspect that is not quite standard is that our NFAs may have multiple start states; this is a useful generalisation that simplifies some aspects of the theory.

These three topics are important elements of the foundations of the discipline of Informatics that we believe all students need to learn at some point during their studies. A benefit of starting with these topics is that they are accessible to all first-year students regardless of their background, including non-Informatics students and students who have no prior exposure to programming. At the same time, they give a glimpse into the intellectual depth of

Informatics, making it clear to beginning students that the journey they are starting is about more than achieving proficiency in technical skills.

By learning functional programming, students discover a way of thinking about programming that will be useful no matter which programming languages they use later. Understanding of at least basic symbolic logic is essential for programming and many other topics in Informatics. Going beyond the basics helps develop clarity of thought as well as giving useful practice in accurately manipulating symbolic expressions. Finite automata have many practical applications in Informatics. Basic automata theory is simple and elegant, and provides a solid basis for future study of theoretical topics.

We choose to teach these three topics together in a single course because they are related in important and interesting ways, including at least the following:

— Symbolic logic may be used to describe properties of finite automata and functional programs, and their computations.
— Functional programming may be used to represent and compute with logical expressions, including checking validity of logical reasoning.
— Functional programming may also be used to represent and compute with finite automata.
— Functional programming may even be used to represent and compute with functional programs!
— There is a deep relationship between types and functional programs on one hand, and logical expressions and their proofs on the other.
— Finite automata and logical expressions are used in the design and construction of both hardware and software for running programs, including functional programs.

In the course of teaching functional programming, symbolic logic, and finite automata, we explain and take advantage of these relationships. Therefore, we cover the material in two intertwined strands—one on functional programming, and one on symbolic logic and finite automata—starting with the basic concepts and building up from there.

We try to keep things simple—even though they won't always feel simple!—so that we can work out clearly what is going on, and get a clear picture of how things work and how they fit together. Later, this provides a basis for working with more complicated systems. It turns out that even simple systems can have complex behaviours.

Prerequisites

No prior background in programming is assumed. Admission to the University of Edinburgh is highly selective and all of our students are intelligent and have good academic records, but they have a very wide range of backgrounds. While some of them have never programmed before, others start university as highly proficient programmers, some having won programming competitions during secondary school. But only a few have been exposed to functional programming. Teaching functional programming in Haskell is an excellent way of levelling out this range of backgrounds. Students who have never programmed before need to work a little harder. But students who have programming experience often have just as much difficulty understanding that Haskell is not just Python with different notation.

Beginning Informatics students generally have a good mathematical background, but few have been exposed to symbolic logic. And they will often lack knowledge of or sufficient practice with some topics in Mathematics that are relevant to Informatics. Some introductory mathematical material is therefore included—Chap. 1 (Sets) and most of Chap. 4 (Venn Diagrams and

Pronunciation of symbols, names, etc. is indicated in marginal notes where it isn't obvious. It's a small thing, but we find that not knowing how to say a formula or piece of code in words can be a barrier to understanding.

Logical Connectives)—that will already be well known to many students but might have been presented differently or using different notation.

Using this Book for Teaching

At the University of Edinburgh, almost all of the material in this book is taught in 11 weeks, with four 50-minute lectures per week—usually two on functional programming and two on logic or automata—plus one optional lecture covering basic concepts in Mathematics.

Since doing exercises is essential for learning this material, we set exercises in functional programming and in logic/automata each week, and expect students to devote considerable time to working on them. Similar exercises are provided at the end of each chapter of this book. In weekly tutorial sessions, the students work on and discuss their solutions to these exercises. To provide some extra help and encouragement for students who are less confident, we have found it useful to offer students a choice between "beginner-friendly" and normal tutorial sessions. Both cover the same ground, but beginner-friendly sessions put more emphasis on ensuring that all students can do the easier exercises before discussing the more challenging ones.

The following diagram gives the dependencies between the chapters of the book. A dashed arrow indicates a weak dependency, meaning that only some of the exercises rely on the content of the indicated chapter.

In Edinburgh, we omit Chap. 32 (Non-Regular Languages). Some of the more advanced material in some chapters is omitted or not covered in full detail, examples being structural induction and mutual recursion. We add a lecture near the end covering the relationship between propositions and types, based on the article ▶ "Propositions as Types" by Philip Wadler in *Communications of the ACM* 58(12):75–84 (2015).

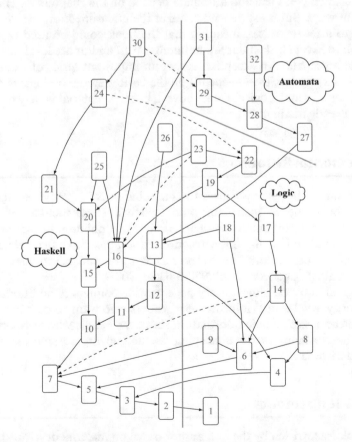

This is a lot of material for an 11-week course. If more time is available, as in universities with 14–15-week semesters, then using the extra time to proceed at a more leisurely pace would probably be better. If less time is available, as in universities with 9-week quarters, then some of the material would need to be cut. Here are some suggestions:

Haskell: Nothing else depends on Chaps. 25 (Search in Trees) or 26 (Combinatorial Algorithms). Only a small part of Chap. 24 (Type Classes) builds on Chap. 21 (Data Abstraction). Nothing else depends on Chap. 30 (Input/Output and Monads) but students need to be exposed to at least the first part of this chapter.

Logic: Nothing else depends on Chaps. 18 (Relations and Quantifiers), 22 (Efficient CNF Conversion), or 23 (Counting Satisfying Valuations). Provided the explanation of CNF at the start of Chap. 17 (Karnaugh Maps) is retained, the rest of the chapter can be omitted. Nothing else depends on Chap. 19 (Checking Satisfiability) but it is a natural complement to the material in Chap. 14 (Sequent Calculus) and demonstrates logic in action.

Automata: Chap. 32 (Non-Regular Languages) can safely be omitted. A course that covers only Haskell and logic, omitting finite automata, would also be possible.

Using this Book for Self-Study

The comments above on using this book for teaching also apply to professionals or non-specialists who wish to use part or all of this book for self-study. The exercises at the ends of each chapter are essential when no instructor's guidance is available.

The chapters on logic and automata depend on the chapters on Haskell in important ways. But most dependencies in the other direction are weak, only relating to a few exercises, meaning that the book could be used by readers who wish to learn Haskell but are unfamiliar with and/or are less interested in logic and automata. An exception is that an important series of examples in Chap. 16 (Expression Trees) depends on the basic material on logic in Chap. 4 (Venn Diagrams and Logical Connectives), but this material is easy to learn if it is not already familiar.

Supplemental Resources

The website ▶ https://www.intro-to-computation.com/ provides resources that will be useful to instructors and students. These include: all of the code in the book, organised by chapters; links to online information about Haskell, including installation instructions; and other links that relate to the content of the book. Solutions to the exercises are available to instructors at ▶ https://link.springer.com/book/978-3-030-76907-9.

Marginal notes on almost every page provide pointers to additional material—mostly articles in Wikipedia—for readers who want to go beyond what is presented. These include information about the people who were originally responsible for many of the concepts presented, giving a human face to the technical material.

Acknowledgements

This book was written by the first author based on material developed for the course *Introduction to Computation* by all of the authors. The treatment of logic and automata draws on a long history of teaching related material in Edinburgh.

Preface

IX

Material in some chapters is partly based on notes on basic set theory produced by John Longley, and lecture notes and tutorials from earlier courses on Computation and Logic developed and delivered by Stuart Anderson, Mary Cryan, Vivek Gore, Martin Grohe, Don Sannella, Rahul Santhanam, Alex Simpson, Ian Stark, and Colin Stirling. The techniques for counting satisfying valuations of a 2-CNF formula described in Chap. 23 were developed by the second author and have not been previously published. Exercise 25.6 was contributed by Moni Sannella.

We are grateful to the teaching assistants who have contributed to running *Introduction to Computation* and its predecessors over many years, and to the material on which this book is based:

Functional Programming: Chris Banks, Roger Burroughes, Ezra Cooper, Stefan Fehrenbach, Willem Heijltjes, DeLesley Hutchins, Laura Hutchins-Korte, Karoliina Lehtinen, Orestis Melkonian, Phil Scott, Irene Vlassi Pandi, Jeremy Yallop, and Dee Yum.

Logic and Automata: Paolo Besana, Claudia Chirita, Dave Cochrane, Thomas French, Mark McConville, Areti Manataki, and Gavin Peng.

Thanks to Claudia Chirita for her assistance with the hard work of moving the course online during the Covid-19 pandemic in 2020/2021, and help with LaTeX. Thanks from Don Sannella to the staff of Station 1A in St Josef-Hospital in Bonn-Beuel who helped him recover from Covid-19 in early February 2021; some of Chap. 32 was written there. Deep thanks from Don to Moni for her love and support; this book is dedicated to her.

Thanks to our colleagues Julian Bradfield, Stephen Gilmore, and Perdita Stevens for detailed comments and suggestions, and to the students in *Introduction to Computation* in 2020/2021 who provided comments and corrections on the first draft, including: Ojaswee Bajracharya, Talha Cheema, Pablo Denis González de Vega, Ruxandra Icleanu, Ignas Kleveckas, Mateusz Lichota, Guifu Liu, Peter Marks, Max Smith, Alexander Strasser, Massimiliano Tamborski, Yuto Takano, and Amy Yin. Plus Hisham Almalki, Haofei Chen, Ol Rushton, and Howard Yates from 2021/2022. Thanks to Matthew Marsland for help with the Edinburgh Haskell Prelude, to Miguel Lerma for an example that was useful in the solution to Exercise 21.7(c), to Marijn in the StackExchange TeX forum for help with LaTeX, and to the publisher's anonymous reviewers for their helpful comments.

Image credits:

- Grain of sand (page 2): ► https://wellcomecollection.org/works/e2ptvq7g (single grain of sand, SEM). Credit: Stefan Eberhard. License: Creative Commons Attribution-NonCommercial 4.0 International (CC BY-NC 4.0).
- Square of opposition (page 74): ► https://commons.wikimedia.org/wiki/File:Square_of_opposition,_set_diagrams.svg. Credit: Tilman Piesk. License: Public domain.
- Karnaugh map in a plane and on a torus (page 166): adapted from ► https://commons.wikimedia.org/wiki/File:Karnaugh6.gif. Credit: Jochen Burghardt. License: Creative Commons Attribution-Share Alike 3.0 Unported (CC BY-SA 3.0).
- Global and local maxima (page 267): ► https://commons.wikimedia.org/wiki/File:Hill_climb.png and ► https://commons.wikimedia.org/wiki/File:Local_maximum.png. Credit: Headlessplatter at English Wikipedia. License: Public domain.

Haskell was originally designed in 1990 by a committee as a standard non-strict purely functional programming language. (Lazy evaluation is a form of non-strict evaluation.) Prominent members of the Haskell committee included Paul Hudak, John Hughes, Simon Peyton Jones, and Philip Wadler. See ▶ https://en.wikipedia.org/wiki/Haskell_(programming_language) and ▶ https://www.haskell.org/.

Type these into a command-line terminal window or "shell", not into a Haskell interactive session!

Haskell

Haskell is a purely functional programming language—meaning that functions have no side effects—with a strong static type system featuring polymorphic types and type classes, and lazy evaluation. A mechanism based on monads (Chap. 30) is used to separate code with side-effects from pure functions. Haskell is used in academia and industry and is supported by a large and active community of developers and users. This book uses the Haskell 2010 language and the Glasgow Haskell Compiler (GHC), which is freely available for most platforms and is the de facto standard implementation of the language.

To get started with Haskell, download the Haskell Platform—GHC together with the main Haskell library modules and tools—from ▶ https://www.haskell.org/platform/, and follow the installation instructions. Once that is done, install QuickCheck by running the commands

```
$ cabal update
$ cabal install QuickCheck
```

To start GHCi, the interactive version of GHC, run the command ghci:

```
$ ghci
GHCi, version 8.0.2: http://www.haskell.org/ghc/   :? for help
Prelude>
```

To exit, type :quit, which can be abbreviated :q.

Although you can experiment with Haskell programming by typing definitions into GHCi, it's better to type your program—known as a **script**, with extension .hs—into a text editor, save it and then load it into GHCi using :load, which can be abbreviated :l, like so:

```
Prelude> :load myprogram.hs
[1 of 1] Compiling Main             ( myprogram.hs, interpreted )
Ok, modules loaded: Main.
*Main>
```

When you subsequently change your program in the text editor, you need to save it and then reload it into GHCi using :reload (:r). Haskell remembers the file name so there is no need to repeat it.

In this book, expressions typed into GHCi will be preceded by a Haskell-style prompt >:

```
> even 42
True
```

to distinguish them from code in a file:

```
even :: Int -> Bool
even n = n `mod` 2 == 0
```

By default, Haskell's prompt starts out as Prelude> and changes when scripts are loaded and modules are imported. Use the command :set prompt "> " if you want to use the simpler prompt >, matching the examples in this book.

Contents

Contents

Sets

Contents

1

Things and Equality of Things

The world is full of **things**: people, buildings, countries, songs, zebras, grains of sand, colours, noodles, words, numbers, …

An important aspect of things is that we're able to tell the difference between one thing and another. Said another way, we can tell when two things are the same, or **equal**. Obviously, it's easy to tell the difference between a person and a noodle. You'd probably find it difficult to tell the difference between two zebras, but zebras can tell the difference. If two things a and b are equal, we write $a = b$; if they're different then we write $a \neq b$.

Sets, Set Membership and Set Equality

One kind of thing is a collection of other things, known as a **set**. Examples are the set of people in your class, the set of bus stops in Edinburgh, and the set of negative integers. One way of describing a set is by writing down a list of the things that it contains—its **elements**—surrounded by curly brackets, so the set of positive odd integers less than 10 is $\{1, 3, 5, 7, 9\}$. Some sets that are commonly used in Mathematics are infinite, like the set of **natural numbers** $\mathbb{N} = \{0, 1, 2, 3, \ldots\}$ and the set of **integers** $\mathbb{Z} = \{\ldots, -3, -2, -1, 0, 1, 2, 3, \ldots\}$, so listing all of their elements explicitly is impossible.

$x \in A$ is pronounced "x is in A" or "x is a member of A".

🪨 is a grain of sand.

A thing is either in a set (we write that using the **set membership** symbol $\in$, like so: $3 \in \{1, 3, 5, 7, 9\}$) or it isn't ($16 \notin \{1, 3, 5, 7, 9\}$). Two sets are equal if they have the same elements. The order doesn't matter, so $\{1, 🪨\} = \{🪨, 1\}$: these sets are equal because both have 1 and 🪨 as elements, and nothing else. A thing can't be in a set more than once, so we can think of sets as unordered collections of things without duplicates. The **empty set** {} with no elements, usually written $\varnothing$, is also a set. A set like $\{7\}$ with only one element is called a **singleton**; note that the set $\{7\}$ is different from the number 7 that it contains. Sets can contain an infinite number of elements, like $\mathbb{N}$ or the set of odd integers. And, since sets are things, sets can contain sets as elements, like this: $\{\varnothing, \{1\}, \{Edinburgh, \{yellow\}\}, 🖼\}$. The size or **cardinality** $|A|$ of a finite set A is the number of elements it contains, for example, $|\{1, 3, 5, 7, 9\}| = 5$ and $|\varnothing| = 0$.

There are different "sizes" of infinity, see ▶ https://en.wikipedia.org/wiki/Cardinality—for instance, the infinite set of real numbers is bigger than the infinite set $\mathbb{Z}$, which is the same size as $\mathbb{N}$—but we won't need to worry about the sizes of infinite sets.

Subset

Suppose that one set B is "bigger" than another set A in the sense that B contains all of A's elements, and maybe more. Then we say that A is a **subset** of B, written $A \subseteq B$. In symbols:

$$A \subseteq B \text{ if } x \in A \text{ implies } x \in B$$

You might see the symbol $\subset$ used elsewhere for subset. We use $A \subseteq B$ to remind ourselves that A and B might actually be equal, and $A \subset B$ to mean $A \subseteq B$ but $A \neq B$.

Here are some examples of the use of subset and set membership:

$$\{a, b, c\} \subseteq \{s, b, a, e, g, i, c\} \qquad \{a, b, j\} \nsubseteq \{s, b, a, e, g, i, c\}$$
$$\{s, b, a, e, g, i, c\} \nsubseteq \{a, b, c\} \qquad \{s, b, a, e, g, i, c\} \subseteq \{s, b, a, e, g, i, c\}$$
$$\{a, \{a\}\} \subseteq \{a, b, \{a\}\} \qquad \{\{a\}\} \subseteq \{a, b, \{a\}\}$$
$$\varnothing \subseteq \{a\} \qquad \varnothing \notin \{a\}$$
$$\{a\} \nsubseteq \{\{a\}\} \qquad \{a\} \in \{\{a\}\}$$

To show that $A = B$, you need to show that $x \in A$ if and only if $x \in B$. Alternatively, you can use the fact that if $A \subseteq B$ and $B \subseteq A$ then $A = B$. This allows you to prove separately that $A \subseteq B$ ($x \in A$ implies $x \in B$) and that $B \subseteq A$ ($x \in B$ implies $x \in A$). That's sometimes easier than giving a single proof of $x \in A$ if and only if $x \in B$.

Set Comprehensions

One way of specifying a set is to list all of its elements, as above. That would take forever, for an infinite set like $\mathbb{N}$!

Another way is to use **set comprehension** notation, selecting the elements of another set that satisfy a given property. For example:

$\{p \in \textit{Students} \mid p$ has red hair$\}$
$\{x \in \mathbb{N} \mid x$ is divisible by 3 and $x > 173\}$
$\{c \in \textit{Cities} \mid (c$ is in Africa and c is south of the equator) or
   $(c$ is in Asia and c is west of Mumbai) or
   $(c$'s name begins with Z) or
   $(c = $ Edinburgh or $c = $ Buenos Aires or $c = $ Seattle)$\}$

The name of the variable ($p \in \textit{Students}$, $x \in \mathbb{N}$, etc.) doesn't matter: $\{p \in \textit{Students} \mid p$ has red hair$\}$ and $\{s \in \textit{Students} \mid s$ has red hair$\}$ are the same set. As the last example above shows, the property can be as complicated as you want, provided it's clear whether an element satisfies it or not.

For now, we'll allow the property to be expressed in English, as in the examples above, but later we'll replace this with logical notation. The problem with English is that it's easy to express properties that aren't precise enough to properly define a set. For example, consider $\textit{TastyFoods} = \{b \in \textit{Foods} \mid b$ is tasty$\}$. Is Brussel sprouts $\in \textit{TastyFoods}$ or not?

Operations on Sets

Another way of forming sets is to combine existing sets using operations on sets.

The **union** $A \cup B$ of two sets A and B is the set that contains all the elements of A as well as all the elements of B. For example,

$\{p \in \textit{Students} \mid p$ has red hair$\} \cup \{p \in \textit{Students} \mid p$ has brown eyes$\}$

is the subset of $\textit{Students}$ having *either* red hair *or* brown eyes, or both. The **intersection** $A \cap B$ of two sets A and B is the set that contains all the things that are elements of *both* A and B. For example,

$\{p \in \textit{Students} \mid p$ has red hair$\} \cap \{s \in \textit{Students} \mid s$ has brown eyes$\}$

is the subset of $\textit{Students}$ having *both* red hair *and* brown eyes.

Both union and intersection are symmetric, or **commutative**: $A \cup B = B \cup A$ and $A \cap B = B \cap A$. They are also **associative**: $(A \cup B) \cup C = A \cup (B \cup C)$ and $(A \cap B) \cap C = A \cap (B \cap C)$.

The **difference** $A - B$ of two sets A and B is the set that contains all the elements of A that are *not* in B. That is, we *subtract* from A all of the elements of B. For example:

$(\{p \in \textit{Students} \mid p$ has red hair$\} \cup \{p \in \textit{Students} \mid p$ has brown eyes$\})$
$-(\{p \in \textit{Students} \mid p$ has red hair$\} \cap \{p \in \textit{Students} \mid p$ has brown eyes$\})$

Set equality can be tricky. For instance, consider the set $\textit{Collatz} = \{n \in \mathbb{N} \mid n$ is a counterexample to the Collatz conjecture$\}$, see ▶ https://en.wikipedia.org/wiki/Collatz_conjecture. As of 2021, nobody knows if $\textit{Collatz} = \varnothing$ or $\textit{Collatz} \neq \varnothing$. But one of these statements is true and the other is false—we just don't know yet which is which!

$\mid$ is pronounced "such that", so $\{p \in \textit{Students} \mid p$ has red hair$\}$ is pronounced "the set of p in $\textit{Students}$ such that p has red hair".

Set difference is often written $A \setminus B$.

is the subset of *Students* having either red hair or brown eyes, but *not* both. Obviously, $A - B \neq B - A$ in general.

The **complement** of a set A is the set $\bar{A}$ of everything that is not in A. Complement only makes sense with respect to some **universe** of elements under consideration, so the universe always needs to be made clear, implicitly or explicitly. For instance, with respect to the universe $\mathbb{N}$ of natural numbers, $\overline{\{0\}}$ is the set of strictly positive natural numbers. With respect to the universe $\mathbb{Z}$ of integers, $\overline{\{0\}}$ also includes negative numbers.

The name "powerset" comes from the fact that $|\wp(A)| = 2^{|A|}$, i.e. "2 raised to the *power* of $|A|$".

The set $\wp(A)$ of all subsets of a set A is called the **powerset** of A. In symbols, $x \in \wp(A)$ if and only if $x \subseteq A$. For example, $\wp(\{1, 2\}) = \{\varnothing, \{1\}, \{2\}, \{1, 2\}\}$. Note that $\wp(A)$ will always include the set A itself as well as the empty set $\varnothing$.

Ordered Pairs and Cartesian Product

Two things x and y can be combined to form an **ordered pair**, written (x, y). In contrast to sets, order matters: $(x, y) = (x', y')$ if and only if $x = x'$ and $y = y'$, so (x, y) and (y, x) are different unless $x = y$. An example of a pair is (Simon, 20), where the first component is a person's name and the second component is a number, perhaps their age. Joining the components of the pair to make a single thing means that the pair can represent an association between the person's name and their age. The same idea generalises to ordered triples, quadruples, etc. which are written using the same notation, with (Introduction to Computation, 1, Sets, 4) being a possible representation of this page in Chap. 1 of this book.

The Cartesian product is named after the French philospher, mathematician, and scientist René Descartes (1596–1650), see ▶ https://en.wikipedia.org/wiki/Ren%C3%A9_Descartes. Cartesian coordinates for points in the plane are ordered pairs in $\mathbb{R} \times \mathbb{R}$, where $\mathbb{R}$ is the set of real numbers.

The **Cartesian product** of two sets A and B is the set $A \times B$ of all ordered pairs (x, y) with $x \in A$ and $y \in B$. For example, {Fiona, Piotr} $\times$ {19, 21} = {(Fiona, 19), (Fiona, 21), (Piotr, 19), (Piotr, 21)}. Notice that $A \times B = \varnothing$ if and only if $A = \varnothing$ or $B = \varnothing$, and $|A \times B| = |A| \times |B|$. The generalisation to the n-fold Cartesian product of n sets, producing a set of ordered n-tuples, is obvious.

Relations

A set of ordered pairs can be regarded as a **relation** in which the first component of each pair stands in some relationship to the second component of that pair. For example, the "less than" relation $<$ on the set $\mathbb{N}$ of natural numbers is a subset of $\mathbb{N} \times \mathbb{N}$ that contains the pairs $(0, 1)$ and $(6, 15)$, because $0 < 1$ and $6 < 15$, but doesn't contain the pair $(10, 3)$ because $10 \not< 3$.

Set comprehension notation is a convenient way of specifying relations, for example:

$\{(p, q) \in Students \times Students \mid p \text{ and } q \text{ are taking the same courses}\}$
$\{(x, y) \in \mathbb{Z} \times \mathbb{Z} \mid y \text{ is divisible by } x\}$
$\{(p, n) \in Students \times \mathbb{N} \mid p \text{ is taking at least } n \text{ courses}\}$

All of these are **binary** relations: sets of pairs. Relations between more than two items can be represented using sets of ordered n-tuples, for example:

$\{(p, q, n) \in Students \times Students \times \mathbb{N} \mid p \text{ and } q \text{ have } n \text{ courses in common}\}$
$\{(a, b, n, c) \in Places \times Places \times Distances \times Places$
       $\mid \text{there is a route from } a \text{ to } b \text{ of length } n \text{ that goes through } c\}$

Functions

Some relations $f \subseteq A \times B$ have the property that the first component uniquely determines the second component: if $(x, y) \in f$ and $(x, y') \in f$ then $y = y'$. Such a relation is called a **function**. We normally write $f : A \to B$ instead of $f \subseteq A \times B$ to emphasise the function's "direction". Given $x \in A$, the notation $f(x)$ is used for the value in B such that $(x, f(x)) \in f$. If there is such a value for every $x \in A$ then f is called a **total** function; otherwise, it is a **partial** function.

$f(x)$ is pronounced "f of x".

None of the examples of relations given above are functions, for example, $< \subseteq \mathbb{N} \times \mathbb{N}$ isn't a function because $0 < 1$ and $0 < 2$ (and $0 < n$ for many other $n \in \mathbb{N}$). Here are some examples of functions:

$\{(m, n) \in \mathbb{N} \times \mathbb{N} \mid n = m + 1\}$
$\{(n, p) \in \mathbb{N} \times \mathbb{N} \mid p$ is the smallest prime such that $p > n\}$
$\{(p, n) \in \textit{Students} \times \mathbb{N} \mid p$ is taking n courses$\}$
$\{(p, n) \in \textit{Students} \times \mathbb{N} \mid p$ got a mark of n in *Advanced Paleontology*$\}$

Notice that it is *not* required that $f(x)$ uniquely determines x, for example, the smallest prime greater than 8 is 11, which is the same as the smallest prime greater than 9.

The first three are total functions, but the last one is partial unless all students have taken *Advanced Paleontology*.

The **composition** of two functions $f : B \to C$ and $g : A \to B$ is the function $f \circ g : A \to C$ defined by $(f \circ g)(x) = f(g(x))$, applying f to the result produced by g. Here's a diagram:

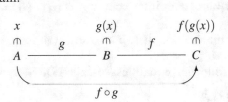

Notice that the order of f and g in the composition $f \circ g$ is the same as it is in the expression $f(g(x))$, but the opposite of the order of application "first apply g to x, then apply f to the result".

For any set A, the **identity function** $id_A : A \to A$ is defined by $id_A(x) = x$ for every $x \in A$. The identity function is the identity element for composition: for every function $f : A \to B$, $f \circ id_A = f = id_B \circ f$. Furthermore, function composition is associative: for all functions $f : A \to B$, $g : B \to C$ and $h : C \to D$, $(h \circ g) \circ f = h \circ (g \circ f)$. But it's not commutative: see Exercise 12 for a counterexample.

A function $f : A_1 \times \cdots \times A_n \to B$ associates an n-tuple $(a_1, \ldots, a_n) \in A_1 \times \cdots \times A_n$ with a uniquely determined $f(a_1, \ldots, a_n) \in B$. It can be regarded as a binary relation $f \subseteq (A_1 \times \cdots \times A_n) \times B$, or equivalently as an $(n + 1)$-ary relation where the first n components of any $(a_1, \ldots, a_n, b) \in f$ uniquely determine the last component. Another example of a partial function is the natural logarithm $\{(a, b) \in \mathbb{R} \times \mathbb{R} \mid a = e^b\}$; this isn't a total function because for any $a \leq 0$ there is no b such that $a = e^b$.

Any subset X of a set A corresponds to a function $1_X : A \to \{0, 1\}$ such that, for any $a \in A$, $1_X(a)$ says whether or not $a \in X$:

$1_X : A \to \{0, 1\}$ is called the **characteristic function** or **indicator function** of X, see ▶ https://en.wikipedia.org/wiki/Indicator_function. Here, 0 denotes false (i.e. the value *is not* in X) and 1 denotes true (i.e. the value *is* in X).

$$1_X(a) = \begin{cases} 1, & \text{if } a \in X \\ 0, & \text{if } a \notin X \end{cases}$$

Using this correspondence, a relation $R \subseteq A \times B$ can be defined by a function $1_R : A \times B \to \{0, 1\}$, where

$$1_R(a, b) = \begin{cases} 1, & \text{if } (a, b) \in R \\ 0, & \text{if } (a, b) \notin R \end{cases}$$

As you will see, this is the usual way of defining relations in Haskell.

1

The question of whether the function *halts* can be computed algorithmically is called the **Halting Problem**, see ▶ https://en.wikipedia.org/wiki/Halting_problem. The British mathematician and logician Alan Turing (1912–1954), see ▶ https://en.wikipedia.org/wiki/Alan_Turing showed that such an algorithm cannot exist.

You will see many examples of functions in the remaining chapters of this book. Most of them will be functions defined in Haskell, meaning that there will an algorithmic description of the way that $f(x)$ is computed from x. But there are functions that can't be defined in Haskell or any other programming language. A famous one is the function *halts* : *Programs* × *Inputs* → $\{0, 1\}$ which, for every program p (say, in Haskell) and input i for p, produces 1 if p eventually halts when given input i and produces 0 otherwise.

Exercises

1. How many elements does the following set have?

   $$\{\{3\}, \{3, 3\}, \{3, 3, 3\}\}$$

2. Show that $A \subseteq B$ and $B \subseteq A$ implies $A = B$.
3. Show that if $A \subseteq B$ and $B \subseteq C$ then $A \subseteq C$.
4. Show that $A - B = B - A$ is false by finding a counterexample. Find examples of A, B for which $A - B = B - A$. Is set difference associative?
5. Show that:

   (a) intersection distributes over union: $A \cap (B \cup C) = (A \cap B) \cup (A \cap C)$
   (b) union distributes over intersection: $A \cup (B \cap C) = (A \cup B) \cap (A \cup C)$

   Does union or intersection distribute over set difference?

6. Show that $\bar{\bar{A}} = A$.

De Morgan's laws arise in logic—see Chap. 14 for details—as well as set theory.

7. Show that the following equalities (so-called **De Morgan's laws**) hold:

   (a) $\overline{A \cup B} = \bar{A} \cap \bar{B}$
   (b) $\overline{A \cap B} = \bar{A} \cup \bar{B}$

   Can you think of a similar law for set difference?

8. Show that $\wp(A \cap B) = \wp(A) \cap \wp(B)$. What about union: does $\wp(A \cup B) = \wp(A) \cup \wp(B)$ hold in general or not?

See ▶ https://en.wikipedia.org/wiki/Russell's_paradox. Bertrand Russell (1872–1970) was a British philosopher, logician, mathematician, political activist, and Nobel laureate.

9. Carelessness with set comprehension notation is dangerous, as shown by Russell's paradox: Let $R = \{x \mid x \notin x\}$, and then show that $R \in R$ if and only if $R \notin R$. That is, the definition of R is meaningless! The mistake with the definition of R is that x isn't specified as being an element of an existing set.

10. Show that $|\wp(A)| = 2^{|A|}$, for every finite set A.

11. Show that:

    (a) $A \subseteq A'$ and $B \subseteq B'$ implies $A \times B \subseteq A' \times B'$
    (b) $(A \cap B) \times (C \cap D) = (A \times C) \cap (B \times D)$

    Use a counterexample to show that $(A \cup B) \times (C \cup D) \neq (A \times C) \cup (B \times D)$.

12. Let

    $$f : \mathbb{N} \to \mathbb{N} = \{(m, n) \in \mathbb{N} \times \mathbb{N} \mid n = m + 1\}$$
    $$g : \mathbb{N} \to \mathbb{N} = \{(n, p) \in \mathbb{N} \times \mathbb{N} \mid p \text{ is the smallest prime such that } p > n\}$$

    What are the functions $f \circ g : \mathbb{N} \to \mathbb{N}$ and $g \circ f : \mathbb{N} \to \mathbb{N}$?

Types

Contents

© The Author(s), under exclusive license to Springer Nature Switzerland AG 2021
D. Sannella et al., *Introduction to Computation*, Undergraduate Topics
in Computer Science, https://doi.org/10.1007/978-3-030-76908-6_2

2

Types are used in logic and Mathematics with a similar meaning but without the computational constraints imposed by their use in programming, see ▶ https://en. wikipedia.org/wiki/Type_theory.

See ▶ https://en.wikipedia.org/wiki/ Zeller's_congruence for one way of defining this function.

$v :: t$ is pronounced "v has type t".

Bool is named after George Boole (1815–1864), see ▶ https://en. wikipedia.org/wiki/George_Boole, and values of type Bool are called **Booleans**.

Don't mix up single quotation marks (') , used for characters, and double quotation marks ("), used for strings! And notice that "a" :: String is different from 'a' :: Char!

Sets Versus Types

A set is a collection of things. In programming, a **type** is also a collection of things. The difference is that a type is a collection of things that the programmer has decided are in some way *related* or *belong together* in some sense that may depend on the situation at hand. A few examples are:

- Weekdays (Monday, Tuesday, … , Sunday)
- Dates (5 Sep 1993, 13 Dec 1996, 28 Jun 1963, …)
- People (Julius Caesar, Ludwig van Beethoven, Elvis Presley, …)
- Integers (… , −2, −1, 0, 1, 2, …)
- Colours (red, orange, yellow, …)

The reason for distinguishing types from sets of unrelated things is that they are useful to classify relationships between kinds of things, including **computations** that produce one kind of thing from another kind of thing. For example, each date falls on one of the days of the week. The association can be computed by a Haskell **function** that takes a date and produces a day of the week. Such a function, which might be called day, would have type Date -> Weekday. That type classifies all of the functions that take a date as input and produce a day of the week as output. Another such function is the one that takes a date and always produces Friday.

Types in Haskell

Types are an important element of the Haskell programming language. In Haskell, we compute with **values**, and every value has a type. This **typing** relationship can be expressed using the notation $v :: t$. So, for example, 17 :: Int where Int is Haskell's built-in type of integers.

When we define things in Haskell, we normally declare their types. For example, if we want to tell Haskell that x is an integer and it has a value of 20 − 3, we write:

```
x :: Int
x = 20 - 3
```

where the **type signature** on the first line declares the type of the value on the second line. The type signature isn't compulsory—Haskell will almost always be able to figure out types on its own, using **type inference**—but it's useful documentation that makes programs easier to understand.

Some other built-in types are:

- Bool, the type of truth values False and True;
- Float, single-precision floating-point numbers, with values like 3.14159 and 6.62607e-34 which is Haskell's notation for 6.62607×10^{-34};
- Double, the type of double-precision floating-point numbers, for applications where more than 6–7 decimal digits of accuracy is required;
- Char, the type of characters, for example 'a', '3', '!' and ' ' (space); and
- String, the type of strings of characters, for example, "a" (a string containing just one character), "this is a string" and "" (the empty string).

We will encounter other built-in types later. All type names in Haskell begin with an upper case letter.

Polymorphic Types

Some functions work on values of lots of different types. A very simple example is the identity function `id`, which is defined so that applying `id` to a value returns the same value. It works for values of any type:

```
> id 3
3
> id "hello"
"hello"
```

This shows that `id :: Int -> Int` and `id :: String -> String`. It even works for functions: `id id` returns `id`.

All of the types that we can write for `id` have the form t -> t. In Haskell, we can give `id` a single **polymorphic** type that stands for all types of this form: `id :: a -> a`, written using the **type variable** `a`, which stands for any type. All of the other types for `id` are obtained by replacing `a` with a type. (Of course, we need to replace both occurrences of `a` with the *same* type!) Type variables always begin with a lower case letter.

Functions with polymorphic types are common in Haskell, and you will see many examples later.

In Haskell, we write "f v" to apply a function f to a value v. In many other programming languages, this is written $f(v)$. Parentheses are used for grouping in Haskell, and grouping things that don't need grouping is allowed, so $f(v)$ is okay too. So is $(f)v$ and $((f)((v)))$.

Functions with polymorphic types are called **generics** in some other languages, including Java, see ▶ https://en.wikipedia.org/wiki/Generics_in_Java.

Equality Testing, `Eq` and `Num`

Two values of the same type can be compared using the **equality testing** operator `==`. For example:

```
> 3 == 3
True
> "hello" == "goodbye"
False
```

Haskell allows you to test equality of the values of most types. But an exception is values of function types, because testing equality of functions would involve testing that equal results are produced for all possible input values. Since that's not possible for functions with infinite domains, Haskell refuses to try, even for functions with finite domains. You'll see later that function types aren't the only exception.

Polymorphic functions sometimes need to be restricted to be used only on types for which equality testing is permitted. This situation arises when they use equality testing, either directly (as part of the function definition) or indirectly (because the function definition uses another function that has such a restriction). The restriction is recorded in the function's type by adding an `Eq` requirement to the front of the type. The equality test `==` itself is such a function: its type signature is

```
(==) :: Eq a => a -> a -> Bool
```

And ditto for **inequality testing**:

```
(/=) :: Eq a => a -> a -> Bool
```

These types for `==` and `/=` mean that a type error will arise when an attempt is made to apply them to values of a type for which equality testing isn't available.

This type is pronounced "a arrow a arrow `Bool`, for any instance a of `Eq`". Notice the *double* arrow `=>` between the requirement `Eq a` and the rest of the type.

2

The same mechanism is used to deal with the fact that there are different kinds of numbers in Haskell—`Int`, `Float`, `Double`, and others—all of which support arithmetic operations like addition and multiplication. Addition has the type signature:

```
(+) :: Num a => a -> a -> a
```

and it works for values of any numerical type:

```
> 3 + 7
10
> 8.5e2 + 17.37
867.37
```

The same goes for a function for squaring a number, which can be defined in terms of multiplication, and has the type signature:

```
square :: Num a => a -> a
```

For example:

```
> square 3
9
> square 3.14
9.8596
```

But square `'w'` would produce a type error.

Types that include `Eq a =>` or `Num a =>` are examples of the use of **type classes**, an important feature of Haskell that will be explained in full detail in Chap. 24.

Defining New Types

We can define new types in Haskell. One very simple way is to enumerate the values of the type, for example:

```
data Weekday = Monday | Tuesday | Wednesday | Thursday
                     | Friday | Saturday | Sunday
```

and then we have `Monday :: Weekday`, etc. The names of the values need to begin with an upper case letter for reasons that will be explained later.

In fact, the built-in type `Bool` is defined this way:

```
data Bool = False | True
```

Another way is to give a new name to an existing type or type expression, for example:

```
type Distance = Float
```

Defining **Distance** as a synonym for **Float**—and other synonyms for **Float**, like **Velocity** and **Acceleration**—might be useful for specifying the types of functions that calculate the movement of objects, as documentation. Another example:

```
type Curve = Float -> Float
```

This type could be used to represent curves in two-dimensional space, with functions like

```
zero :: Curve -> Float
```

for computing the point at which a curve crosses the x-axis.

An example of a polymorphic type definition is the type of binary relations over a given type

```
type Relation a = a -> a -> Bool
```

with values like

```
isPrefixOf :: Relation String
```

in Haskell's **Data.List** library module, which checks whether or not one string is the first part of another string. Since **Relation** a is a synonym for a -> a -> Bool, this is the same as

```
isPrefixOf :: String -> String -> Bool
```

To use **isPrefixOf** to find out if s is a prefix of t, we write **isPrefixOf** s t. For instance:

```
> isPrefixOf "pre" "prefix"
True
> isPrefixOf "hello" "hell"
False
```

We will come to other ways of defining types later, as are required for types like **String**.

Types Are Your Friend!

Types are useful for keeping things organised. Knowing a value's type tells us what kinds of computations can be performed with it. If we have a value d :: **Date**, we can apply day to it to produce a value day d :: **Weekday**. Obviously, applying day to a value like **True** :: **Bool** would be meaningless, and a mistake: truth values aren't associated with days of the week. An algorithm for calculating the day of the week that corresponds to a date wouldn't produce any useful result when applied to **True**, and in the worst case it might even cause some kind of catastrophic failure.

Luckily, we know already from the type of **True** that it's pointless to apply day to it. We can use the type mismatch to avoid simple mistakes like this one, and similar but much more subtle and complicated mistakes in our programs. For this reason, Haskell **typechecks** your code before running it. It will refuse point-blank to run any program that fails the typecheck until you fix the problem. In the case of

```
day True
```

Notice the difference between **data** (used to define a new type) and **type** (used to give a new name to an existing type)!

This uses the idea of defining a relation $R \subseteq A \times A$ via its characteristic function $1_R : A \times A \rightarrow \{0, 1\}$, see page 5, with Bool standing in for $\{0, 1\}$.

Use **import Data.List** to get access to **isPrefixOf** and the other items in that library module. Hoogle (▶ https://hoogle.haskell.org/) is a good tool for exploring the contents of Haskell's library.

The reason why the type of **isPrefixOf** is written **String -> String -> Bool** and not **String * String -> Bool**, as in some other programming languages, will be explained later. For now, just note that most functions in Haskell that take multiple inputs have types like this.

2

Polymorphic types and the algorithm used in Haskell and other languages to check and infer polymorphic types are due to British logician J. Roger Hindley (1939–) and British computer scientist and 1991 Turing award winner Robin Milner (1934–2010), see ▶ https://en. wikipedia.org/wiki/Hindley-Milner_type_system.

NASA lost a spacecraft at the end of its 461-million-mile flight to Mars because of an imperial/metric unit mismatch in the code, see ▶ https:// en.wikipedia.org/wiki/ Mars_Climate_Orbiter. See ▶ https://hackage.haskell.org/ package/uom-plugin for a Haskell library that adds support for typechecking units of measure. The F# functional language supports typechecking units directly.

it will produce an error message like this:

```
program.hs:3:5: error:
```

- Couldn't match expected type 'Date' with actual type 'Bool'
- In the first argument of 'day', namely 'True'
  In the expression: day True
  In an equation for 'it': it = day True

Yikes! This is a little complicated, and some of the details might be difficult to understand at this point, but the problem is clear: there is a type mismatch between **Date** and **Bool** in the "argument" (that is, the input value) **True** of the function day.

If Haskell's typechecker reports a type error in your program, then there is something that you need to fix. The error message might not lead you directly to the source of the problem, but it reports an inconsistency that will be helpful in tracking down your mistake. And, if it doesn't report a type error, then certain kinds of failure are guaranteed not to arise when your program is run. It will definitely not crash due to an attempt to apply a function to a value of the wrong type, but (for example) typechecking won't detect potential division by zero.

When you are writing Haskell programs and the typechecker signals a type error, you may regard its complaint as an irritating obstacle that is interfering with your progress. You can't even test to see if your program works until the typechecker stops getting in your way! But it's much better for Haskell to catch mistakes in your program automatically at this stage than requiring you to find them later manually using testing—or even worse, *failing* to find one of them using testing, and having it cause a serious failure later.

As you'll see later, Haskell's type system is flexible and expressive, so defining types that fit the problem you are trying to solve is normally straightforward. And this often helps you to understand the problem domain by giving you notation for defining important collections of things that are relevant to your problem.

Exercises

1. The Haskell command :t can be used to ask about the types of expressions:

```
> :t id
id :: a -> a
> :t id 'w'
id 'w' :: Char
```

Soon you'll learn about *lists*, which is one of the most important types in Haskell. Lists are polymorphic: [1,3,0] is a list of integers and [True,False] is a list of Booleans. Here are some functions on lists: head, tail, replicate, take, drop, reverse, elem.

Use :t to learn the types of these functions. Try applying them to some values and use :t to see the types of the results. Include some examples that combine these functions. If you run into a type error, use :t to figure out what went wrong.

2. Consider the following description of some students:

   ▶▶ Ashley, Eric, and Yihan are students. Ashley lives in a room in Pollock Halls for which she pays rent of £600 per month. All three are taking Informatics 1A, and Eric and Yihan are also taking Archeology 1A. Ashley and Eric were both born in 2003, and Yihan was born in 2002. Ashley listens to hip hop music while Eric prefers folk music and Yihan likes C-pop. Eric plays table tennis every Wednesday with Yihan.

Give names (but not definitions) of some types that might be used to model these students and what is described about them, together with names (but not definitions) of relevant values/functions and their types.

3. Try to guess the types of the following functions from their names and the clues given:

   - `isDigit` (for example, `isDigit '4'`);
   - `intToDigit` (yields a result like `'5'`);
   - `gcd` (greatest common divisor of two integers);
   - `studentSurname` (for application to a value of type **Student**);
   - `yearOfStudy` (ditto);
   - `scale` (for shrinking or enlarging an **Image** by a given scale factor);
   - `negate` (for example, `negate (-3.2)` and `negate 3`);
   - `howManyEqual` (applied to three values, computes how many of them are the same).

4. Think of some examples of sets that would probably not constitute a type.

Simple Computations

Contents

© The Author(s), under exclusive license to Springer Nature Switzerland AG 2021
D. Sannella et al., *Introduction to Computation*, Undergraduate Topics
in Computer Science, https://doi.org/10.1007/978-3-030-76908-6_3

3

Arithmetic Expressions

Now that we have some values, classified into types, we can start to compute with them. Starting with something familiar, let's use Haskell to do some simple arithmetic calculations:

```
> 1 + 2 ^ 3 * 4
33
> (1 + 2) ^ 3 * 4
108
> (1 + 2) ^ (3 * 4)
531441
```

Watch out when using negation! 4 - -3 will give a syntax error, and Haskell will understand function application f -3 as an attempt to subtract 3 from f. So use parentheses: 4 - (-3) and f (-3).

Here you see that Haskell obeys the rules of arithmetic operator precedence that you learned in school—do multiplication and division before addition and subtraction, but after exponentiation (^)—and you need to use parentheses if you want something different.

A difference to normal mathematical notation is that you always need to write multiplication explicitly, for instance, 3 * (1+2) instead of 3(1+2). Another difference is that Haskell uses only parentheses for grouping, with other kinds of brackets used for other things, so instead of

```
> [(1 + 2) * 3] - (4 - 5)
<interactive>:1:1: error:
    • Non type-variable argument in the constraint: Num [t]
      (Use FlexibleContexts to permit this)
    • When checking the inferred type
        it :: forall t. (Num [t], Num t) => [t]
```

The error message is Haskell's way of saying that your use of mixed brackets has confused it. Don't try to understand the details at this point! But it's handy to know that 1:1 near the beginning of the error message says that the problem starts on line **1**, column **1**: the square bracket.

you need to write

```
> ((1 + 2) * 3) - (4 - 5)
10
```

Int and Float

Because division has a different meaning for integers and for other numerical types like **Float**, there are different operators for integer division (div) and for normal division (/):

```
> 53634 / 17
3154.9411764705883
> div 53634 17
3154
```

Actually, div and mod have a more general type, of which Int -> Int -> Int is an instance. The function name mod is short for "modulo", referring to modular arithmetic, see ▶ https://en.wikipedia.org/wiki/Modular_arithmetic.

There's also an operator to compute the remainder after integer division (mod). The operators div and mod are functions, with type Int -> Int -> Int. The last example above shows how a function with this type can be applied.

Operators like +, -, *, /, and ^ are functions too. Because they are symbols rather than having an alphabetic name like div, they are applied using **infix** notation. You can surround div and mod with "backticks" to use them as infix operators:

```
> 53634 `div` 17
3154
```

To use / in **prefix** style, which is the default for `div` and `mod`, you need to enclose it in parentheses:

```
> (/) 53634 17
3154.9411764705883
```

At this point that might seem like a pretty odd thing to want to do, but you'll see later why it might be useful.

Notice that the backticks in `div` are different from the single quotation marks used for characters, as in `'a'`!

Function Definitions

You might find it handy to use Haskell to do simple arithmetic calculations, if you can't find the calculator app on your phone. But of course the useful thing about Haskell or any other programming language is the way that you can write general recipes for computation that can be applied to lots of different values, by defining functions. You can turn arithmetic expressions like the ones above into function definitions by replacing parts of them with **variables**:

```
square :: Int -> Int
square n = n * n

pyth :: Int -> Int -> Int
pyth x y = square x + square y
```

When defining functions, it's good practice to give them meaningful names and to give their type signatures. Type signatures are optional but giving them makes it easier to track down type errors; without them, a type error may be reported in one function when the actual mistake is in another function, far away.

Function names always begin with a lower case letter, except that symbols like `**` and `<+>` are also allowed:

```
(**) :: Int -> Int -> Int
m ** n = m ^ n + n ^ m
```

In a function definition like

```
square n = n * n
```

the variable `n` is called the **formal parameter**. It stands for the value that is supplied when the function is applied, which is called the **actual parameter**. You can use any variable name you like, of course, but it's good practice to use a name that hints at its type, where the names `m` and `n` used here suggest the type `Int`. Variable names like `x'` and `x''` are okay too. Variable names always begin with a lower case letter, the same name needs to be used consistently throughout a function definition, and you can't use the same variable name in a single function for different formal parameters. So all of these are wrong:

`x'` is pronounced "ex-prime" in American English and "ex-dash" in British English.

```
square N = N * N
square m = n * n
pyth x x = square x + square x
```

The right-hand side of a function definition is called the **function body**.

Note that the type of `pyth` is `Int -> Int -> Int`. Application of `pyth` to actual parameters *a* and *b* is written `pyth` *a b* and not (as in many other programming languages) `pyth`(*a*,*b*). The reason for this will become clear later.

3

This function is already supplied in Haskell.

Make sure you understand the difference between equality used in a **definition**, written using =, and an equality **test**, written using ==.

This function is also supplied in Haskell.

Here's a function that produces a result of type `Bool`.

```
even :: Int -> Bool
even n = n `mod` 2 == 0
```

This function definition uses an **equality test**, written using the operator ==. Here we're testing whether two values of type `Int` are equal.

Case Analysis

The result returned by a function will often need to be different depending on some condition on its input(s). A simple example is the absolute value function on `Int`.

```
abs :: Int -> Int
abs n = if n<0 then -n else n
```

Here we use a **conditional expression**, written using the syntax `if` *exp1* `then` *exp2* `else` *exp3*. The **condition** *exp1* needs to be an expression producing a value of type `Bool`, and the result expressions *exp2* and *exp3* need to produce values of the same type. All of these expressions can be as complicated as you like. You might be familiar with conditional *statements* in other programming languages, where you can omit the `else` part to say that if the condition is false, do nothing. In Haskell, you always need to include the `else` part; otherwise, the expression will have no value when the condition is false.

An alternative way of writing such a function definition is to explicitly split it into cases using **guards**. Here's an example, for a function to compute the maximum of three integers:

```
max3 :: Int -> Int -> Int -> Int
max3 a b c
  | a>=b && a>=c = a
  | b>=a && b>=c = b
  | otherwise    = c        -- here, c>=a and c>=b
```

Function definitions using guards often end with an `otherwise` case, but that's not required.

Each guard is an expression that returns `True` or `False`. When `max3` is applied to values, each of the guards is evaluated in order, from top to bottom, until one of them produces `True`; then the result of the application is the value of the corresponding result expression on the right-hand side of the equals sign. The expression `otherwise` is just another name for `True`, so the last result expression will be the result if no previous guard has produced `True`.

The first two guards in this example use the **conjunction** function && for combining two expressions of type `Bool`. The result is `True` only when both of the expressions have value `True`. In other examples, the **disjunction** function || is useful. The result is `True` when either or both of the expressions have value `True`. Negation (not) is also useful, on its own or in combination with conjunction and/or disjunction.

x && y is pronounced "x and y", and x || y is pronounced "x or y".

The last line of the definition of `max3` includes a **comment**, starting with the symbol -- and continuing until the end of the line. You can write anything you want in a comment, but normally they are used to explain something about the code that a reader might find helpful. That includes things that you might yourself forget, if you come back to look at your code next month or next year!

Defining Functions by Cases

We have seen two ways to distinguish cases in a function definition. A third way is to give a separate definition for each case. This is possible when the input has a type that has been defined by cases, such as

```
data Bool = False | True
```

and the case analysis is according to the values of the type. We will see as we go along that this situation arises very frequently.

The built-in negation function is defined using two separate equations:

```
not :: Bool -> Bool
not False = True
not True  = False
```

The built-in conjunction function can be defined as follows:

```
(&&) :: Bool -> Bool -> Bool
True && y  = y
False && y = False
```

or, equivalently:

```
(&&) :: Bool -> Bool -> Bool
True && True  = True
_    && _     = False
```

The first definition of conjunction shows how a variable can be used for a parameter on which no case analysis is required. The second definition shows how case analysis can be done on multiple parameters. The second line uses the **wildcard** _, which is useful when the corresponding value isn't needed in the body: it saves you thinking of a name for it, as well as telling the reader that its value won't be used. Each use of _ is independent, so using it twice in the last line above doesn't break the "don't use the same variable name for different formal parameters" rule, and the matching actual parameters can be different.

_ is pronounced "underscore".

Dependencies and Scope

All of the function definitions above have been self-contained, apart from their references to Haskell's built-in functions. Writing more complicated function definitions often requires auxiliary definitions. You might need to define variables to keep track of intermediate values, and/or **helper functions** that are used to define the function you really want.

Here's an example, for computing the angle between two 2-dimensional vectors as $\cos^{-1}$ (in Haskell, acos) of the quotient of their dot product by the product of their lengths:

```
angleVectors :: Float -> Float -> Float -> Float -> Float
angleVectors a b a' b' = acos phi
  where phi = (dotProduct a b a' b')
                / (lengthVector a b * lengthVector a' b')

dotProduct :: Float -> Float -> Float -> Float -> Float
dotProduct x y x' y' = (x * x') + (y * y')

lengthVector :: Float -> Float -> Float
lengthVector x y = sqrt (dotProduct x y x y)
```

In Chap. 5, you will see that a 2-dimensional vector can be represented by a single value of type (Float,Float).

3

The definition of `angleVectors` uses a **where** clause to define a variable locally to the function definition—that is, `phi` is only visible in the body of `angleVectors`.

The functions `dotProduct` and `lengthVector` are helper functions that are required for computing the result of `angleVectors`. If we judge that `dotProduct` and/or `lengthVector` will only ever be needed inside the definition of `angleVectors`, we can define one or both of them locally inside that definition as well, by including them in the **where** clause. This avoids distracting attention from `angleVectors`, which is the main point of this sequence of definitions:

```
angleVectors :: Float -> Float -> Float -> Float -> Float
angleVectors a b a' b' = acos phi
  where phi = (dotProduct a b a' b')
                 / (lengthVector a b * lengthVector a' b')

        dotProduct ::
          Float -> Float -> Float -> Float -> Float
        dotProduct x y x' y' = (x * x') + (y * y')

        lengthVector :: Float -> Float -> Float
        lengthVector x y = sqrt (dotProduct x y x y)
```

You can include type signatures for functions defined inside a **where** clause, as in this example: just put the type signature on a separate line, preferably just before the function definition, like a type signature outside a **where** clause.

While there is a choice whether or not to make `dotProduct` and/or `lengthVector` local to the definition of `angleVectors`, there is no such choice for the variable `phi`: its definition depends on the parameters `a`, `b`, `a'` and `b'`. Suppose we try to define `phi` separately, outside the definition of `angleVectors`:

```
angleVectors :: Float -> Float -> Float -> Float -> Float
angleVectors a b a' b' = acos phi

phi = (dotProduct a b a' b')
        / (lengthVector a b * lengthVector a' b')
```

Haskell will produce a sequence of error messages, starting with these two:

```
vectors.hs:4:19: error: Variable not in scope: a :: Float
```

```
vectors.hs:4:21: error: Variable not in scope: b :: Float
```

What Haskell is complaining about is that `a` and `b` are undefined in the definition of `phi`. They are defined inside the body of `angleVectors`, but are meaningless outside that definition.

The technical term for an association between a name and a value is a **binding**. This includes function definitions, which bind names to functional values, as well as local variable/function definitions. And function application creates a temporary binding between its formal parameters and the actual parameter values. The part of a program in which the name is then defined (or **bound**) is called its **scope**. The variables `a` and `b` (and `a'` and `b'`) are formal parameters of `angleVectors`, and so their scope is its body. The scope of `angleVectors` itself is the whole program. In the first definition of `angleVectors`, that was also the case for `dotProduct` and `lengthVector`, but in the second definition their scope was restricted to the body of `angleVectors`.

Notice that the order of definitions doesn't matter! In the first definition of `angleVectors`, the scope of `dotProduct` and `lengthVector` include the definition of `angleVectors`, where they are used, even though their definitions came later.

Exercises

21   **3**

Indentation and Layout

You might have noticed that some care has been taken in the examples above to indent code so that things line up nicely. This is done mainly in order to make it easy to read, but in Haskell layout actually matters, and sloppy indentation can lead to syntax errors. Haskell's **offside rule** requires, among other things, that the definitions in a **where** clause are lined up vertically. And parts of a phrase should be indented further than the start of the phrase. So

```
angleVectors a b a' b' = acos phi
  where phi = (dotProduct a b a' b')
              / (lengthVector a b * lengthVector a' b')
        dotProduct x y x' y' = (x * x') + (y * y')   -- fail
        lengthVector x y = sqrt (dotProduct x y x y)
```

and

```
angleVectors a b a' b' = acos phi
  where phi = (dotProduct a b a' b')
              / (lengthVector a b * lengthVector a' b')
       dotProduct x y x' y' = (x * x') + (y * y') -- fail
        lengthVector x y = sqrt (dotProduct x y x y)
```

and

```
angleVectors a b a' b' = acos phi
  where phi = (dotProduct a b a' b')
        / (lengthVector a b * lengthVector a' b')   -- fail
        dotProduct x y x' y' = (x * x') + (y * y')
        lengthVector x y = sqrt (dotProduct x y x y)
```

will all produce syntax errors. (The failing line has a comment in each case.)

It's possible to play football without paying much attention to the details of the offside rule, and the same goes for Haskell programming. If you take a little bit of care to make your programs readable by lining things up in the obvious way, you should have no trouble.

The tab character has different behaviour in different text editors; some editors will replace it by spaces and some will include the tab character in your program. The latter is a potential source of great confusion, since Haskell may interpret your program differently from the way it looks on your screen. So it's better to avoid tabs completely.

By now you will probably have also noticed that different fonts are used when displaying Haskell programs: boldface for type names, italics for comments, etc. This is **syntax highlighting**, used to improve readability. You don't need to take it into account when you enter your programs. Colour is often used in syntax highlighting but here we stick with boldface and italics.

See ▶ https://en.wikibooks.org/wiki/Haskell/Indentation for more information on indentation in Haskell.

Exercises

1. Define a function root :: `Float -> Float -> Float -> Float` that takes the coefficients a, b, c of a quadratic equation $ax^2 + bx + c = 0$ as input and computes one of its solutions using the quadratic formula

$$\frac{-b \pm \sqrt{b^2 - 4ac}}{2a}$$

   If the equation has no real solutions because $b^2 < 4ac$ then root will return NaN, which means "not a number".

Use sqrt :: `Float -> Float` to compute the square root.

2. Define a function hour :: `Int -> Int` that takes as input a number of minutes and calculates what hour it is after that time has passed, starting at 1 o'clock and using a 12-hour clock. For example, hour 50 should be 1, hour 60 should be 2, and hour 2435 should be 5.
3. Define a function between : `Int -> Int -> Int -> Int` that computes which of its three inputs is between the other two. For example, between 3 5 4 should be 4 and between 3 5 3 should be 3.
4. Define a "exclusive or" function xor :: `Bool -> Bool -> Bool` that returns True if *either* of its inputs is True, but not *both*.

3

Haskell also provides a type `Integer` of unbounded-size integers.

5. Give three definitions of the disjunction function (`||`) `:: Bool -> Bool -> Bool`: one using `if-then-else`, one using guards, and one using definition by cases. Which definition do you think is most elegant or easiest to understand? Why?

6. Sometimes you need to convert a `Float` to an `Int`. Haskell provides three functions that do the conversion: `ceiling`, `floor`, and `round`. Investigate their behaviour, including what they do on negative numbers, numbers like 4.5 and −4.5 that are halfway between adjacent integers, and values of type `Float` that are larger than the maximum (`maxBound :: Int`) and smaller than the minimum (`minBound :: Int`) values of type `Int`.

Venn Diagrams and Logical Connectives

Contents

© The Author(s), under exclusive license to Springer Nature Switzerland AG 2021
D. Sannella et al., *Introduction to Computation*, Undergraduate Topics
in Computer Science, https://doi.org/10.1007/978-3-030-76908-6_4

Visualising Sets

John Venn (1834–1923) was an English mathematician, logician, and philosopher, see ▶ https://en. wikipedia.org/wiki/John_Venn.

4

It's possible to work with sets by listing all of their elements—provided they are finite sets, of course—and then comparing sets by comparing the elements in these lists. But a visual representation provides much better support for human intuition. You've probably already seen **Venn diagrams** used to represent and relate sets. Here's a quick reminder of how to draw them and what they mean.

Let's start with some small sets of numbers:

$S = \{1, 2, 3, 4, 5, 6, 7, 8, 9, 10, 11, 12\}$
$A = \{x \in S \mid x \text{ is divisible by } 2\}$
$B = \{x \in S \mid x \text{ is divisible by } 3\}$

For the sake of this example, we'll regard S as the **universe** of elements that we're interested in, with A and B being subsets of that universe. We can draw that situation like this:

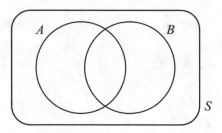

The fact that the circles corresponding to A and B are completely within S means that they are subsets of S. The fact that the circles overlap, but only partially, gives a place (in the middle) to put elements that belong to both A and B, as well as places to put elements that belong to each set but not the other.

We can fill in the diagram by testing all of the elements in S against the properties in the definitions of A and B to decide where they belong. For example, 1 isn't divisible by 2 and it also isn't divisible by 3, so it doesn't belong to either A or B. We therefore put it outside of both circles:

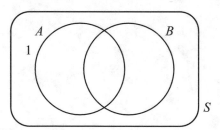

2 is divisible by 2, obviously, but it isn't divisible by 3, so it belongs to A but not B. So we put it into the region of the circle for A that doesn't overlap with B:

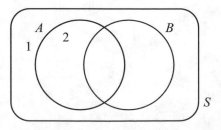

And so on. 6 is divisible by both 2 and 3, so it belongs to both sets. We therefore put it into the middle region, where the circles overlap.

Here's the completed diagram with all of the elements of S where they belong:

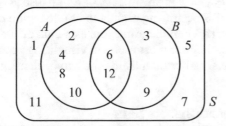

Now, let's consider a different set A, while keeping S and B the same:

$S = \{1, 2, 3, 4, 5, 6, 7, 8, 9, 10, 11, 12\}$
$A = \{x \in S \mid x \text{ is divisible by } 5\}$
$B = \{x \in S \mid x \text{ is divisible by } 3\}$

That gives the following Venn diagram:

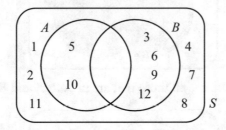

Finally, let's consider yet another choice of the set A:

$S = \{1, 2, 3, 4, 5, 6, 7, 8, 9, 10, 11, 12\}$
$A = \{x \in S \mid x \text{ is divisible by } 6\}$
$B = \{x \in S \mid x \text{ is divisible by } 3\}$

That gives the following Venn diagram:

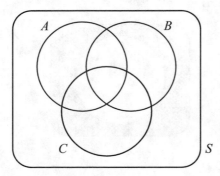

The examples above involve just two sets, A and B (and a universe S). Venn diagrams involving three sets have eight regions:

Venn diagrams involving more than three sets are possible but are harder to draw, see ▶ https://en.wikipedia.org/wiki/Venn_diagram.

We have already seen how subset and set membership are reflected in a Venn diagram. Since $A = B$ means that $A \subseteq B$ and also $B \subseteq A$, set equality is represented by A and B occupying exactly the same region of the diagram.

When we work with Venn diagrams, the elements of the sets aren't usually filled in. As we'll see, they are most useful in visualising potential relationships between arbitrary sets, rather than listing the actual elements of specific sets.

Visualising Operations on Sets

Venn diagrams can be used to represent operations on sets, using shading of regions. The elements of $A \cup B$ are all those that are in either A or B, or in both. Thus, $A \cup B$ is the shaded region in this diagram:

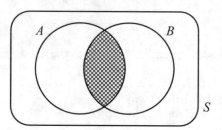

The elements of $A \cap B$ are all those that are in both A and B:

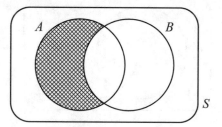

The elements in $A - B$ are those that are in A but not in B:

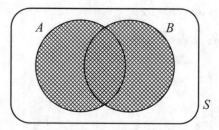

and the elements in the complement $\bar{A}$ of A (with respect to the universe S) are those that are not in A:

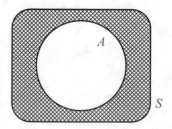

We can do proofs of set-theoretic equalities involving these operations by drawing diagrams and comparing them. For example, to prove $A \cap (B \cup C) = (A \cap B) \cup (A \cap C)$ we can draw a diagram of $A \cap (B \cup C)$:

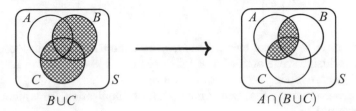

$B \cup C$ $A \cap (B \cup C)$

and a diagram of $(A \cap B) \cup (A \cap C)$:

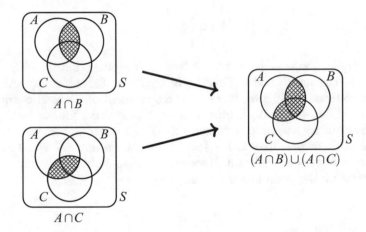

$A \cap B$

$A \cap C$ $(A \cap B) \cup (A \cap C)$

The fact that the shaded region in both diagrams are the same means that $A \cap (B \cup C) = (A \cap B) \cup (A \cap C)$.

Diagrammatic proofs are often more intuitively appealing than symbolic proofs. Compare this proof of $A \cap (B \cup C) = (A \cap B) \cup (A \cap C)$ with your solution to Exercise 1.5(a) and see what you think.

Logical Connectives

We have already seen some Haskell functions that are used to express guards and conditions in conditional expressions:

```
not  :: Bool -> Bool              -- negation
(&&) :: Bool -> Bool -> Bool      -- conjunction
(||) :: Bool -> Bool -> Bool      -- disjunction
```

These functions, called **logical connectives**, are fundamental building blocks of logic. In logic and Mathematics, we use the symbol $\neg$ for not, the symbol $\wedge$ for &&, the symbol $\vee$ for ||, 0 for **False**, and 1 for **True**.

The meanings of the logical connectives are defined by **truth tables** which show what output is produced when they are applied to each of the possible combinations of inputs.

Negation is easy since it takes just one input. The table says that if a is false then $\neg a$ is true, and vice versa.

The pronunciation of these symbols is the same as for the Haskell versions: "not" for $\neg$, "and" for $\wedge$, and "or" for $\vee$. T or $\top$ is used instead of 1 in some books, and F or $\bot$ instead of 0.

a	$\neg a$
0	1
1	0

Conjunction and disjunction have two inputs each, so their truth tables have four rows. For $a \wedge b$ to be true, both a and b must be true, while for $a \vee b$ to be true, at least one of a or b must be true.

a	b	$a \wedge b$
0	0	0
0	1	0
1	0	0
1	1	1

a	b	$a \vee b$
0	0	0
0	1	1
1	0	1
1	1	1

An alternative is **exclusive or**, written $\oplus$. The or function in Haskell corresponds to $\vee$.

The connective $\vee$ is sometimes called *inclusive* disjunction since it doesn't capture the exclusive sense of "or" in English sentences like "You can have soup or salad", where having both is not an option.

Another connective that will come up later is **implication**, for which we will use the symbol $\rightarrow$.

$\rightarrow$ is pronounced "implies". Some books use $\Rightarrow$ or $\supset$ instead of $\rightarrow$. Don't confuse it with Haskell's function arrow ->!

a	b	$a \rightarrow b$
0	0	1
0	1	1
1	0	0
1	1	1

In logical systems that allow more values than 0 and 1, these problems can be partly resolved by using a third value meaning "undefined" for some of these combinations.

The implication $a \rightarrow b$, which is intended to capture "if a then b", is false only if a is true and b is false. This can be confusing for at least two reasons. First, in contrast to English, there is no requirement of a causal connection between a and b. "If the sky is blue then $1 + 1 = 2$" doesn't make much sense in English, but it's true in logic because $1 + 1 = 2$, whether the sky is blue or not. Second, sentences like "If $1 + 1 = 3$ then grass is orange" where the first statement is false don't make much sense in English, but in logic they are true no matter what the second statement is.

Truth Tables

We have just seen truth tables used to define the meanings of the connectives. The same idea can be used to work out the values of complex logical expressions that involve combinations of the connectives.

Here's a simple example, for the expression $(a \vee b) \wedge (a \vee c)$. We need a column for each of the variables in the expression (a, b, c), each of its sub-expressions ($a \vee b$ and $a \vee c$), and for the expression $(a \vee b) \wedge (a \vee c)$ itself. And we need a row for each combination of values of the variables:

Writing down the values of the variables in the order given—or in some other systematic order—rather than writing down different combinations until you can't think of any more, is strongly recommended!

a	b	c	$a \vee b$	$a \vee c$	$(a \vee b) \wedge (a \vee c)$
0	0	0			
0	0	1			
0	1	0			
0	1	1			
1	0	0			
1	0	1			
1	1	0			
1	1	1			

We can fill in a row by working from left to right: the value of each expression is determined by values that are already to its left in the table. Taking the second row, where $a = b = 0$ and $c = 1$, because it's a little more interesting than the first row: we can easily see that $a \vee b = 0$ and $a \vee c = 1$ because the values of a, b, and c are available in the first three columns. Then we can use the values of $a \vee b$ and $a \vee c$ to calculate that $(a \vee b) \wedge (a \vee c) = 0$:

a	b	c	$a \vee b$	$a \vee c$	$(a \vee b) \wedge (a \vee c)$
0	0	0			
0	0	1	0	1	0
0	1	0			
0	1	1			
1	0	0			
1	0	1			
1	1	0			
1	1	1			

Proceeding in the same way to fill in the rest of the table gives:

a	b	c	$a \vee b$	$a \vee c$	$(a \vee b) \wedge (a \vee c)$
0	0	0	0	0	0
0	0	1	0	1	0
0	1	0	1	0	0
0	1	1	1	1	1
1	0	0	1	1	1
1	0	1	1	1	1
1	1	0	1	1	1
1	1	1	1	1	1

The columns in the middle of the table—between the values of the variables and the final result—are only used to hold intermediate values, so once the table is finished they can be left out:

a	b	c	$(a \vee b) \wedge (a \vee c)$
0	0	0	0
0	0	1	0
0	1	0	0
0	1	1	1
1	0	0	1
1	0	1	1
1	1	0	1
1	1	1	1

One reason to build a truth table for an expression is in order to find out if it has certain properties. An expression that is always true—that is, where the final column contains only 1s—is called a **tautology**. An example of a tautology is the expression $a \vee \neg a$.

On the other hand, a **contradiction** is an expression that is always false, so the final column contains only 0s. An example is $a \wedge \neg a$. Complex expressions can be simplified by replacing tautologies by the expression 1 and contradictions by the expression 0.

Finally, an expression that is true for *at least one* combination of values of variables—that is, where the final column of the truth table contains at least one 1—is called **satisfiable**. The truth table above shows that $(a \vee b) \wedge (a \vee c)$ is satisfiable but is neither a tautology nor a contradiction.

Once you get used to building truth tables, you might be able to leave out some of the columns in more complicated examples, to avoid excessive numbers of columns. For example, here is a truth table for the expression $(a \wedge \neg b \wedge (c \vee (d \wedge b)) \vee (\neg b \wedge \neg a)) \wedge c$, in which columns for the sub-expressions $\neg a$, $\neg b$, $d \wedge b$, and $a \wedge \neg b \wedge (c \vee (d \wedge b)) \vee (\neg b \wedge \neg a)$ are omitted:

4

a	b	c	d	$c \vee (d \wedge b)$	$a \wedge \neg b \wedge (c \vee (d \wedge b))$	$\neg b \wedge \neg a$	$(a \wedge \neg b \wedge (c \vee (d \wedge b)) \vee (\neg b \wedge \neg a)) \wedge c$
0	0	0	0	0	0	1	0
0	0	0	1	0	0	1	0
0	0	1	0	1	0	1	1
0	0	1	1	1	0	1	1
0	1	0	0	0	0	0	0
0	1	0	1	1	0	0	0
0	1	1	0	1	0	0	0
0	1	1	1	1	0	0	0
1	0	0	0	0	0	0	0
1	0	0	1	0	0	0	0
1	0	1	0	1	1	0	1
1	0	1	1	1	1	0	1
1	1	0	0	0	0	0	0
1	1	0	1	1	0	0	0
1	1	1	0	1	0	0	0
1	1	1	1	1	0	0	0

Or you can use Haskell to produce the entries in the first place, if you prefer typing to thinking!

It's easy to make mistakes in complicated examples like this. Since the logical connectives are Haskell functions, you can use Haskell to check that entries are correct. For example, to check that the last entry in the 11th row above is correct, we can define:

```
complicated :: Bool -> Bool -> Bool -> Bool -> Bool
complicated a b c d =
  (a && not b && (c || (d && b)) || (not b && not a)) && c
```

and then

```
> complicated True False True False
True
```

There are many truth table generators available on the web, for example, ▶ https://web.stanford.edu/class/cs103/tools/truth-table-tool/.

The truth table for an expression with n variables will have 2^n rows, so writing out truth tables by hand is infeasible for really complex logical expressions.

Exercises

1. Draw a three-set Venn diagram for the following sets:

   $$S = \{1, 2, 3, 4, 5, 6, 7, 8, 9, 10, 11, 12\}$$
   $$A = \{x \in S \mid x > 6\}$$
   $$B = \{x \in S \mid x < 5\}$$
   $$C = \{x \in S \mid x \text{ is divisible by 3}\}$$

   where S is taken to be the universe.

2. Use Venn diagrams to prove:

   (a) union distributes over intersection: $A \cup (B \cap C) = (A \cup B) \cap (A \cup C)$
   (b) associativity of union and intersection
   (c) De Morgan's laws: $\overline{A \cup B} = \bar{A} \cap \bar{B}$ and $\overline{A \cap B} = \bar{A} \cup \bar{B}$

3. Use Venn diagrams to prove that $(A \cap B) - (A \cap C) \subseteq B - C$.

4. Produce truth tables for the following expressions, and check whether they are tautologies, or contradictions, or satisfiable.

   (a) $(a \vee b) \wedge (\neg a \wedge \neg b)$
   (b) $a \rightarrow ((b \rightarrow c) \vee (b \rightarrow \neg c))$
   (c) $((a \wedge \neg b) \vee c \vee (\neg d \wedge b) \vee a) \wedge \neg c$

5. There is a close relationship between operations on sets and the logical connectives. For example, union corresponds to disjunction: $x \in A \cup B$ iff $x \in A \vee x \in B$.

It follows from this relationship that proofs of equalities on sets correspond to proofs of equalities between logical expressions. For example, suppose we prove that union is commutative: $A \cup B = B \cup A$. Then $x \in A \vee x \in B$ iff $x \in A \cup B$ iff $x \in B \cup A$ iff $x \in B \vee x \in A$. Once we introduce **predicates** in Chap. 6 and explain their relationship to subsets of the universe in Chap. 8, this will turn out to amount to a proof of $a \vee b = b \vee a$.

(a) What (combinations of) connectives correspond to intersection, difference, and complement?

(b) What equalities between logical expressions correspond to the equalities on sets proven in Exercise 2?

Lists and Comprehensions

Contents

© The Author(s), under exclusive license to Springer Nature Switzerland AG 2021
D. Sannella et al., *Introduction to Computation*, Undergraduate Topics
in Computer Science, https://doi.org/10.1007/978-3-030-76908-6_5

5

[1,3,2] is pronounced "the list containing 1, 3 and 2" and [Int] is pronounced "list of Int". [[Int]] is pronounced "list of list of Int".

[] is pronounced "nil".

In the error message, Haskell says that [1,2] has type [Integer]. It also has types [Int], [Float], [Double], etc.

: is pronounced "cons", which is short for "construct".

Lists

So far, our computations have involved simple "atomic" data values, having types like Int and Bool. More interesting programs involve the use of "compound" **data structures**, which group together a number of data values and allow them to be handled as a unit. The most important data structure in Haskell and similar programming languages is the **list**, which is a sequence of values of the same type. Most of the programs you will write in Haskell will involve lists, and Haskell provides several special notations to make such programs easy to write.

Here are some examples of lists, including their types:

```
someNumbers :: [Int]
someNumbers = [1,3,2,1,7]

someLists :: [[Int]]
someLists = [[1], [2,4,2], [], [3,5]]

someFunctions :: [Bool -> Bool -> Bool]
someFunctions = [(&&), (||)]
```

A list is written using square brackets, with commas between its elements. Order matters: [1,2] and [2,1] are different lists. The type of a list is written [t], where t is the type of the elements in the list. As the examples above demonstrate, t can be any type, including a list type [s] (so that the elements in the list are themselves lists, containing elements of type s) or a type of functions. The lists in a list of type [[s]] can have different lengths, as in someLists. And the empty list [], which contains no elements and has the polymorphic type [a] (because it is the empty list of integers, and also the empty list of strings, etc.) is also a list.

Here is an example of a mistaken attempt to define a list:

```
> someStuff = [True,[1,2],3]
<interactive>:1:19: error:
    • Couldn't match expected type 'Bool' with actual type '[Integer]'
    • In the expression: [1, 2]
      In the expression: [True, [1, 2], 3]
      In an equation for 'someStuff': someStuff = [True, [1, 2], 3]
```

Remember, all of the values in a list are required to be of the same type.

Functions on Lists

Haskell provides many functions for operating on lists. Most important are the functions for building a list by adding an element to the front of an existing list, and for taking apart a list by returning its first element (its **head**) and the rest (its **tail**). Other functions are for computing the length of a list, for testing whether or not a list is empty, for reversing the order of the elements in a list, for joining two lists together, etc.

```
> 1:[3,2,1,7]
[1,3,2,1,7]
> head [1,3,2,1,7]
1
> tail [1,3,2,1,7]
[3,2,1,7]
> tail []
```

```
*** Exception: Prelude.tail: empty list
> length [1,3,2,1,7]
5
> null [1,3,2,1,7]
False
> null []
True
> reverse [1,3,2,1,7]
[7,1,2,3,1]
> [1,3,2,1,7] ++ reverse [1,3,2,1,7]
[1,3,2,1,7,7,1,2,3,1]
```

Many functions on lists have polymorphic types:

- `(:) :: a -> [a] -> [a]`
- `head :: [a] -> a`
- `tail :: [a] -> [a]`
- `length :: [a] -> Int`
- `null :: [a] -> Bool`
- `reverse :: [a] -> [a]`
- `(++) :: [a] -> [a] -> [a]`

The list notation `[1,3,2,1,7]` is shorthand for the expression `1:(3:(2:(1:(7:[]))))` which can be written without parentheses `1:3:2:1:7:[]` because `:` is **right associative**, meaning that Haskell reads $x:y:z$ as $x:(y:z)$ rather than $(x:y):z$.

Note that `:` can only be used to add an element to the **front** of a list! So the following doesn't work:

```
> [1,3,2,1]:7
<interactive>:1:1: error:
    • Non type-variable argument in the constraint: Num [[t]]
      (Use FlexibleContexts to permit this)
    • When checking the inferred type
        it :: forall t. (Num [[t]], Num t) => [[t]]
```

You can use `++` to add an element to the end of a list, but—because it operates on two lists, rather than an element and a list—you need to first put the element into a singleton list:

`++` is pronounced "append".

```
> [1,3,2,1] ++ [7]
[1,3,2,1,7]
```

Built-in functions like `tail` are defined in the **Prelude**, a part of Haskell's library that is loaded automatically when it starts. There are thousands of other modules in Haskell's library, each providing a group of related definitions. To use a function that is defined there, you need to make it clear where in the library the definition is located. For example, the function `transpose :: [[a]] -> [[a]]`, which interchanges the rows and columns of an $m \times n$ matrix represented as a list of lists, is in the **Data.List** library module along with many other functions for operating on lists. You can load the module using an `import` declaration:

Reminder: Hoogle (▶ https://hoogle.haskell.org/) is a good tool for exploring the contents of Haskell's library.

```
> import Data.List
> transpose [[1,2,3],[4,5,6]]
[[1,4],[2,5],[3,6]]
```

Alternatively, you can import just the `transpose` function by writing `import Data.List (transpose)` or import everything *except* `transpose` by writing `import Data.List hiding (transpose)`.

Of course, you can use all of the notation introduced earlier to write function definitions on lists. For instance, this function definition uses guards:

```
headEven :: [Int] -> Bool
headEven xs | not (null xs) = even (head xs)
            | otherwise     = False
```

Alternatively, you can give separate equations to define the function for the empty list and for non-empty lists:

```
headEven :: [Int] -> Bool
headEven []     = False
headEven (x:xs) = even x
```

Variables with names like xs, pronounced "exes", are often used for lists. Then the names of variables for list elements are chosen to match, in this case x.

Literal value—"literal" for short—is the terminology for a value that doesn't need to be evaluated, like 42, True or 'z'.

Look carefully at the second equation: the **pattern** x:xs will only match a non-empty list, with the variables x and xs giving access to its head and tail in the body of the function. Since there happens to be no use for xs in the body in this particular case, you can replace it with a wildcard if you like.

List patterns are only allowed to contain variables (with no repeated variable names), wildcards, [], : and literal values. Expressions that are equivalent to such patterns, like [a,2,_] (shorthand for a:2:_:[]), are also allowed. Here's a function for returning the square root of the second element in a list, or -1.0 if there is no second element:

```
sqrtSecond :: [Float] -> Float
sqrtSecond []      = -1.0
sqrtSecond (_:[])  = -1.0
sqrtSecond (_:a:_) = sqrt a
```

The second equation can also be written

```
sqrtSecond [_]     = -1.0
```

Other ways of defining functions on lists will be introduced in later chapters.

Strings

Recall that **Char** is Haskell's type of characters such as 'w'. Haskell provides special notation for lists of characters: the type **String** is another name for the type [**Char**], and the notation "string" is shorthand for the list of characters ['s','t','r','i','n','g']. Accordingly, all of the functions on lists work on strings:

```
> 's':"tring"
"string"
> head "string"
's'
> tail "string"
"tring"
> length "string"
6
> reverse "string"
"gnirts"
```

Equality testing (==) and the order relations (<, >, <= and >=) work on strings in the way you would expect, with (<) :: **String** -> **String** -> **Bool** giving the familiar dictionary ordering:

```
> "app" < "apple" && "apple" < "banana"
True
> "apple" < "app"
False
```

And you can define functions on strings using patterns:

```
import Data.Char
capitalise :: String -> String
capitalise ""       = ""
capitalise (c:cs) = (toUpper c) : cs
```

where the function toUpper :: **Char -> Char**, in the **Data.Char** library module, converts a character to upper case.

Tuples

Tuples are another kind of compound data structure. A tuple is a fixed length sequence of data values where the components of the tuple can have different types. For example:

A 2-tuple is sometimes called a **pair**, a 3-tuple is sometimes called a **triple**, etc.

```
coordinates3D :: (Float,Float,Float)
coordinates3D = (1.2, -3.42, 2.7)

friends :: [(String,Int)]
friends = [("Hamish",21), ("Siobhan",19), ("Xiaoyu",21)]
```

("Siobhan",19) is pronounced "the pair (or tuple) containing the string "Siobhan" and the integer 19", or sometimes just "Siobhan, 19".

A tuple is written with parentheses, with its elements separated by commas, and a tuple type is written using the same notation.

A tuple is fixed-length in the sense that adding more components yields a tuple of a different type. So its length is determined by its type. This is in contrast to a list, where a list of type [t] may have any length. But as with lists, the order of the components of a tuple matters: (1,2) and (2,1) are different tuples.

Here are two functions using tuples as parameter or result:

```
metresToFtAndIn :: Float -> (Int,Int)
metresToFtAndIn metres = (feet,inches)
  where feet = floor (metres * 3.28084)
        inches = round (metres * 39.37008) - 12 * feet

nameAge :: (String,Int) -> String
nameAge (s,n) = s ++ "(" ++ show n ++ ")"
```

The style of name used in metresToFtAndIn, where a sequence of words is squashed together into a single word with capitalisation indicating the word boundaries, is called "camel case" because of the "humps" caused by the protruding capital letters.

The definition of nameAge uses a pattern to extract the components of the actual parameter pair, and the Prelude function show to convert an **Int** to a **String**. An equivalent definition uses the **selector functions** fst :: (a,b) -> a and snd :: (a,b) -> b instead of a pattern, as follows:

show can be used to convert values of many types to **String**—more later.

```
nameAge :: (String,Int) -> String
nameAge person = fst person ++ "(" ++ show (snd person) ++ ")"
```

fst is pronounced "first" and snd is pronounced "second".

but function definitions using patterns are usually shorter and easier to read, as in this case.

The singleton list [7] is different from the value 7 that it contains. In contrast, there is no difference between the 1-tuple (7) and the value 7, or between the types (Int) and Int. However, there is such a thing as a 0-tuple, namely, the value (), which has type (). Since there is just one 0-tuple, such a value carries no information! But 0-tuples can still be useful, as you will see later on.

List Comprehensions

List comprehensions in Haskell are a powerful and convenient notation for defining computations on lists, inspired by set comprehension notation (see page 3). Some simple examples will help to introduce the main ideas.

[n*n | n <- [1,2,3]] is pronounced "the list of n*n where n is drawn from [1,2,3]".

```
> [ n*n | n <- [1,2,3] ]
[1,4,9]

> [ toLower c | c <- "Hello, World!" ]
"hello, world!"

> [ (n, even n) | n <- [1,2,3] ]
[(1,False),(2,True),(3,False)]

> [ s++t | s <- ["fuzz","bizz"], t <- ["boom","whiz","bop"] ]
["fuzzboom","fuzzwhiz","fuzzbop","bizzboom","bizzwhiz","bizzbop"]
```

List comprehensions are written using the notation $[\cdots|\cdots]$. The second part of the comprehension, after the vertical bar |, usually includes one or more **generators**, each of which binds a local variable to consecutive elements of the indicated list, using the notation *var <- list*. The part before the vertical bar is an expression producing an element of the resulting list, given values of those variables. The examples show that elements are selected by the generators in the same order as they appear in the list, and what happens in the case of multiple generators.

Note the direction of the arrow in a generator! A good way to remember is to notice that <- looks a little like ∈ (set membership): n <- [1,2,3] versus $n \in \{1, 2, 3\}$. The other kind of arrow, ->, is used for function types and other things in Haskell.

Guards may be added to the second part of comprehensions, after a comma, to specify which of the values selected by the generators are to be included when returning results.

[n*n | n <- 1, odd n, n>0] is pronounced "the list of n*n where n is drawn from 1 such that n is odd and n is greater than 0".

```
> [ n*n | n <- [-3,-2,0,1,2,3,4,5], odd n, n>0 ]
[1,9,25]

> [ s++t | s <- ["fuzz","bizz"], t <- ["boom","whiz","bop"], s<t ]
["fuzzwhiz","bizzboom","bizzwhiz","bizzbop"]
```

Notice the different uses of the conditions n<0 and odd n in the following example:

```
> [ if n<0 then "neg" else "pos" | n <- [3,2,-1,5,-2], odd n ]
["pos","neg","pos"]
```

The condition odd n is a guard used in the second part of the comprehension to select the odd elements from the list [3,2,-1,5,-2]. On the other hand, the condition n<0 is used in a conditional expression in the first part of the comprehension to do different things with the elements that are selected.

List comprehensions can be used anywhere that a list is required, including in function definitions.

```
squares :: [Int] -> [Int]
squares ns = [ n*n | n <- ns ]

odds :: [Int] -> [Int]
odds ns = [ n | n <- ns, odd n ]

sumSqOdds :: [Int] -> Int
sumSqOdds ns = sum [ n*n | n <- ns, odd n ]
```

In sumSqOdds, the Prelude function sum :: [Int] -> Int is used to add together all of the numbers in the list defined by the comprehension. Applying sum to [] gives 0. There is also a function product :: [Int] -> Int for multiplying together all of the numbers in a list, but product [] gives 1. Similar functions that work on lists of Boolean values are and :: [Bool] -> Bool (conjunction) and or :: [Bool] -> Bool (disjunction), where and [] is True and or [] is False. (Why do you think those are the right results for []?)

sum and product also work on lists of Float, Integer, Double, etc.

One of the things that makes set comprehensions so powerful is the way that they define operations on entire lists. This lifts the conceptual level of programming from consideration of single values to transformations on whole data structures, all at once. The change in perspective is analogous to the way that arithmetic operations in computing conceptually operate on numbers rather than on their binary representations. Of course, there is an underlying computation on list elements—or on binary representations of numbers, in the case of arithmetic—but most of the time you can think at the level of the whole list.

Enumeration Expressions

Haskell makes it easy to work with lists of consecutive values like [0,1,2,3,4, 5,6,7,8,9,10] by providing a notation that allows you to just indicate the endpoints in the sequence, in this case [0..10]. You can get it to count down, or to use steps different from 1, by providing more information, and it works for other types too:

```
> [10,9..0]
[10,9,8,7,6,5,4,3,2,1,0]
> [10,8..0]
[10,8,6,4,2,0]
> ['a'..'p']
"abcdefghijklmnop"
```

and such expressions are also useful in function definitions:

```
isPrime :: Int -> Bool
isPrime n = null [ n | x <- [2..n-1], n `mod` x == 0 ]

pythagoreanTriples :: [(Int,Int,Int)]
pythagoreanTriples =
  [ (a,b,c) | a <- [1..10], b <- [1..10], c <- [1..10],
              a^2 + b^2 == c^2 ]
```

You can even leave off the second endpoint to get an **infinite list**! The expression [0..] gives [0,1,2,3,4,5,6,7,8,9,10,11,...]. If you try typing it into GHCi, you will need to interrupt the computation unless your idea of a good time is to spend all day and all night watching your screen fill with increasingly large numbers.

Typing Control-C is the usual way to interrupt an infinite computation.

Although they might not seem very useful right now, we will see later that Haskell's ability to compute with infinite lists like [0..] using a computation strategy called **lazy evaluation**—which does just enough computation to produce a result—makes it possible to write elegant programs. And because of lazy evaluation, you don't always need to interrupt them:

```
> head (tail [0..])
1
```

Lists and Sets

intersect, union, and set difference (\\) are in the **Data.List** library module.

Sets are conceptually important for describing the world and the relationships between things in the world. In Haskell, we can use lists for most of the things that make sets useful, and to make that easier we can use functions on lists that mimic set operations. For example:

```
intersect :: Eq a => [a] -> [a] -> [a]
s `intersect` t = [ x | x <- s, x `elem` t ]
```

where `elem` is a function in the Prelude that checks list membership for lists over any type on which equality testing works:

```
elem :: Eq a => a -> [a] -> Bool
```

But when using lists to represent sets, we need to be aware of some differences between lists and sets.

- All of the elements of a list have the same type, so we can't represent sets that have elements of different types, unless we define a type that includes all the kinds of elements we need. That is sometimes possible, as you'll see later.
- The order of the elements in a list is significant, and lists may contain repeated elements, unlike sets. But we can mimic the "no repeated elements" property by using lists without repetitions. If the type of elements has a natural order, like < on Int, then we can mimic the "order doesn't matter" property as well, by keeping the elements in ascending or descending order: then both {1, 3, 2} and {3, 1, 2}—which are the same set, just written differently— are represented by the same list, [1,2,3].
- Because the order of the elements in a list is significant, we can obtain the first element of a list (its head) and the rest (its tail), and this is one of the most important ways of computing with lists. That doesn't make sense with sets.

Exercises

1. Write a version of the function `angleVectors` from Chap. 3 that represents two-dimensional vectors using the type (**Float,Float**).
2. Let's represent a line with equation $y = ax + b$ using the pair (a,b).

   ```
   type Line = (Float,Float)
   ```

   Write a function

   ```
   intersect :: Line -> Line -> (Float,Float)
   ```

Note that (Float,Float) in the result represents the Cartesian coordinates of a point, not a line.

   to find the coordinates of the intersection between two lines. Don't worry about the case where there is no intersection.
3. Using list comprehension, write a function

   ```
   halveEvens :: [Int] -> [Int]
   ```

   that returns half of each even number in a list. For example,

   ```
   halveEvens [0,2,1,7,8,56,17,18] == [0,1,4,28,9]
   ```
4. Using list comprehension, write a function

   ```
   inRange :: Int -> Int -> [Int] -> [Int]
   ```

   that returns all numbers in the input list that fall within the range given by the first two parameters (inclusive). For example,

   ```
   inRange 5 10 [1..15] == [5,6,7,8,9,10]
   ```

5. Using list comprehension, write a function `countPositives` to count the number of positive numbers in a list. For example,

   ```
   countPositives [0,1,-3,-2,8,-1,6] == 4
   ```

   You will probably want to use the `length` function in your definition. (Why do you think it's not possible to write `countPositives` using only list comprehension, without use of a function like `length`?)

   We will consider 0 to be a positive number.

6. Using list comprehension, write a function

   ```
   multDigits :: String -> Int
   ```

   that returns the product of all the digits in the input string. If there are no digits, your function should return 1. For example,

   ```
   multDigits "The time is 4:25" == 40
   multDigits "No digits here!"  ==  1
   ```

   You'll need a library function to determine if a character is a digit, one to convert a digit to an integer, and one to do the multiplication.

7. Using list comprehension, write an improved definition of the function

   ```
   capitalise :: String -> String
   ```

   that converts the first character in a string to upper case *and converts the rest to lower case*. For example, `capitalise "edINBurgH" == "Edinburgh"`.

8. Use a list comprehension to check whether the expression $(a \wedge \neg b \wedge (c \vee (d \wedge b)) \vee (\neg b \wedge \neg a)) \wedge c$ (see page 29) is a tautology or not. You will need to use the function `and :: [Bool] -> Bool`.

9. Dame Curious is a crossword enthusiast. She has a list of words that might appear in a crossword puzzle, but she has trouble finding the ones that fit a slot. Using list comprehension, write a function

   ```
   crosswordFind :: Char -> Int -> Int -> [String] -> [String]
   ```

   to help her. The expression

   ```
   crosswordFind letter pos len words
   ```

   should return all the items from `words` which (a) are of the given length `len` and (b) have `letter` in position `pos`, starting counting with position 0. For example, if Curious is looking for seven-letter words that have `'k'` in position 1, she can evaluate the expression

   ```
   crosswordFind 'k' 1 7
   ["baklava", "knocked", "icky", "ukelele"]
   ```

   to get `["ukelele"]`. You'll need a library function that returns the nth element of a list, for a given n, and the function `length`.

10. Consider the following definition of the infinite list of all Pythagorean triples:

    ```
    pythagoreanTriples :: [(Int,Int,Int)]
    pythagoreanTriples =
       [ (a,b,c) | a <- [1..], b <- [1..], c <- [1..],
                 a^2 + b^2 == c^2 ]
    ```

    What is the problem with this definition? How might you improve it?

Features and Predicates

Contents

© The Author(s), under exclusive license to Springer Nature Switzerland AG 2021
D. Sannella et al., *Introduction to Computation*, Undergraduate Topics
in Computer Science, https://doi.org/10.1007/978-3-030-76908-6_6

6

An example of a tricky English sentence is "All that glitters is not gold" (originally from *The Merchant of Venice* by Shakespeare) which appears to mean "Everything that glitters isn't gold" but really means "Not all that glitters is gold".

Our logical language is simple enough that we'll be able to decide *mechanically* whether any statement is true or false. That is, all statements will be **decidable**. More complicated logical languages don't have that property. See ▶ https://en.wikipedia.org/wiki/Decidability_(logic).

In **fuzzy logic**, a real number between 0 (completely false) and 1 (completely true) is used to express the "degree of truth" of a statement. See ▶ https://en.wikipedia.org/wiki/Fuzzy_logic.

Logic

Human language is often ambiguous, verbose and imprecise, and its structure is complex. Its ability to express layers and shades of meaning makes it well-suited to writing poetry, but not so good for applications that demand precision or when simplicity is important.

You're going to learn to use a language based on **propositional logic** for describing features of things, making statements about them, and deciding whether or not statements are true or false. Propositional logic is a very simple form of logic where the focus is on ways of building up complex statements from simpler ones using **logical connectives** including conjunction ($\wedge$, or && in Haskell), disjunction ($\vee$, or ||), and negation ($\neg$, or not).

We'll start with a simple world containing a fixed set of things and a fixed vocabulary of features of those things, and where we know which things have which features. Every statement that we'll be able to make using our logical language will have a precise meaning and will be either true or false.

As you learn how to write statements in logic, you'll also learn how to represent the world and statements about the world in Haskell. This will allow you to **compute** whether a statement is true or false, saving you the trouble of working it out by hand—or at least, giving you an easy way to check the results of your hand calculations—which can be hard work once things get complicated. It also demonstrates how logic can be mechanised, which is essential for many applications of logic in Informatics.

Later, we'll look at aspects of logic that relate to all worlds, or to sets of worlds that have something in common, rather than just to a particular world. We'll also "abstract away" from specific statements about the world, and focus on ways of combining statements and on reasoning about the truth of complex statements using knowledge of the truth of simpler statements. This gets to the heart of logic as the science of **pure reasoning**: reasoning that doesn't depend on the subject matter.

Logic is a big subject that has many applications in Informatics, especially in Artificial Intelligence. Sometimes the applications require a logical language that's much more complex than the simple one we're going to study here, but our language is the basis of all of these more complex systems. An example of an aspect that could be added is the ability to reason about sequences of events and about events that are *possible* (might happen) versus events that are *necessary* (will definitely happen). And some varieties of logic involve more truth values than true and false: for example, there could be a third value meaning "unknown". You'll learn about some of these things later, in courses covering topics where logic is used.

Our Universe of Discourse

To start with, we need to know something about the domain that we're going to be talking about: what are the things? and what features of those things do we want to talk about? In logic, this is called the **universe of discourse**. The word "universe" suggest that it contains *everything*, but the universe of discourse is limited to just the things *that are of interest at the moment*. The idea is to give meaning to words like "everything" and "something": "everything has property P" means that all of the things in the universe of discourse have property P, and "something has property P" means that there is at least one thing in the universe of discourse that has property P. This is the same idea as the universe that we use when taking the complement $\bar{A}$ of a set A. Without knowing what universe we have in mind, we can't form the set of everything that's not in A.

An example of a universe of discourse would be all of the people who are matriculated students at the University of Edinburgh on 30 September 2021, together with certain features of those people: age, eye colour, etc. We're only going to look at finite universes of discourse in this book. Our first universe of discourse will be very small, just nine things, all of which are shapes, either triangles or discs. Each of these shapes is either small or big, and is either white or black or grey. Here it is:

An example of an infinite universe of discourse is the set of real numbers. A feature of things in this universe is whether they are rational or irrational. An example of a large but finite universe is the set of all people in China who were alive at 00:01 CST on 1 January 2020, with features relating to their past and present health.

We can make statements about the things in the universe of discourse. Some are true, for example:

- Every white triangle is small.
- Some big triangle is grey.
- No white thing is black.

and some are false, for example:

- Every small triangle is white.
- Some small disc is black.

For this small universe, it's easy to work out which statements are true, just by looking at the picture. For bigger and more complicated universes, it might be much harder.

We're now going to look at how to represent our universe of discourse in Haskell and how to turn such statements into Haskell code that will produce the correct answer. The methods we use will work for any other finite universe, so they'll scale up to universes of any size and complexity.

Representing the Universe

The first step is to give names to the things in the universe. The names we use don't matter, but we need them to be able to refer to the things in Haskell. Here goes:

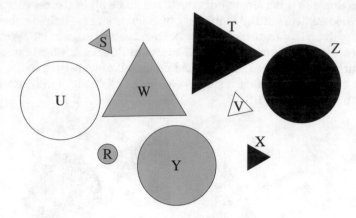

6

Using these names, we can define a Haskell type of things, and a list of all the things in the universe:

```
data Thing = R | S | T | U | V | W | X | Y | Z
  deriving (Eq,Show)
things :: [Thing]
things = [R, S, T, U, V, W, X, Y, Z]
```

We'd actually like to use sets rather than lists for representing the universe of discourse, since the order of the things in our lists isn't significant. But Haskell's features for computing with lists are very convenient, so we'll go ahead and use them.

One way of describing the features of things in Haskell is by creating types for the features—colours, shapes, sizes—and then defining functions that say which things are triangles, which are grey, etc.

```
data Colour = White | Black | Grey
data Shape = Disc | Triangle
data Size = Big | Small

colour :: Thing -> Colour
colour R = Grey
...

shape :: Thing -> Shape
shape R = Disc
...

size :: Thing -> Size
size R = Small
...
```

You need to include "deriving (Eq,Show)" in the type definition in order to use elem in the definitions below, and to get the results of expressions involving values of type Thing to print out. What this means and why you need it will be explained later.

This is an accurate way of representing features, but it's hard to work with because the features have different types. This makes the representation highly dependent on what the features in the universe of discourse happen to be. So let's instead use **predicates**: Bool-valued functions for each value of each feature, saying whether each thing has that value of that feature or not.

It's convenient to start with a definition of Predicate as an abbreviation for the type Thing -> Bool, but we'll need predicates for other kinds of things later, so we'll define

```
type Predicate u = u -> Bool
```

instead. And then we can define the predicates:

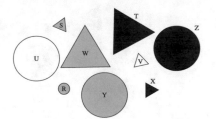

```
isSmall :: Predicate Thing
isSmall x = x `elem` [R, S, V, X]

isWhite :: Predicate Thing
isWhite x = x `elem` [U, V]

...
```

There are seven of these predicates (`isDisc`, `isTriangle`, `isWhite`, `isBlack`, `isGrey`, `isBig`, `isSmall`). They all have the same type, so this representation is uniform and independent of the subject matter.

We can use Haskell's logical connectives

```
not  :: Bool -> Bool
(&&) :: Bool -> Bool -> Bool
(||) :: Bool -> Bool -> Bool
```

to help define the predicates. Once we have the definition of `isSmall` above, we can define

```
isBig :: Predicate Thing
isBig x = not (isSmall x)
```

and once we've defined `isWhite` and `isBlack`, we can define

```
isGrey :: Predicate Thing
isGrey x = not (isWhite x) && not (isBlack x)
```

When applying a predicate to a thing *x* produces **True**, we say that *x* **satisfies** the predicate.

Things Having More Complex Properties

We can now use list comprehensions together with the definitions above to compute lists of things in our universe that have properties involving combinations of features. For example, here's the list of small triangles:

```
> [ x | x <- things, isSmall x, isTriangle x ]
[S,V,X]
```

or equivalently, using `&&` to combine the two guards:

```
> [ x | x <- things, isSmall x && isTriangle x ]
[S,V,X]
```

and here's the list of grey discs:

```
> [ x | x <- things, isGrey x && isDisc x ]
[R,Y]
```

We can also handle other ways of combining features, not just conjunction. For example, here's the list of things that are either big or triangles, or both:

```
> [ x | x <- things, isBig x || isTriangle x ]
[S,T,U,V,W,X,Y,Z]
```

which is different from the list of big triangles:

```
> [ x | x <- things, isBig x && isTriangle x ]
[T,W]
```

And here's the list of discs that aren't grey:

```
> [ x | x <- things, isDisc x && not (isGrey x) ]
[U,Z]
```

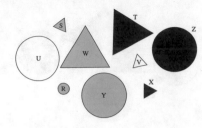

6

In English, use "every" or "all" and be very careful with "any" whose meaning depends on context: compare "Can you do anything?" with "I can do anything!"

It's easy to see that conjunction of features amounts to intersection, for example, the list of small triangles

```
> [ x | x <- things, isSmall x && isTriangle x ]
[S,V,X]
```

is the intersection of the list of small things

```
> [ x | x <- things, isSmall x ]
[R,S,V,X]
```

and the list of triangles

```
> [ x | x <- things, isTriangle x ]
[S,T,V,W,X]
```

In the same way, disjunction of features corresponds to union.

Checking Which Statements Hold

We can now check statements about our universe of discourse to see whether they are true or false.

To check that a property holds for *everything*, we need to check that it holds for all of the things in the universe. We can check each of them individually, and then take the *conjunction* of the results.

For example, consider the statement "Every small triangle is white". We compute the list of small triangles

```
> [ x | x <- things, isSmall x && isTriangle x ]
[S,V,X]
```

and then we look to see whether or not they are all white:

```
> [ (x, isWhite x) | x <- things, isSmall x && isTriangle x ]
[(S,False),(V,True),(X,False)]
```

We discover that S and X are small triangles that aren't white, so the statement is false. If we don't care about the specific counterexamples, that conclusion can be obtained more directly using the function and :: [Bool] -> Bool, like so:

```
> and [ isWhite x | x <- things, isSmall x && isTriangle x ]
False
```

On the other hand, every white triangle is small:

```
> and [ isSmall x | x <- things, isWhite x && isTriangle x ]
True
```

Similarly, we check that a property holds for *something* by checking the things in the universe and taking the *disjunction* of the results.

For example, consider the statement "Some big triangle is grey". We compute the list of big triangles

```
> [ x | x <- things, isBig x && isTriangle x ]
[T,W]
```

and then we look to see which ones are grey:

```
> [ (x,isGrey x) | x <- things, isBig x && isTriangle x ]
[(T,False),(W,True)]
```

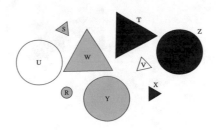

We see that `W` is a big grey triangle, so the statement is true. And again, we can reach that conclusion more directly using the function `or :: [Bool] -> Bool`:

```
> or [ isGrey x | x <- things, isBig x && isTriangle x ]
True
```

But it's not the case that some small disc is black:

```
> or [ isBlack x | x <- things, isSmall x && isDisc x ]
False
```

Sequents

Here is another way of saying that "Every white triangle is small" is true:

`isWhite, isTriangle ⊨ isSmall`

This is an example of a **sequent**. The **antecedents** are listed before ⊨, and the **succedents** are listed after ⊨. A sequent is **valid** if everything in the universe of discourse that satisfies all of the antecedents satisfies at least one of the succedents. We'll start with examples where there's just one succedent, which makes things a little easier to understand.

If a sequent is invalid, we write it using the symbol ⊭. For example:

`isSmall, isTriangle ⊭ isWhite`

This says that "Every small triangle is white" is false. That is, it's *not* the case that everything in the universe of discourse that satisfies both `isSmall` and `isTriangle` will satisfy `isWhite`. Or equivalently, there's something that satisfies both `isSmall` and `isTriangle` but doesn't satisfy `isWhite`.

Any or all of the antecedents and succedents can be negated. For example:

`isWhite ⊨ ¬ isBlack`

says that "Every white thing is not black" (or equivalently, "No white thing is black") is true. We use the mathematical symbol ¬, rather than `not`, because with sequents we're in the world of Mathematics rather than in Haskell. Also because `not isBlack` would give a type error: `not :: Bool -> Bool` while `isBlack :: Predicate Thing`.

There's a subtle but very important difference between negation of predicates and a sequent that's invalid, written using ⊭. A sequent that's valid expresses a statement of the form "Every X is Y". If Y is of the form $¬Z$, then we have "Every X is not Z", as in the last example, "Every white thing is not black".

On the other hand, a sequent being invalid means that there's at least one thing that satisfies the antecedents but doesn't satisfy any of the succedents: "It's not the case that every X is Y".

We can put negation of predicates and ⊭ together to express statements of the form "Some X is Y". For example, recall the statement "Some big triangle is grey":

```
> or [ isGrey x | x <- things, isBig x && isTriangle x ]
True
```

That statement is equivalent to the statement "It's not the case that all big triangles are not grey", which can be expressed:

`isBig, isTriangle ⊭ ¬ isGrey`

`isWhite, isTriangle ⊨ isSmall` is pronounced "isWhite and isTriangle **satisfies** isSmall". The symbol ⊨ is called a **double turnstile**.

⊭ is pronounced "does not satisfy".

Think about this and make sure that you understand it! It's a little counterintuitive that expressing the "positive" statement "Some X is Y" requires two forms of negation. We'll need this later in Chap. 9, where there is some more explanation, and thereafter.

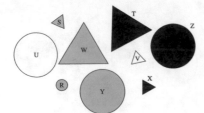

6

Exercises

1. Express the following statements in Haskell using the logical operations `&&`, `||` and/or `not` to combine the predicates defined above. Give the values of the Haskell expressions and check that they are correct according to the diagram.

   - Every small disc is white.
   - Some grey things are not discs.
   - Every big triangle is either grey or black.
   - Some grey triangle is not big.

2. The statement "No white thing is black" doesn't fit either of the patterns "Every X is Y" or "Some X is Y". But it's equivalent to a statement of the form "It's not the case that some X is Y" and also to a statement of the form "Every X is not Y". Give those two equivalent statements and express them in Haskell using the logical operations `&&`, `||` and/or `not` to combine the predicates defined above.

3. Most uses of the words "and", "or" and "not" in English correspond pretty closely to the meanings of the corresponding logical operations. For instance, "R is grey and small" corresponds to `isGrey R && isSmall R`, and "T is white or black" corresponds to `isWhite T || isBlack T`. But there are some subtleties. Consider the sentence "T is not a black disc". Express all of the readings you can think of in Haskell, using the logical operations `&&`, `||` and/or `not` to combine the predicates defined above.

4. Explain why our method of translating sentences into Haskell gives the correct answer in the following cases involving the empty list of big white triangles:

   - All big white triangles are black.
   - No big white triangles are black.
   - Some big white triangle is black.

5. Find two more statements of the form "Every X Y is Z" that are true, and two more statements of the form "Some X Y is Z" that are false, for X, Y, and Z being combinations of predicates or negated predicates.

6. Express each of the following using a sequent:

   - "Every big grey thing is a triangle" is false.
   - "Something small is black" is true.
   - "Some small disc is black" is false.

Testing Your Programs

Contents

© The Author(s), under exclusive license to Springer Nature Switzerland AG 2021
D. Sannella et al., *Introduction to Computation*, Undergraduate Topics
in Computer Science, https://doi.org/10.1007/978-3-030-76908-6_7

Making Mistakes

As you will have discovered by now, it's easy to make mistakes in your function definitions. Sometimes you won't be paying enough attention. Sometimes your fingers will type something different from what your brain intended. Nobody's perfect!

Fortunately, Haskell will detect some of your mistakes automatically. These include syntax and type errors as well as typos, like misspelled function names.

Should be "n <- ns", not "ns -> n".

```
sumSqOdds :: [Int] -> Int
sumSqOdds ns = sum [ n*n | ns -> n, odd n ]
```

```
sumSqOdd.hs:2:31: error: parse error on input '->'
Failed, modules loaded: none.
```

Should be "odd n", not "n odd".

```
sumSqOdds :: [Int] -> Int
sumSqOdds ns = sum [ n*n | n <- ns, n odd ]
```

```
sumSqOdd.hs:2:37: error:
    • Couldn't match expected type '(Integer -> Bool) -> Bool'
                   with actual type 'Int'
    • The function 'n' is applied to one argument,
      but its type 'Int' has none
      In the expression: n odd
      In a stmt of a list comprehension: n odd
Failed, modules loaded: none.
```

Typo in "odd n".

```
sumSqOdds :: [Int] -> Int
sumSqOdds ns = sum [ n*n | n <- ns, odf n ]
```

```
sumSqOdd.hs:2:37: error:
    • Variable not in scope: odf :: Int -> Bool
    • Perhaps you meant 'odd' (imported from Prelude)
Failed, modules loaded: none.
```

You still have to figure out what the error message means, locate the mistake in your code, and fix it. But having it pointed out to you automatically is a big help.

According to the *Oxford English Dictionary*, "thinko" is not an actual word. But it should be.

Much more challenging is finding and fixing "**thinkos**", like this one:

```
allDifferent :: Int -> Int -> Int -> Bool
allDifferent a b c | a/=b && b/=c = True
                   | otherwise    = False
```

This function is supposed to return **True** when it is given three different numbers as input. The problem here is that, although *equality* is transitive (that is, if $x==y$ and $y==z$ then $x==z$) *inequality* is not, meaning that the reasoning used when writing the guard is wrong. So allDifferent will sometimes incorrectly return **True** when a==c.

Sometimes these are dumb mistakes, but ones that Haskell can't detect automatically. Sometimes they are more subtle, for example, when you misunderstand some aspect of the problem you are trying to solve or you forget to consider certain cases.

Finding Mistakes Using Testing

One way of finding thinkos is to **test** your function with sample inputs to see if it does what you intended.

```
> allDifferent 1 2 3
True
> allDifferent 0 0 0
False
> allDifferent 1 1 0
False
> allDifferent 0 1 1
False
> allDifferent 1 0 1
True
```

The last test reveals that there is a mistake in the definition of `allDifferent`.

Sometimes you will be provided with information about how your function should behave in response to certain sample inputs, and those inputs are then obvious choices of test cases.

In general, a range of different test cases—rather than just a few tests that are all similar to each other—is the best way to find mistakes. Make sure to test boundary cases (the empty list, for functions on lists; 0, for functions on `Int`) and cases that you might not have considered as the primary ones (e.g. negative numbers, strings that contain funny symbols instead of just letters).

If your problem involves searching for something in a data structure, say, then try test cases where it is absent and cases where it is present more than once, rather than just cases where it is present exactly once. If your function produces a result of type `Bool`, then test inputs for which it should produce the result `False`, not just inputs for which it should produce `True`. And if you can manage it, try some very large test values.

It's good practice to test your functions right away, as soon as you have written them. That way, when another function that builds on previously defined functions gives an unexpected result, you know to look for the mistake in the new code rather than in the previous function definitions. It's a lot easier to find mistakes when you can focus attention on a small chunk of code. It's also easier to remember what you were thinking when the code is fresh in your mind. On the other hand, sometimes you won't be able to find a mistake after staring at the code for ages. And looking for a bug shortly after you created it means that you might still be suffering from the same misapprehension that led you to make the mistake. In that case it might be best to come back later for a fresh look after a break.

Finally, when you change the definition of a function that you have already tested, it's a good idea to re-run the previous tests to make sure that your changes haven't broken anything.

Test-driven development is a popular software engineering approach that advocates writing down the test cases even *before* you write the code! See ▶ https://en.wikipedia.org/wiki/Test-driven_development.

Re-running tests to ensure that previously tested code still works after a change is called **regression testing**.

Triple modular redundancy—where three sub-systems independently compute a result and the final result is produced by majority voting—is a technique used in fault-tolerant systems to deal with potential sub-system failure, see ▶ https://en. wikipedia.org/wiki/ Triple_modular_redundancy. Using just two doesn't work because you don't know which one is wrong in case the results differ. Sometimes more than three sub-systems are used: Charles Darwin's ship HMS Beagle carried 22 chronometers!

Testing Multiple Versions Against Each Other

One way to achieve considerable confidence that your code is correct is to write two versions—perhaps using two different algorithms, or produced by different people—and then test that both versions produce the same result. If there is a discrepancy, then at least one of the versions is wrong.

Writing two versions of every function definition doubles the work so it is not very practical as a general rule. One case where it is practical is in a situation where a particularly efficient implementation of some function is required. As you will see later, efficiency in software often comes at the cost of simplicity. So more efficient code is likely to be more complicated and harder to understand, and it is more likely to contain mistakes.

❯❯ There are two ways of constructing a software design: One way is to make it so simple that there are *obviously* no deficiencies and the other way is to make it so complicated that there are no *obvious* deficiencies.
C.A.R. Hoare, 1980 Turing Award winner

In the rare situation where efficiency is important, a good way to achieve *both* correct *and* efficient code is to proceed as follows:

1. Write the simplest possible definition of the function;
2. Check that it is correct. Simple code is easier to understand and usually has fewer cases that need to be checked separately;
3. Write a more efficient version of the same function; and
4. Test that the more efficient version produces the same results as the simple version.

Property-Based Testing

A way to achieve some of the advantages of testing multiple versions of a function definition against each other without doing everything twice is to instead think of **properties** that your function should satisfy, and use testing to check whether or not they hold.

One example of a property of sumSqOdds above is that its result should always be less than or equal to the sum of the squares of the numbers in its input list, not just the odd ones. Another is that its result will be odd (resp. even) if an odd (resp. even) number of the values in its input list are odd. Finally, suppose that we do have a different version of sumSqOdds, for example:

```
sumSqOdds' :: [Int] -> Int
sumSqOdds' ns = sum (squares (odds ns))
```

then always producing the same result as sumSqOdds' is an important property of sumSqOdds.

Thinking about properties—*what* the function should compute—is quite a different activity from thinking about *how* to compute it. Thinking of the same problem from a different angle, as with testing multiple versions of a function definition against each other, is a good way of exposing gaps in your thinking.

If you code the properties as Haskell functions, you can test them, rather than just thinking about them. Here are Haskell versions of the properties above:

```
sumSqOdds_prop1 :: [Int] -> Bool
sumSqOdds_prop1 ns = sumSqOdds ns <= sum (squares ns)

sumSqOdds_prop2 :: [Int] -> Bool
sumSqOdds_prop2 ns = odd (length (odds ns)) == odd (sumSqOdds ns)
```

```
sumSqOdds_prop3 :: [Int] -> Bool
sumSqOdds_prop3 ns = sumSqOdds ns == sumSqOdds' ns
```

If you compare two functions and discover that they are different, you need to figure out which of the two functions is wrong. Or if you test a property and discover that it doesn't hold, you need to figure out whether the mistake is in the code or in the property.

An example of a mistake in a property is the following version of sumSqOdds_prop1, where "less than or equal" has been incorrectly coded using the < relation:

```
sumSqOdds_prop1' :: [Int] -> Bool
sumSqOdds_prop1' ns = sumSqOdds ns < sum (squares ns)
```

This property will fail for a correct implementation of sumSqOdds on any list containing no non-zero even numbers, such as the empty list.

Automated Testing Using `QuickCheck`

If you test a few cases, you gain some confidence that your code is correct. Testing more cases gives you more confidence. If you could test *all* cases then you could be *very* confident, but unfortunately that is usually impossible unless the domain of possible inputs is finite and relatively small. Still, more testing is better than less testing, and if possible it should be done on as many different kinds of inputs as possible.

Fortunately, Haskell makes it easy to do lots of testing. Haskell's QuickCheck library module provides tools for testing properties that have been coded as Haskell functions on 100 automatically generated random inputs.

After importing QuickCheck

```
import Test.QuickCheck
```

we can test the above properties of sumSqOdds using the function quickCheck:

```
> quickCheck sumSqOdds_prop1
+++ OK, passed 100 tests.
> quickCheck sumSqOdds_prop2
+++ OK, passed 100 tests.
> quickCheck sumSqOdds_prop3
+++ OK, passed 100 tests.
> quickCheck sumSqOdds_prop1'
*** Failed! Falsifiable (after 1 test):
[]
```

To do more tests, run quickCheck repeatedly, or increase the number of tests it does:

```
> quickCheck sumSqOdds_prop3
+++ OK, passed 100 tests.
> quickCheck sumSqOdds_prop3
+++ OK, passed 100 tests.
> quickCheck sumSqOdds_prop3
+++ OK, passed 100 tests.
> quickCheck (withMaxSuccess 1000000 sumSqOdds_prop3)
+++ OK, passed 1000000 tests.
```

When quickCheck encounters a failing test case, it reports the inputs on which the test fails:

Values of type Int occupy 32 or 64 bits, so any function on Int has a finite domain of possible inputs. But since $2^{32} = 4294967296$ and $2^{64} = 18446744073709551616$, it is infeasible to test them all.

Notice that the library module is called QuickCheck, with an upper case Q, but the function is called quickCheck!

```
allDifferent_prop :: Int -> Int -> Int -> Bool
allDifferent_prop a b c
  | allDifferent a b c
              = a/=b && a/=c && b/=a && b/=c && c/=a && c/=b
  | otherwise = a==b || a==c || b==a || b==c || c==a || c==b
```

```
> quickCheck allDifferent_prop
*** Failed! Falsifiable (after 31 tests and 2 shrinks):
27
0
27
```

Before reporting a failing test case, quickCheck will first try to "shrink" it to find a similar but smaller failing test case. This is useful because smaller and simpler counterexamples are simpler to analyse than the possibly more complicated counterexample that quickCheck happened to encounter.

Without doing exhaustive testing of all input values, testing can only ever reveal mistakes, never guarantee correctness. So, quickCheck not finding a mistake doesn't mean that there aren't any.

```
> quickCheck allDifferent_prop
*** Failed! Falsifiable (after 5 tests and 1 shrink):
-3
0
-3
> quickCheck allDifferent_prop
+++ OK, passed 100 tests.
```

In the second run of quickCheck, we were lucky (that is, unlucky): it didn't happen to generate any failing test cases.

Conditional Tests

Sometimes you need to restrict the test cases that quickCheck generates to a subset of the values of the parameter type. This might be necessary to avoid testing your function on inputs for which it was not intended to produce a sensible result.

For example, given a number n and an approximation r to $\sqrt{n}$, the Newton–Raphson method computes a closer approximation:

```
newton :: Float -> Float -> Float
newton n r = (r + (n / r)) / 2.0
```

Iterating will eventually converge to $\sqrt{n}$.

Suppose that we want to test that newton $n\ r$ is indeed closer to $\sqrt{n}$ than r is. To avoid attempting to find the square root of a negative number, we need to restrict attention to $n \geq 0$, and then it is sensible to restrict to $r > 0$ as well, excluding $r = 0$ to avoid division by zero. We can write this **conditional test** using the operator ==>:

```
newton_prop :: Float -> Float -> Property
newton_prop n r =
  n>=0 && r>0 ==> distance n (newton n r) <= distance n r
    where distance n root = abs (root^2 - n)
```

Testing this property using quickCheck easily finds cases where it isn't satisfied, since the Newton–Raphson method performs poorly when the chosen approximation to the square root is too far from the actual square root:

```
> quickCheck newton_prop
*** Failed! Falsifiable (after 32 tests and 6 shrinks):
3.0
0.1
```

The margin notes:

Haskell's **SmallCheck** library module (▶ https://hackage.haskell.org/package/smallcheck) does exhaustive testing of properties for all test cases up to a given size.

The Newton–Raphson method is named after the British mathematician and scientist Isaac Newton (1642–1727) and the British mathematician Joseph Raphson (c. 1648–c. 1715), but its essence is much older, see ▶ https://en.wikipedia.org/wiki/Newton's_method.

The symbol ==> is meant to suggest logical implication.

Note that the result of a conditional test has type `Property`, rather than `Bool`, but `quickCheck` works for that type too.

Test Case Generation

`QuickCheck` knows how to automatically generate random test values for Haskell's built-in types. You need to tell it how to generate test values for types that you define yourself.

For example, consider this type definition from Chap. 2:

```
data Weekday = Monday | Tuesday | Wednesday | Thursday
                     | Friday | Saturday | Sunday
  deriving Show
```

You need to include "`deriving Show`" to allow `quickCheck` to print out failing test cases.

Here is an example of a function that takes an input of type `Weekday`:

```
isSaturday :: Weekday -> Bool
isSaturday Saturday = True
isSaturday _        = False
```

and a simple test for isSaturday:

```
isSaturday_prop :: Weekday -> Bool
isSaturday_prop d = not (isSaturday d)
```

Attempting to run this test using `quickCheck` fails:

```
> quickCheck isSaturday_prop
<interactive>:2:1: error:
    • No instance for (Arbitrary Weekday)
        arising from a use of 'quickCheck'
    • In the expression: quickCheck isSaturday_prop
      In an equation for 'it': it = quickCheck isSaturday_prop
```

If we first declare a `QuickCheck` generator for `Weekday`:

```
instance Arbitrary Weekday where
    arbitrary = elements [Monday, Tuesday, Wednesday,
                              Thursday, Friday, Saturday, Sunday]
```

Don't try to understand this declaration at this point! It uses a feature of Haskell ("type classes") that won't be covered until Chap. 24. But feel free to follow the same scheme to get a `QuickCheck` generator for other types that are defined by enumerating their values.

then `quickCheck` runs the test, but (of course) finds an input on which it fails:

```
> quickCheck isSaturday_prop
*** Failed! Falsifiable (after 2 tests):
Saturday
```

Later on you'll learn how to define new types in Haskell that are more complicated than `Weekday`. Getting `QuickCheck` to generate tests for these goes into territory that is beyond the scope of this book. See the `QuickCheck` manual (▶ http://www.cse.chalmers.se/~rjmh/QuickCheck/manual_body.html#13) and the extensive comments in the `QuickCheck` code (▶ https://hackage.haskell.org/package/QuickCheck-2.14/docs/src/Test.QuickCheck.Arbitrary.html) for relevant information.

Another useful source of information on `QuickCheck` is the blog post in ▶ https://begriffs.com/posts/2017-01-14-design-use-quickcheck.html.

It's also possible to tweak the way that `QuickCheck` generates test cases for built-in types, for example, to change the distribution of long inputs versus short inputs. The way that `QuickCheck` attempts to shrink failing test cases is also customisable.

7

Testing Polymorphic Properties

Testing polymorphic properties with `QuickCheck` is an important special case. Consider commutativity of the polymorphic function `++`, which obviously doesn't hold:

```
append_prop :: Eq a => [a] -> [a] -> Bool
append_prop xs ys = xs ++ ys == ys ++ xs
```

But `QuickCheck` doesn't catch the problem:

```
> quickCheck append_prop
+++ OK, passed 100 tests.
```

This shows that, although `QuickCheck` is able to test polymorphic properties, the test values it generates are insufficient to catch even obvious mistakes.

You might think that polymorphic properties should be tested for values of many different types, in order to catch errors that might arise for one type but not another. But since polymorphic functions behave uniformly for all type instances, there's no need to test them on different types. It suffices to test polymorphic properties for `Int` or some other infinite type:

Testing with types having small numbers of values is not appropriate. For example, `append_prop :: [()] -> [()] -> Bool` holds, where `()` is the type of 0-tuples, having the single value `()`.

```
append_prop' :: [Int] -> [Int] -> Bool
append_prop' xs ys = xs ++ ys == ys ++ xs
```

QuickCheck quickly finds a counterexample:

```
> quickCheck append_prop'
*** Failed! Falsified (after 3 tests and 1 shrink):
[0]
[1]
```

Exercises

1. If you can remember any of the mistakes you made when solving exercises in previous chapters, formulate some test cases that would have revealed them.
2. Write and run tests to check that reversing a list of type `[Int]` twice produces the same list. Think of a property that relates `reverse` and `++`, and test that it holds for lists of type `[Int]`.
3. Recall the function `max3` from Chap. 3:

```
max3 :: Int -> Int -> Int -> Int
max3 a b c
   | a>=b && a>=c = a
   | b>=a && b>=c = b
   | otherwise    = c
```

Write and run a test to check that the result of `max3` is always greater than or equal to all of its inputs. Then replace all occurrences of `>=` in `max3` with `>` and re-run your test to see if that change introduces a problem.
4. Write and run tests to check whether Haskell's integer addition and multiplication are commutative and associative, and whether integer arithmetic satisfies $(m + n) - n = m$ and $(m * n)/n = m$. Check which of these tests fail for floating-point arithmetic.
5. Exercise 5.2 asks you to write a function

See ► https://en.wikipedia.org/wiki/Floating-point_arithmetic for information about accuracy problems with floating-point arithmetic.

```
intersect :: Line -> Line -> (Float,Float)
```

to compute the coordinates of the intersection between two lines. Write a test to check that the intersection of two lines is on both lines. Use `quickCheck` to check if your solution for intersect passes these tests. Does anything go wrong? Can you improve your test to avoid the problem?

6. The following function from Chap. 5 capitalises a word:

```
import Data.Char
capitalise :: String -> String
capitalise ""     = ""
capitalise (c:cs) = (toUpper c) : cs
```

Write tests to check that:

- The first character of the result is upper case, using isUpper from the Data.Char library module.
- The length of the input is the same as the length of the result.
- The characters in the input are the same as the characters in the result, if case is disregarded. (You will probably want to use the list indexing function !!.)

You may need to use conditional tests. (For which properties?)

Use quickCheck to check if capitalise passes these tests. Does anything go wrong? Can you improve your tests to avoid the problem?

In Exercise 5.7, you wrote an improved version of capitalise. Use the same tests to check that version. Any new problems?

Patterns of Reasoning

Contents

Aristotle (384–322 BC) was the founder of the western tradition of logic. The study of logic started earlier in India, see ▶ https://en. wikipedia.org/wiki/Indian_logic, and its study in China started around the same time as Aristotle, see ▶ https:// en.wikipedia.org/wiki/ Logic_in_China.

8

Syllogisms

Now that you know how to use logic for making statements about things and how to check whether or not a given statement is true, the next step is to study patterns of reasoning that allow true statements to be combined to give other statements that are guaranteed to be true. We'll start with ideas that go all the way back to Aristotle, the founder of logic, who looked at simple patterns of logical argument with two premises and a conclusion, called **syllogisms**. The study of the sound syllogisms and the relationships between them dominated the subject for the next 2000 years.

Nowadays, using the modern notation of **symbolic logic**, we can express what scholars were studying for all this time in a much simpler way, and see it as a straightforward combination of a couple of simple ideas. Since we're not historians, the point of looking at syllogisms isn't to learn about Aristotle and the history of logic. Rather, it's to learn those simple ideas and how they fit together, as a starting point for studying the rest of symbolic logic, and to demonstrate the power of a good notation.

Relationships Between Predicates

Recall that a predicate is a `Bool`-valued function p that says whether any given thing x in the universe has the given property ($p\ x$ is `True`) or not ($p\ x$ is `False`). An example in Chap. 6 was `isGrey`, which is `True` for all grey things.

Each predicate p determines a subset of the universe of discourse containing all of the things in the universe that satisfy the predicate, $\{x \in Universe \mid p\ x\}$. (To avoid boring clutter we'll write this as $\{x \mid p\ x\}$, leaving the requirement that $x \in Universe$ implicit.)

We draw these sets in a Venn diagram as circles, labelled with the name of the predicate, to show relationships between predicates. The different regions in the diagram represent subsets where different combinations of predicates are true. In a Venn diagram with two circles, we can depict four combinations: a and b are both true; a is true but b is false; a is false but b is true; a and b are both false.

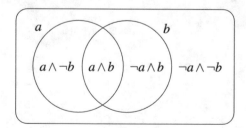

(The region that's outside both of the circles represents the subset of the universe in which a and b are both false.) With three circles we get eight combinations:

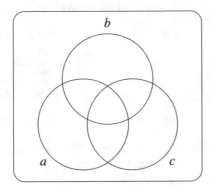

An **Euler diagram** is a similar way of drawing relationships between predicates, where regions that are empty are omitted. For example:

Euler is pronounced "oiler". Leonhard Euler (1707–1783) was the most prolific mathematician in history. See ▶ https://en.wikipedia. org/wiki/Leonhard_Euler.

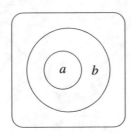

This diagram says that everything that satisfies a also satisfies b. In other words, every a is b.

We can represent the same situation with a Venn diagram, using the convention that a region that's coloured grey is empty. For example:

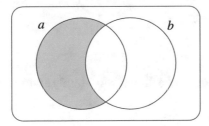

This diagram says that there's nothing in the universe that satisfies a that doesn't also satisfy b, since the region where a is true and b is false is empty.

The same thing can be written in symbols using sequent notation (Chap. 6) as $a \vDash b$: everything in the universe of discourse that satisfies the antecedent a will satisfy the succedent b. Good diagrams are intuitively appealing, but symbolic representations are powerful because they can easily be manipulated using standard techniques like substitution for variables.

A Deductive Argument

Let's look at a simple deductive argument. Suppose that every a is b, and every b is c. Can we then conclude that every a is c? Here's a concrete example. Suppose that every professor is a genius, and that every genius is arrogant. Can we conclude that every professor is arrogant?

We can represent the assumptions as Venn diagrams like this:

Laying those two diagrams on top of each other, and then focusing on the part with just a and c, we can see that every a must be c because the $a \land \lnot c$ region of the diagram is coloured grey:

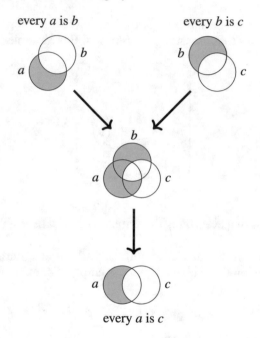

The same argument can be drawn very simply as an Euler diagram like this:

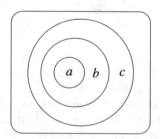

This diagram can be translated into symbols as follows:

$$\{x \mid a\ x\} \subseteq \{x \mid b\ x\} \subseteq \{x \mid c\ x\} \text{ implies } \{x \mid a\ x\} \subseteq \{x \mid c\ x\}$$

which reveals the argument as following directly from transitivity of the subset relation.

It can also be written symbolically using sequents as a **rule**, like so:

$$\frac{a \vDash b \qquad b \vDash c}{a \vDash c} \text{ Barbara}$$

In any **sound** rule, if all of the statements above the line (the **premises**) are true, for any predicates a, b, and c and any universe, then the statement below the line (the **conclusion**) is true as well. ("Barbara" on the right is just the name of the rule.) This rule is indeed sound, as the explanations via Venn and Euler diagrams have shown. It can therefore be used to validate arguments in any universe—for example, one in which all professors are geniuses and all geniuses are arrogant—and for any choice of predicates. In that sense, it's an abstract pattern of reasoning that's independent of the subject matter.

Since rules that are unsound aren't useful, writing a rule is generally intended as an assertion that it's sound. As you'll see, more complicated arguments can be constructed by putting together such rules.

Medieval monks studying logic called this rule "Barbara", not in commemoration of Saint Barbara's logical skills but as part of a mnemonic scheme for keeping track of syllogisms.

Negated Predicates

The fact that negation of a predicate corresponds to the complement of the set that it determines is built into Venn diagram and Euler diagram notation. We can prove the same fact in symbols, once we define the negation of a predicate a as $(\neg a)\, x = \neg(a\, x)$:

$$\{x \mid (\neg a)\, x\} = \{x \mid \neg(a\, x)\} = \overline{\{x \mid a\, x\}}$$

Let's now consider the Barbara rule with a negated predicate $\neg c$ in place of c:

$$\frac{a \vDash b \qquad b \vDash \neg c}{a \vDash \neg c}$$

An example in English is the following. Suppose that every snake is a reptile, and that every reptile has no fur (or, equivalently, that no reptile has fur). We can then conclude that every snake has no fur (or, equivalently, that no snake has fur).

This rule is sound because the Barbara rule was sound for any choice of c. The negation of a predicate is also a predicate, and is therefore a perfectly cromulent choice. You can also see that it's sound from the corresponding Euler diagram:

The monks called this rule "Celarent". (You don't need to remember these rule names, except for Barbara, which is the simplest one.)

There is often more than one way to say things. "Every x is not y" has the same meaning as "No x is y".

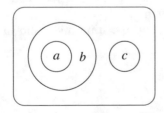

or using Venn diagrams:

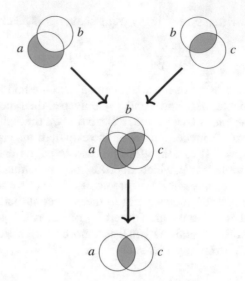

but since it's just an instance of another sound rule, no separate explanation is required.

This is a case where the notation makes things easier. For Aristotle and medieval scholars, these were two completely different syllogisms, because they used different words to write them down—every *b* is *c* versus no *b* is *c*—but for us, they're the application of the same rule to different predicates. It's obvious that substituting $\neg c$ for *c* in a sound rule gives another sound rule, but the explanations of soundness via Euler and Venn diagrams both look quite different from the corresponding explanations for the original rule.

Contraposition and Double Negation

Another important rule of logic is the **contraposition** rule:

$$\frac{a \vDash b}{\neg b \vDash \neg a} \text{ contraposition}$$

An English example is the following. Suppose that every human is a mammal. We can then conclude that every non-mammal is a non-human (or, equivalently, that no non-mammal is a human). The sequent $\neg b \vDash \neg a$ is called the **contra-positive** of the sequent $a \vDash b$.

The soundness of the contraposition rule is shown using Venn diagrams as follows:

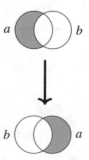

The second diagram is the mirror image of the first diagram. It says that every-thing that's not *b*—that is, everything in the universe that isn't in the circle for *b*—is not *a*, since the part that's in the circle for *a* is empty.

Now, we've already seen that negation of a predicate corresponds to the complement of the set that it determines. Since $\bar{\bar{A}} = A$, it follows that double negation of a predicate gives the same predicate. This is the **double negation** law:

$$\neg\neg a = a$$

Using the double negation law, we can show that turning the contraposition rule upside-down also gives a sound rule:

$$\frac{\dfrac{\neg b \vDash \neg a}{\neg\neg a \vDash \neg\neg b}}{a \vDash b} \begin{array}{l}\text{contraposition}\\[1ex]\text{double negation, twice}\end{array}$$

When a rule is sound in both directions—that is, the conclusion follows from the premise, and the premise also follows from the conclusion—the rule is called an **equivalence** and is written using a double bar, like so:

$$\frac{a \vDash b}{\neg b \vDash \neg a} \text{ contraposition}$$

Intuitionistic logic is a weaker system of logic than **classical logic**, which is the system we are studying, in which neither the double negation law nor the method of proof by contradiction is accepted, see ▶ https://en.wikipedia.org/wiki/Intuitionistic_logic.

More Rules

Other sound rules can be obtained from the contraposition rule by substituting negated predicates and using the double negation law. For example, replacing the predicate *b* in the contraposition rule by its negation $\neg b$ and then applying the double negation law gives:

$$\frac{\dfrac{a \vDash \neg b}{\neg\neg b \vDash \neg a}}{b \vDash \neg a} \begin{array}{l}\text{contraposition}\\[1ex]\text{double negation}\end{array} .$$

In English, this says that if no *a* is *b*, then no *b* is *a*. (Concrete example: Suppose that no cats are green. Then nothing green is a cat.) And this rule is also sound in both directions:

$$\frac{a \vDash \neg b}{b \vDash \neg a}$$

Here's another example, obtained from the Celarent rule:

$$\frac{a \vDash b \quad \dfrac{\dfrac{c \vDash \neg b}{\neg\neg b \vDash \neg c}}{b \vDash \neg c}}{a \vDash \neg c} \begin{array}{l}\text{contraposition}\\[1ex]\text{double negation}\\[1ex]\text{Celarent}\end{array}$$

and a concrete example is: Suppose that all humans are mammals, and that no reptiles are mammals. Then no humans are reptiles.

A valid combination of sound rules, which can be as complicated as you want, is called a **proof**. It shows that the conclusion at the "bottom" of the proof follows from the set of premises along the "top" of the proof. This can be expressed as a new rule, removing all of the detail of the proof between the premises and the conclusion. The proof above yields the rule

$$\frac{a \vDash b \qquad c \vDash \neg b}{a \vDash \neg c}$$

The monks called this rule "Cesare".

Any combination of sound rules will yield a rule that's guaranteed to be sound, so no explanation of soundness via Venn diagrams or Euler diagrams is required. If all of the rules involved in the proof are equivalences, then the result is an equivalence.

Other rules can be derived by combining the ones above in other ways.

Exercises

1. Consider the following Euler diagram, where every region is non-empty:

    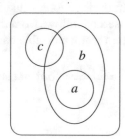

    Which of the following sequents is valid: $a \vDash b$, $b \vDash c$, $b \vDash a$, $a \vDash c$, $c \vDash a$, $a \vDash \neg b$, $b \vDash \neg c$, $b \vDash \neg a$, $a \vDash \neg c$, $c \vDash \neg a$.

2. Use the contraposition rule and the double negation law to show that the following rule is sound:

    $$\frac{\neg a \vDash b}{\neg b \vDash a}$$

    Use Venn diagrams to show that this rule is sound, and give an English example of its application.

3. Derive the following rules from the Cesare rule using contraposition and double negation, and give English examples of their application.

    $$\frac{a \vDash b \quad c \vDash \neg b}{c \vDash \neg a} \text{ Camestres}$$

    $$\frac{a \vDash b \quad b \vDash \neg c}{c \vDash \neg a} \text{ Calemes}$$

4. Give counterexamples to show that the following three rules are unsound. (**Hint:** There are counterexamples for all of these in the universe of discourse of Chap. 6.)

    $$\frac{a \vDash b}{b \vDash a}$$

    $$\frac{a \vDash b \qquad a \vDash c}{a \vDash \neg b}$$

    $$\frac{a \vDash c \qquad b \vDash c}{c \vDash b}$$

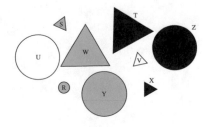

5. (This exercise and the next one use the Haskell implementation of the universe of discourse of Chap. 6, which defines the list `things :: Thing` and predicates like `isGrey :: Predicate Thing`.)

Define an infix function

```
(|=) :: Predicate Thing -> Predicate Thing -> Bool
```

for testing whether a sequent involving one antecedent and one succedent is valid or invalid. Use it to check that isWhite ⊭ isBlack.
Define another infix function

```
(||=) :: [Predicate Thing] -> Predicate Thing -> Bool
```

for testing whether a sequent involving a list of antecedents and one succedent is valid or invalid. Use it to check that isWhite, isTriangle ⊨ isSmall and that isSmall, isTriangle ⊭ isWhite.

6. Recall that the type **Predicate** u is defined as u -> Bool. The following function negates a predicate:

```
neg :: Predicate u -> Predicate u
(neg a) x = not (a x)
```

For example, (neg isWhite) R = not (isWhite R) = True. And isWhite |= neg isBlack produces True.
Define functions

```
(|:|) :: Predicate u -> Predicate u -> Predicate u
(&:&) :: Predicate u -> Predicate u -> Predicate u
```

that compute the disjunction and conjunction of two predicates. Which of the following produce **True**?

- isBig &:& isWhite |= isDisc
- isBig &:& isDisc |= isWhite
- isSmall &:& neg isGrey |= neg isDisc
- isBig |:| isGrey |= neg isTriangle
- neg (isTriangle |:| isGrey) |= isDisc
- neg isTriangle &:& neg isGrey |= isDisc

More Patterns of Reasoning

Contents

Denying the Conclusion

So far you've seen how to use rules to combine valid sequents into simple deductive arguments about statements of the form "every *a* is *b*". Sometimes the antecedents and/or consequent of a sequent involved negation, to allow statements like "every *a* is not *b*". You're now going to learn how to build arguments that also take account of *invalid* sequents, which relies on a different kind of negation. This provides new ways of reasoning, as well as allowing arguments about *existence* of things that satisfy predicates ("some *a* is *b*").

Let's start by looking at a simple example of commonsense reasoning. We'll start by making an underlying assumption that Fiona is old enough to buy alcohol. Here's a sound deduction, according to current laws in Scotland relating to buying alcohol in a supermarket, expressed in the form of a rule:

The relevant law is the Licensing (Scotland) Act 2005, see ▶ http://www.legislation.gov.uk/asp/2005/16/contents.

$$\frac{\text{Fiona is in Scotland} \qquad \text{The time is between 10am and 10pm}}{\text{Fiona can legally buy alcohol}}$$

It's legal to buy alcohol at any time of the day in some countries, but not in Scotland. And there are other countries where the sale of alcohol is illegal at any time.

Now, suppose we know that Fiona is in Scotland and that Fiona can't legally buy alcohol. What can we conclude from that? Well, then it must not be between 10am and 10pm: if it were between those times, the rule would apply and then the conclusion would contradict the fact that Fiona can't legally buy alcohol.

On the other hand, suppose that the time is between 10am and 10pm and that nevertheless Fiona can't legally buy alcohol. From that, it follows by the same reasoning that Fiona must not be in Scotland.

This is also known as **proof by contradiction**.

So from a deduction that we know to be sound, we can get two more sound deductions, by **denying the conclusion**: one using the first premise and one using the second premise. In this case, we get the two deductions

$$\frac{\text{Fiona is in Scotland} \qquad \text{Fiona can't legally buy alcohol}}{\text{The time isn't between 10am and 10pm}}$$

$$\frac{\text{The time is between 10am and 10pm} \qquad \text{Fiona can't legally buy alcohol}}{\text{Fiona isn't in Scotland}}$$

Now, let's look at the same pattern but using rules with sequents written using predicates.

We already know that the following rule is sound:

$$\frac{a \vDash b \qquad b \vDash c}{a \vDash c} \text{ Barbara}$$

Suppose that we know that $a \vDash b$ ("every *a* is *b*") and $a \nvDash c$ ("some *a* is not *c*"). Since the rule is sound, one of the premises is valid and the conclusion is invalid, we know that the other premise must be invalid, so $b \nvDash c$ ("some *b* is not *c*"). Here it is as a rule:

$$\frac{a \vDash b \qquad a \nvDash c}{b \nvDash c}$$

The same thing works applied to the other premise of the rule: if $b \vDash c$ and $a \nvDash c$, then $a \nvDash b$.

$$\frac{b \vDash c \qquad a \nvDash c}{a \nvDash b}$$

What we're doing here is related to the way that the contraposition rule

$$\frac{a \vDash b}{\neg b \vDash \neg a} \text{ contraposition}$$

was used earlier, but at a different level: instead of negating and switching antecedents and succedents of *sequents*, we're negating (i.e. asserting as invalid) and switching assumptions and conclusions of *rules*.

Venn Diagrams with Inhabited Regions

We've already seen how to draw a Venn diagram to represent $a \vDash b$ (every a is b):

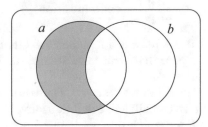

where regions are coloured grey to say that they are empty.

We'll put an x in a region to indicate that it contains at least one thing (that is, it's **inhabited**). For example:

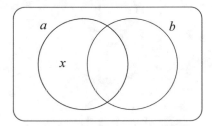

This diagram says that there's at least one thing in the universe that satisfies a but not b. That is, $a \nvDash b$ (some a is not b). A region is inhabited if and only if it's not empty, which explains why this diagram says that the sequent represented by the previous diagram is invalid: it's not the case that every a is b.

Using the same ideas, we can represent $a \vDash \neg b$ (no a is b):

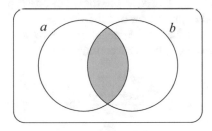

Since the diagram is symmetric, it also represents $b \vDash \neg a$ (no b is a).

and $a \nvDash \neg b$ (some a is b):

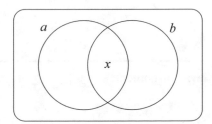

And this symmetric diagram also represents $b \nvDash \neg a$ (some b is a).

These four sequents are Aristotle's *categorical propositions*. They are traditionally arranged in the following *square of opposition* with contradictory sequents along the diagonals:

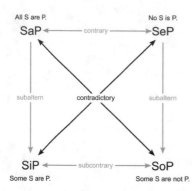

See ▶ https://en.wikipedia.org/wiki/ Square_of_opposition.

Again, comparing with the previous diagram, this one says that the sequent represented by that diagram is invalid: it's not the case that no *a* is *b*.

Contraposition Again

The **contraposition** rule that we saw earlier also holds when the sequent is invalid:

$$\frac{a \nvDash b}{\neg b \nvDash \neg a} \text{ contraposition}$$

In English, this says that if some *a* is not *b*, then something that's not *b* is *a*. A concrete example is: Suppose that some mice are not small. Then some things that are not small are mice.

Let's think about whether this makes sense. The assumption says that there's at least one mouse that isn't small: let's call it Squeaky. Now, since Squeaky is a mouse, and isn't small, the conclusion does indeed hold: something that isn't small—namely, Squeaky—is a mouse.

The soundness of this version of the contraposition rule can be shown using Venn diagrams:

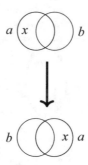

The first diagram represents $a \nvDash b$: some *a* is not *b*. As in the proof of soundness of the original version of contraposition, the second diagram is just the mirror image of the first diagram. It says that the subset of the universe that's not in *b* but is in *a* is not empty. If it were empty, then we would have $\neg b \vDash \neg a$, so instead we have $\neg b \nvDash \neg a$. The *x* in the diagram corresponds to Squeaky the mouse in our informal explanation of the rule.

As before, we can use the double negation law to show that contraposition for invalid sequents works in both directions:

$$\frac{\dfrac{\neg b \nvDash \neg a}{\neg\neg a \nvDash \neg\neg b}}{a \nvDash b} \begin{array}{l}\text{contraposition}\\ \text{double negation, twice}\end{array}$$

and so we can write it using a double bar:

$$\frac{a \nvDash b}{\neg b \nvDash \neg a} \text{ contraposition}$$

Checking Syllogisms

Let's consider the following argument:

- No *b* is *c*.
- Some *c* is *a*.
- Therefore, some *a* is not *b*.

Checking Syllogisms

An example in English is:

- No dogs are insects.
- Some insects are yucky.
- Therefore, there is something yucky that is not a dog.

Is this a valid argument?

Let's start by writing the argument in the form of a rule, using sequents:

$$\frac{b \vDash \neg c \qquad c \nvDash \neg a}{a \nvDash b}$$

We can try to show that this rule is sound, meaning that the argument is valid, using Venn diagrams. The procedure is similar to what we did before, but the presence of invalid sequents, corresponding to Venn diagrams with inhabited regions, makes things a little more complicated.

As before, we can represent the assumptions as Venn diagrams:

and then laying those two diagrams on top of each other gives the following:

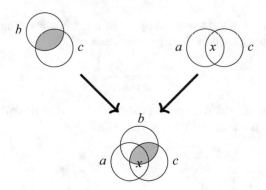

The inhabitant can't be in the part that we know is empty, so it must be in the other part.

Finally, focusing on the part of this diagram with just a and b, we can see that the conclusion holds, because the $a \wedge \neg b$ region of the diagram is inhabited:

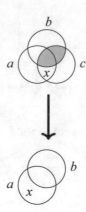

So the rule is sound.

Finding Counterexamples

Let's consider another argument:

In fact, no plants are fungi, see
► https://en.wikipedia.org/wiki/
Fungus and ► https://en.wikipedia.
org/wiki/Plant. But that doesn't
matter: we're studying the *form* of
logical arguments, which doesn't
depend on their subject matter or the
accuracy of their assumptions.

- All plants are fungi.
- Some flowers are not plants.
- Therefore, some flowers are not fungi.

If we let a stand for plants, b for fungi, and c for flowers, we can express the pattern of reasoning used in this argument as a rule:

$$\frac{a \vDash b \qquad c \nvDash a}{c \nvDash b}$$

And we can then use Venn diagrams to try to show that the rule is sound.

We represent the assumptions as Venn diagrams:

and then laying those two diagrams on top of each other gives the following:

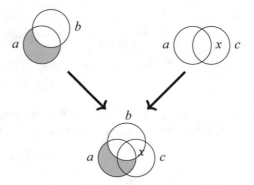

The inhabitant could be in either of the two sub-regions, or both. In order for the conclusion to follow, the inhabitant needs to be in the $c \wedge \neg b$ region of the diagram, as in the second alternative below:

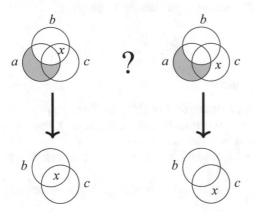

There's no guarantee that this is the case, so the rule isn't sound!

$$\frac{a \vDash b \qquad c \nvDash a}{c \nvDash b}$$

We can derive a counterexample from the failed attempt to show that the rule is sound. According to the diagram, a counterexample would be any universe containing predicates a, b, and c, in which the $\neg a \wedge b \wedge c$ region is inhabited but the $\neg a \wedge \neg b \wedge c$ region is empty.

One such universe would contain a single thing—let's call it x—such that $a\,x$ is false and $b\,x$ and $c\,x$ are both true, and nothing else. In our English version of the syllogism, that would be a universe containing only one thing that's both a fungus and a flower, and nothing else.

In that universe, $a \vDash b$ (because nothing satisfies a, which fits the diagram for the first premise) and $c \nvDash a$ (because x satisfies c but not a, which fits the diagram for the second premise). However, it's not the case that $c \nvDash b$ (because everything that satisfies c, namely x, also satisfies b, meaning that the part of the diagram for the conclusion that's required to be inhabited is empty).

If we know that a region in a Venn diagram is empty, and we split that region into sub-regions, then they must *all* be empty. On the other hand, if we know that a region is inhabited, and we split that region into sub-regions, all we know is that *at least one* of them is inhabited. And, to *show* that a region in a Venn diagram is empty, we need to show that *all* of its sub-regions are empty. But to show that a region is inhabited, we only need to show that *at least one* of its sub-regions is inhabited.

Symbolic Proofs of Soundness

Pictures involving Venn diagrams with grey regions and xs are a good way to explain why a rule is sound. The same thing can be done by giving names to the regions, like so:

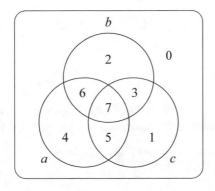

and then giving the soundness proof using the region names. This is just the same proof but the names allow it to be written down more concisely. Going back to the dogs/insects/yucky example on page 75:

$$\frac{b \models \neg c \qquad c \not\models \neg a}{a \not\models b}$$

we can reason as follows:

- $b \models \neg c$ (no b is c) means that both 3 and 7 are empty.
- $c \not\models \neg a$ (some c is a) means that at least one of 7 and 5 is inhabited. Since 7 is empty, 5 must be inhabited.
- Therefore, $a \not\models b$ (some a is not b) since at least one of 4 and 5 is inhabited.

Or, looking at the unsound plants/fungi/flowers example:

$$\frac{a \models b \quad\overset{\text{x}}{}\quad \overset{\text{x}}{c \not\models a}}{c \not\models b}$$

we can reason:

- $a \models b$ (every a is b) means that both 4 and 5 are empty.
- $c \not\models a$ (some c is not a) means that at least one of 1 and 3 must be inhabited.
- In order to show that $c \not\models b$ (some c is not b) we need to show that at least one of 1 and 5 is inhabited. Since 5 is empty, we need to show that 1 is inhabited. But we only know that *at least one of 1 and 3* is inhabited. So 1 could be empty with 3 being inhabited.

Deriving All of the Sound Syllogisms

Sound rules can be obtained from other sound rules using a few simple algebraic manipulations that we have already encountered:
- Denying the conclusion: for example, since

$$\frac{b \models \neg c \qquad c \not\models \neg a}{a \not\models b}$$

is sound, both of the following rules are sound:

$$\frac{c \not\models \neg a \qquad a \models b}{b \not\models \neg c} \qquad\qquad \frac{b \models \neg c \qquad a \models b}{c \models \neg a}$$

The first is obtained by denying the conclusion with the first premise and the second is obtained by denying the conclusion with the second premise.
- Substituting for predicates: for example, we can substitute $\neg b$ for b in

$$\frac{c \not\models \neg a \qquad a \models b}{b \not\models \neg c}$$

to get

$$\frac{c \not\models \neg a \qquad a \models \neg b}{\neg b \not\models \neg c}$$

- Applying contraposition and the double negation law: for example, we can apply contraposition to the conclusion of the last rule to get

$$\frac{c \not\models \neg a \qquad a \models \neg b}{c \not\models b}$$

Exercises

and then apply contraposition to the first premise to get

$$\frac{\neg\neg a \nvDash \neg c \qquad a \vDash \neg b}{c \nvDash b}$$

which can be simplified using the double negation law to get

$$\frac{a \nvDash \neg c \qquad a \vDash \neg b}{c \nvDash b}$$

Since all of these preserve soundness of rules, there's no need to check soundness separately using Venn diagrams.

All of the following rules can be derived, starting from the rule on the top left (Barbara), by applying one or more of these manipulations:

$$\frac{a \vDash b \qquad b \vDash c}{a \vDash c} \qquad \frac{a \vDash b \qquad a \nvDash c}{b \nvDash c} \qquad \frac{b \vDash c \qquad a \nvDash c}{a \nvDash b}$$

$$\frac{a \vDash b \qquad b \vDash \neg c}{a \vDash \neg c} \qquad \frac{a \vDash b \qquad a \nvDash \neg c}{b \nvDash \neg c} \qquad \frac{b \vDash \neg c \qquad a \nvDash \neg c}{a \nvDash b}$$

$$\frac{a \vDash b \qquad c \vDash \neg b}{a \vDash \neg c} \qquad \frac{a \vDash b \qquad a \nvDash \neg c}{c \nvDash \neg b} \qquad \frac{c \vDash \neg b \qquad a \nvDash \neg c}{a \nvDash b}$$

$$\frac{a \vDash b \qquad c \vDash \neg b}{c \vDash \neg a} \qquad \frac{a \vDash \neg b \qquad a \nvDash \neg c}{c \nvDash b} \qquad \frac{b \vDash \neg c \qquad c \nvDash \neg a}{a \nvDash b}$$

$$\frac{a \vDash b \qquad b \vDash \neg c}{c \vDash \neg a} \qquad \frac{a \vDash b \qquad c \nvDash \neg a}{c \nvDash \neg b} \qquad \frac{c \vDash b \qquad a \nvDash \neg c}{b \nvDash \neg a}$$

These are all of the sound syllogisms. Aristotle had nine others that are only sound provided $a \vDash b$ also requires that there's at least one thing that satisfies a, see ▶ https://en.wikipedia.org/wiki/Syllogism.

For example, the second rule in the first row can be obtained from Barbara by denying the conclusion with the second premise, and the second rule in the first column can be obtained from Barbara by substituting $\neg c$ for c. See Exercise 6 for the others.

Exercises

1. Find unambiguous sentences in a language that you're fluent in, other than English, that accurately capture the meaning of: $a \vDash b$ (every a is b); $a \vDash \neg b$ (no a is b); $a \nvDash \neg b$ (some a is b); and $a \nvDash b$ (some a is not b).

2. Use Venn diagrams to show that the following rules are sound:

$$\frac{a \vDash b \qquad a \nvDash c}{b \nvDash c} \qquad \frac{b \vDash c \qquad a \nvDash c}{a \nvDash b}$$

3. Use the contraposition rule for invalid sequents and the double negation law to show that the following rules are sound:

$$\frac{a \nvDash \neg b}{b \nvDash \neg a} \qquad \frac{\neg a \nvDash b}{\neg b \nvDash a}$$

Use Venn diagrams to show that these rules are sound, and give English examples of their application.

4. Consider the following arguments:

- Some cats have no tails. All cats are mammals. Therefore, some mammals have no tails.
- All informative things are useful. Some websites are not useful. Therefore, some websites are not informative.

- All rabbits have fur. Some pets are rabbits. Therefore, some pets have fur.
- No homework is fun. Some reading is homework. Therefore, some reading is not fun.

Formulate them as syllogisms, then use Venn diagrams to show that they are all sound, meaning that they are valid arguments. For the soundness proofs, use either pictures involving Venn diagrams or arguments referring to the region names given above.

5. Consider the following syllogism:

$$\frac{a \nvDash \neg b \qquad b \nvDash \neg c}{a \nvDash \neg c}$$

Use a Venn diagram (either a picture or an argument referring to region names) to show that it's unsound, and give a counterexample.

6. Starting from the rule on the top left (Barbara), show how each of the following rules can be derived from the one above it and the one to its left.

$$\frac{a \vDash b \qquad b \vDash c}{a \vDash c} \qquad \frac{a \vDash b \qquad a \nvDash c}{b \nvDash c} \qquad \frac{b \vDash c \qquad a \nvDash c}{a \nvDash b}$$

$$\frac{a \vDash b \qquad b \vDash \neg c}{a \vDash \neg c} \qquad \frac{a \vDash b \qquad a \nvDash \neg c}{b \nvDash \neg c} \qquad \frac{b \vDash \neg c \qquad a \nvDash \neg c}{a \nvDash b}$$

$$\frac{a \vDash b \qquad c \vDash \neg b}{a \vDash \neg c} \qquad \frac{a \vDash b \qquad a \nvDash \neg c}{c \nvDash \neg b} \qquad \frac{c \vDash \neg b \qquad a \nvDash \neg c}{a \nvDash b}$$

$$\frac{a \vDash b \qquad c \vDash \neg b}{c \vDash \neg a} \qquad \frac{a \vDash \neg b \qquad a \nvDash \neg c}{c \nvDash b} \qquad \frac{b \vDash \neg c \qquad c \nvDash \neg a}{a \nvDash b}$$

$$\frac{a \vDash b \qquad b \vDash \neg c}{c \vDash \neg a} \qquad \frac{a \vDash b \qquad c \nvDash \neg a}{c \nvDash \neg b} \qquad \frac{c \vDash b \qquad a \nvDash \neg c}{b \nvDash \neg a}$$

7. Show how each of the syllogisms in Exercise 4 can be derived from the Barbara rule using denying the conclusion, substituting for predicates, contraposition, and the double negation law.

Lists and Recursion

Contents

© The Author(s), under exclusive license to Springer Nature Switzerland AG 2021
D. Sannella et al., *Introduction to Computation*, Undergraduate Topics
in Computer Science, https://doi.org/10.1007/978-3-030-76908-6_10

10

Well, one way provided you—like Haskell—regard 4:(6:(8:[])) and 4:6:8:[] as the same expression.

Remember that patterns can't contain repeated variables.

We've also seen tuple patterns, where there is just one constructor, namely the notation (.., ..).

Examples of non-well-founded self-reference are "Brexit means Brexit" (Theresa May) and "A rose is a rose is a rose" (Gertrude Stein).

Infinite lists are defined differently, requiring methods that go beyond the scope of this book. See ▶ https://en.wikipedia.org/wiki/Corecursion if you're curious.

This computes the same function as squares in Chap. 5. We call it squaresRec to avoid confusing the two.

Building Lists

Recall that the list notation [4,6,8] is just shorthand for the expression 4:(6:(8:[])). So, every list can be written using : and []. In fact, every list can be written using : and [] *in just one way*. That fact is what makes pattern matching work on lists.

Given a list *l* and a pattern *p* built using variables, [], : and literals, there is at most one way to match *p* against *l*. That match gives values to the variables in *p*. It might not match, if *p* requires a list that's longer than *l*, or if *p* contains a literal that doesn't match the value in the corresponding position in *l*.

Because of this special feature of : and [], they are called **constructors**. The reason why : and [] are special in this sense comes from the following way of defining the type [*t*] of lists with elements of type *t*:

Definition. A **list** of type [*t*] is either

1. *empty*, written [], or
2. *constructed*, written *x*:*xs*, with *head x* (an element of type *t*) and *tail xs* (a list of type [*t*]).

But wait a minute: this definition is self-referential! The second case of the explanation of what a list is, refers to a list!

The kind of self-reference used in this definition of lists is okay. It's called **recursion**, and the definition of lists is a **recursive definition**. It's meaningful because the self-reference is **well-founded**: it defines a complicated list (*x*:*xs*) in terms of a simpler list (*xs*), and ultimately in terms of the simplest list of all, [].

Here's how it explains that 4:(6:(8:[])) is a list of type [Int]:

- [] is a list of type [Int], by case (1);
- and so 8:[] is a list of type [Int], with head 8 and tail [], by case (2);
- and so 6:(8:[]) is a list of type [Int], with head 6 and tail 8:[], by case (2);
- and so 4:(6:(8:[])) is a list of type [Int], with head 4 and tail 6:(8:[]), by case (2).

The same process works for any finite list.

Recursive Function Definitions

We can also write recursive definitions of functions on lists. Here's a simple example of a function that squares each of the elements in a list of integers:

```
squaresRec :: [Int] -> [Int]
squaresRec []     = []
squaresRec (x:xs) = x*x : squaresRec xs
```

Again, the definition of squaresRec is self-referential: the second equation defines squaresRec (x:xs) in terms of squaresRec xs. And again, the self-reference is well-founded, because we are defining squaresRec applied to a complicated list (x:xs) in terms of squaresRec applied to a simpler list (xs), and ultimately in terms of squaresRec applied to the simplest list of all, []. The first equation, which defines squaresRec [], is called the **base case** of the recursion.

If you're not used to recursion, then this definition may look pretty confusing at first. The best way to understand it is by looking at an example of how it can be used to compute the result of applying squaresRec to a specific list, say

[1,2,3], step by step. At each step, we'll expand the underlined part of the expression, usually using the equations in the definition of `squaresRec`. Here goes!

```
squaresRec [1,2,3]
        Expanding list notation
= squaresRec (1 : (2 : (3 : [])))
        Applying second equation, with x = 1 and xs = 2 : (3 : [])
= 1*1 : (squaresRec (2 : (3 : [])))
        Applying second equation, with x = 2 and xs = 3 : []
= 1*1 : (2*2 : (squaresRec (3 : [])))
        Applying second equation, with x = 3 and xs = []
= 1*1 : (2*2 : (3*3 : (squaresRec [])))
        Applying first equation
= 1*1 : (2*2 : (3*3 : []))
        Doing the multiplications
= 1 : (4 : (9 : []))
        Using list notation
= [1,4,9]
```

We could have done the multiplications earlier.

It takes a little while to get used to recursion, but once you do, it is simple and elegant. But if you're not yet used to recursion, and pattern matching still seems a little mysterious, then it might help you to study the following version of `squaresRec` before going any further. It uses recursion too, but pattern matching is replaced by a conditional expression for case analysis and `head`/`tail` for extracting the components of a list.

If you are familiar with other programming languages, what is done in them using iteration—such as `while` loops and `for` loops—is done in Haskell with recursion.

```
squaresCond :: [Int] -> [Int]
squaresCond ws =
  if null ws then []
  else x*x : squaresCond xs
    where x  = head ws
          xs = tail ws
```

Every function definition using pattern matching and/or guards can be rewritten into a definition in this form. But Haskell programmers prefer pattern matching.

The sequence of computation steps above for computing the result of applying `squaresRec` to [1,2,3] works for this version too, where the first equation refers to the `then` case of the conditional, and the second equation refers to the `else` case.

Going back to the definition of `squaresRec`, it fits perfectly with the recursive definition of lists:

- There are two cases, one for the empty list [] and one for the non-empty list `x:xs`;
- The body of the second case refers to the function being defined, but the body of the first case doesn't;
- The self-reference is well-founded; and
- The computation reduces eventually to the base case `squaresRec []` in the same way as the formation of any finite list reduces eventually to the empty list [].

Later, you'll see recursive definitions of functions on lists that don't fit quite so perfectly with the definition of lists. For example, some recursive function definitions have a separate base case for singleton lists, with the definition of `f (x:x':xs)` referring to `f xs` or even to both `f (x':xs)` and `f xs`. But we'll stick to simple examples like `squaresRec` for now.

Before continuing: we now have three functions that we claim compute the same thing, namely `squares` (from Chap. 5), `squaresRec`, and `squaresCond`. Let's test that they are in fact the same:

```
squares_prop :: [Int] -> Bool
squares_prop ns =
    squares ns == squaresRec ns && squares ns == squaresCond ns

> quickCheck squares_prop
+++ OK, passed 100 tests.
```

More Recursive Function Definitions

Here's another example of a recursive function definition. This one selects all of the odd numbers in a list.

This computes the same function as odds in Chap. 5.

```
oddsRec :: [Int] -> [Int]
oddsRec []              = []
oddsRec (x:xs) | odd x     = x : oddsRec xs
               | otherwise = oddsRec xs
```

This definition uses guards as well as recursion. The recursive aspect of `oddsRec` is exactly the same as it was in `squaresRec`: the first equation is the base case, the second and third equations are self-referential, and the self-reference is well-founded.

Here's a step-by-step computation of the result of applying `oddsRec` to [1,2,3].

```
oddsRec [1,2,3]
        Expanding list notation
= oddsRec (1 : (2 : (3 : [])))
        Applying second equation, with x = 1 and xs = 2 : (3 : [])
        since odd x == True
= 1 : (oddsRec (2 : (3 : [])))
        Applying third equation, with x = 2 and xs = 3 : []
        since odd x == False
= 1 : (oddsRec (3 : []))
        Applying second equation, with x = 3 and xs = []
        since odd x == True
= 1 : (3 : (oddsRec []))
        Applying first equation
= 1 : (3 : [])
        Using list notation
= [1,3]
```

Here are four more examples, defining Haskell's Prelude functions for computing the sum and product of a list of integers and conjunction/disjunction of a list of Booleans. These functions are examples of **accumulators**: they collect together information from a list into a single result.

It's probably clear why sum [] = 0 is the right base case, but why is product [] = 1? And why is and [] = **True** but or [] = **False**? To understand this, write out a step-by-step computation of product [2], which will involve the result of product [], and similarly for and and or.

```
sum :: [Int] -> Int
sum []     = 0
sum (x:xs) = x + sum xs

product :: [Int] -> Int
product []     = 1
product (x:xs) = x * product xs

and :: [Bool] -> Bool
and []     = True
and (x:xs) = x && and xs
```

```
or :: [Bool] -> Bool
or []       = False
or (b:bs) = b || or bs
```

Finally, here is a function that combines the computations in squaresRec, oddsRec, and sum to compute the sum of the squares of the odd numbers in a list:

```
sumSqOddsRec :: [Int] -> Int
sumSqOddsRec []                 = 0
sumSqOddsRec (x:xs) | odd x     = x*x + sumSqOddsRec xs
                    | otherwise = sumSqOddsRec xs
```

This computes the same function as sumSqOdds in Chap. 5.

It would be a good idea at this point to use QuickCheck to check that oddsRec and sumSqOddsRec produce the same results as odds and sumSqOdds in Chap. 5, in the same way as we checked squaresRec and squaresCond against squares above. (Do it!)

Sorting a List

Now you're ready for some more interesting examples of recursive function definitions on lists. We'll look at two algorithms for **sorting** a list of integers into ascending order.

The first algorithm is called **insertion sort**. The idea is this: to sort a list of integers, we **insert** each of the elements in turn into another list, that is initially empty, taking care when doing the insertion to keep the list in ascending order. Once we've inserted all of the elements, we're done: the result list contains everything that was in the original list, and it's in ascending order.

Of course, sorting into *descending* order is exactly the same—just replace <= by >= everywhere.

We need two functions. Here's the first one, a function for doing the insertion:

```
-- parameter list is in ascending order; same for result list
insert :: Int -> [Int] -> [Int]
insert m []                 = [m]
insert m (n:ns) | m <= n    = m:n:ns
                | otherwise = n : insert m ns
```

When inserting a number into a non-empty list, there are two cases. The first is where the number we are inserting belongs at the beginning of the list, because it's less than or equal to the head of the list (and is therefore less than or equal to all of the elements in the tail). Otherwise, it belongs somewhere later in the list and a recursive call of insert is used to put it in the right place.

The main function insertionSort builds up the result list starting from the empty list, using insert to do the insertion:

```
insertionSort :: [Int] -> [Int]
insertionSort []     = []
insertionSort (n:ns) = insert n (insertionSort ns)
```

That was short and sweet: just five lines of code plus type signatures and a comment! Let's check that it works:

```
> insertionSort [4,8,2,1,7,17,2,3]
[1,2,2,3,4,7,8,17]
```

In the worst case—when the list to be sorted is already in ascending or descending order, meaning that either `less` or `more` is empty—Quicksort is no faster than insertion sort. If the lengths of `less` and `more` are more balanced, which is usually the case, then Quicksort is faster.

See ▶ https://en.wikipedia.org/wiki/ Quicksort for more about Quicksort. This includes an animated visualisation but where the *last* element in the list is used as the "pivot" value, rather than the *first* element as is done here.

10

Here's another algorithm, called **Quicksort**, for doing the same thing. As the name suggests, it's usually faster than insertion sort.

Given a list `m:ns`, quicksort works by splitting `ns` into two sublists: one (call it `less`) containing all of the elements of `ns` that are *less than* `m`; and the other (call it `more`) containing all of the elements of `ns` that are *greater than or equal to* `m`. Then `less` and `more` are sorted, using recursive calls of `quicksort`, and those results are appended, with `[m]` in between, to give the final result. Here's the code:

```
quicksort :: [Int] -> [Int]
quicksort []     = []
quicksort (m:ns) = quicksort less ++ [m] ++ quicksort more
                   where less = [ n | n <- ns, n < m ]
                         more = [ n | n <- ns, n >= m ]
```

Let's look at `quicksort [4,8,2,1,7,17,2,3]` to see how this works:

- `m = 4`, so `less = [2,1,2,3]` and `more = [8,7,17]`
- `quicksort less = [1,2,2,3]` and `quicksort more = [7,8,17]`
- the result is `[1,2,2,3] ++ [4] ++ [7,8,17] = [1,2,2,3,4,7,8,17]`

Now that we have two ways of doing sorting, we can check that they produce the same results:

```
sort_prop :: [Int] -> Bool
sort_prop ns = insertionSort ns == quicksort ns
```

```
> quickCheck sort_prop
+++ OK, passed 100 tests.
```

Now that we know how to sort lists of integers, what about lists of strings?

All of the function definitions above work for lists of strings as well. The only functions required on `Int`—apart from list operations like `:` and `++`— are the order relations `<=` (in `insert`), and `<` and `>=` (in `quicksort`). Those relations are available for `String` as well, where they give the dictionary ordering, and for many other types including `Float` and (maybe surprising) `Bool`:

```
> False < True
True
```

Recall that a polymorphic function can be restricted to types for which equality testing is available by adding an `Eq` requirement to its type:

```
elem :: Eq a => a -> [a] -> Bool
```

An `Ord` requirement includes an `Eq` requirement, see Chap. 24; that makes sense because of **antisymmetry**: $x \mathrel{<=} y$ and $y \mathrel{<=} x$ imply $x \mathrel{==} y$.

In the same way, an `Ord` requirement can be added to a polymorphic type to restrict to types for which the order relations are available, like so:

```
insert :: Ord a => a -> [a] -> [a]
insertionSort :: Ord a => [a] -> [a]
quicksort :: Ord a => [a] -> [a]
```

With that change, we can use `insertionSort` and `quicksort` to sort lists of strings into alphabetical order:

```
> insertionSort ["elephant","zebra","gnu","buffalo","impala"]
["buffalo","elephant","gnu","impala","zebra"]
> quicksort ["hippopotamus","giraffe","hippo","lion","leopard"]
["giraffe","hippo","hippopotamus","leopard","lion"]
```

Recursion Versus List Comprehension

You've seen that, with some functions (squares/squaresRec, odds/oddsRec, sumSqOdds/sumSqOddsRec), we have a choice whether to write the definition using list comprehension or recursion. Comparing the definitions—here are squares and squaresRec again:

```
squares :: [Int] -> [Int]
squares ns = [ n*n | n <- ns ]

squaresRec :: [Int] -> [Int]
squaresRec []     = []
squaresRec (x:xs) = x*x : squaresRec xs
```

—it seems clear that list comprehension is preferable, since it yields definitions that are shorter and simpler.

But some functions can't be written using comprehension. Examples are sum and product: list comprehensions always produce a list, and these functions don't have lists as results.

There are other cases, like sorting, where list comprehension might be possible but wouldn't be natural. One reason is that list comprehension works on single list elements at a time, and sorting involves comparison of different list elements with each other. Our definition of quicksort uses a mixture: list comprehension to compute intermediate results, with recursion for the overall algorithm structure.

Recursion is about breaking up problems into smaller sub-problems that are easier to solve than the original problem, and then using the solutions to the sub-problems to get a solution to the original problem. This general strategy, known as **divide and conquer**, is one of the most fundamental techniques in Informatics. Sometimes, as in most of our examples so far, there's one sub-problem, and it's just one element smaller than the original problem. Sometimes, as in quicksort, there are *two* sub-problems, and—as usually happens in quicksort—they are about half the size of the original problem. Two small sub-problems are better than one big sub-problem, especially since the same decomposition strategy applies again to each of them. As a result, the overall computation is much more efficient, as you'll see in a later chapter.

For more on divide and conquer, see
► https://en.wikipedia.org/wiki/Divide-and-conquer_algorithm.

Since most recursive function definitions on lists use recursion on the tail, one way to approach the problem of writing a recursive definition is to assume that this is the case, write the skeleton for such a recursive definition, and then fill in the blanks. For a function f of type [a] -> a -> [a], for instance:

```
f :: [a] -> a -> [a]
f [] y     = ... y ...
f (x:xs) y = ... x ... f xs ... y ...
```

Now you just need to fill in the . . . s.

Base case: It's usually easy to fill this in.

Recursive case: Write down a couple of concrete examples of the required results for f (x:xs) y and f xs y. What do you need to do to f xs y, using x and y, to get f (x:xs) y?

If this doesn't work, you need to consider variations: two base cases? a case split? two recursive function calls? If all else fails, you may have a case like quicksort which requires a different problem decomposition.

Exercises

1. Write out the step-by-step computation of `sum [1,2,3]`, `product [1,2,3]`, and `sumSqOddsRec [1,2,3]`. Then do the same for `insert 2 [1,3]`, and finally for `insertionSort [2,1,3]`.

2. Write a function `halveEvensRec :: [Int] -> [Int]` that returns half of each even number in a list. For example,

   ```
   halveEvensRec [0,2,1,7,8,56,17,18] == [0,1,4,28,9]
   ```

   Use recursion, not list comprehension.

   Use `QuickCheck` to test that `halveEvensRec` returns the same result as `halveEvens` in Exercise 5.3.

3. Write a function `inRangeRec :: Int -> Int -> [Int] -> [Int]` that returns all numbers in the input list that fall within the range given by the first two parameters (inclusive). For example,

   ```
   inRangeRec 5 10 [1..15] == [5,6,7,8,9,10]
   ```

   Use recursion, not list comprehension.

   Use `QuickCheck` to test that `inRangeRec` returns the same result as `inRange` in Exercise 5.4.

4. Write a function `countPositivesRec` to count the number of positive numbers in a list. For example,

   ```
   countPositivesRec [0,1,-3,-2,8,-1,6] == 4
   ```

   Use recursion, not list comprehension.

   Use `QuickCheck` to test that `countPositivesRec` returns the same result as `countPositives` in Exercise 5.5.

5. Write a function `multDigitsRec :: String -> Int` that returns the product of all the digits in the input string. If there are no digits, your function should return 1. For example,

   ```
   multDigitsRec "The time is 4:25" == 40
   multDigitsRec "No digits here!"  ==  1
   ```

   Use recursion, not list comprehension. You'll need a library function to determine if a character is a digit and one to convert a digit to an integer.

   Use `QuickCheck` to test that `multDigitsRec` returns the same result as `multDigits` in Exercise 5.6.

6. What's wrong with the following version of `insertionSort`?

   ```
   insertionSort :: Ord a => [a] -> [a]
   insertionSort []     = []
   insertionSort (n:ns) = insertionSort (insert n ns)
   ```

7. Write and run tests to check that:

   - The elements in `quicksort ns` are in ascending order.
   - `quicksort ns` contains the same elements as `ns`.

8. **Merge sort** is another sorting algorithm whose efficiency comes from the fact that no rearrangement of elements is required to merge (interleave) two ordered lists to give an ordered list containing all of the elements of both lists. Given a list of integers, it proceeds as follows:

   - Split the list into two sublists: `front`, containing the first half of the elements; and `back`, containing the rest.
   - Sort `front` and `back` and merge the results.

   Implement merge sort. You will need two functions: one to `merge` two lists that are in ascending order; and the main `mergesort` function that does the split, uses `mergesort` recursively to sort the two sublists, and then uses `merge` to combine them into a sorted list. (**Hint:** To do the split, consider using the Prelude functions `take` and `drop`.)

More Fun with Recursion

Contents

© The Author(s), under exclusive license to Springer Nature Switzerland AG 2021
D. Sannella et al., *Introduction to Computation*, Undergraduate Topics
in Computer Science, https://doi.org/10.1007/978-3-030-76908-6_11

Counting

We'll now look at more examples of recursively defined functions which demonstrate some points that didn't arise in earlier examples. To start, recall the notation [1..10].

Underlying this notation is the following Prelude function, where [*m*..*n*] stands for enumFromTo *m n*:

```
enumFromTo :: Int -> Int -> [Int]
enumFromTo m n | m > n  = []
               | m <= n = m : enumFromTo (m+1) n
```

Here, the recursion is on integers rather than lists.

We've learned how important it is that recursion is well-founded: it's okay to define the result of applying a function to a value in terms of its application to a *smaller* value. But here, we're defining enumFromTo m n in terms of enumFromTo (m+1) n! Of course, m+1 is *larger* than m; how can that be right? And the first equation must be the base case, since there is no recursion, but it looks different from all of the previous examples, so what's going on there?

To understand this definition, let's look at the step-by-step computation of enumFromTo 1 3, expanding the underlined part of the expression at each step:

enumFromTo 1 3

        Applying second equation, with m = 1 and n = 3
        since m <= n == **True**

= 1 : enumFromTo 2 3

        Applying second equation, with m = 2 and n = 3
        since m <= n == **True**

= 1 : 2 : enumFromTo 3 3

        Applying second equation, with m = 3 and n = 3
        since m <= n == **True**

= 1 : 2 : 3 : enumFromTo 4 3

        Applying first equation, with m = 4 and n = 3
        since m > n == **True**

= 1 : 2 : 3 : []

        Using list notation

= [1,2,3]

We see from this example that the recursion in enumFromTo is well-founded too, because the *difference between* m *and* n decreases with each recursive function application! The crucial thing is that *something* gets smaller. And the base case fits with this, since it kicks in as soon as the difference becomes negative.

Here's a similar function that multiplies the numbers in a range together rather than making them into a list:

```
prodFromTo :: Int -> Int -> Int
prodFromTo m n | m > n  = 1
               | m <= n = m * prodFromTo (m+1) n
```

n! is the mathematical notation for factorial *n*.

and then we can use prodFromTo to define the **factorial** function:

```
factorial :: Int -> Int
factorial n = prodFromTo 1 n
```

Notice that the definition of prodFromTo is exactly the same as enumFromTo, where : is replaced by * and [] is replaced by 1. So, prodFromTo *m n* yields

$$m * (m+1) * \cdots * n * 1$$

instead of

$$m : (m + 1) : \cdots : n : [\,]$$

And the recursion in `prodFromTo` is well-founded for the same reason as it is in `enumFromTo`.

Infinite Lists and Lazy Evaluation

Remember that Haskell can compute with infinite lists like `[0..]`. Here's a function, analogous to `enumFromTo`, for producing the infinite list of integers starting from `m`:

```
enumFrom :: Int -> [Int]
enumFrom m = m : enumFrom (m+1)
```

The recursion in this definition isn't well-founded, and there's no base case, so evaluation won't terminate. For example:

$\underline{\text{enumFrom 1}}$
    Applying equation, with $m = 1$
$= 1 : \underline{\text{enumFrom 2}}$
    Applying equation, with $m = 2$
$= 1 : 2 : \underline{\text{enumFromTo 3}}$
    Applying equation, with $m = 3$
$= \ldots$

It's nevertheless possible to compute with such a definition. Given the following definitions of the `head` and `tail` functions:

```
head :: [a] -> a
head []      = error "empty list"
head (x : _) = x

tail :: [a] -> [a]
tail []        = error "empty list"
tail (_ : xs) = xs
```

> The Prelude function **error** is used when you want to stop computation and produce an error message instead of a value.

here is how Haskell computes `head (tail (enumFrom 0))`:

$\text{head (tail (}\underline{\text{enumFrom 0}}\text{))}$
    Applying equation for `enumFrom`, with $m = 0$
$= \text{head (}\underline{\text{tail (0 : enumFrom 1)}}\text{)}$
    Applying equation for `tail`, with $xs = \text{enumFrom 1}$
$= \text{head (}\underline{\text{enumFrom 1}}\text{)}$
    Applying equation for `enumFrom`, with $m = 1$
$= \underline{\text{head (1 : enumFrom 2)}}$
    Applying equation for `head`, with $x = 1$
$= 1$

This gives a hint of how lazy evaluation operates. Expressions are held in unevaluated form until their values are needed. In order to get the value of `head (tail (enumFrom 0))`, we need the value of `tail (enumFrom 0)`. In order to get that value, we need to know whether `enumFrom 0` is empty or not, and if it is non-empty then we need to know its tail. After one step of evaluation, we discover that `enumFrom 0 = 0 : enumFrom 1` and that is enough to compute that `tail (enumFrom 0) = enumFrom 1`. And so on.

Here's a more interesting example:

> There is more to lazy evaluation than this, see ▶ https://en.wikipedia.org/wiki/Lazy_evaluation.

```
primes :: [Int]
primes = [ p | p <- [2..], isPrime p ]

upto :: Int -> [Int] -> [Int]
upto bound []                     = []
upto bound (x:xs) | x > bound = []
                  | otherwise = x : upto bound xs
```

and then

```
> upto 30 primes
[2,3,5,7,11,13,17,19,23,29]
```

Because of the application of upto, none of the primes that are greater than 30 will be computed: those values aren't needed to produce the result of the computation.

Zip and Search

Now we'll look at a frequently-used function from Haskell's Prelude, called zip. It takes two lists—possibly with different types of elements—and produces a single list of pairs, with the first component taken from the first list and the second component taken from the second list. Obviously, the name is intended to suggest a zipper, but in Haskell the "teeth" of the zipper are not interleaved but paired.

The definition of zip uses **simultaneous recursion** on both lists.

```
zip :: [a] -> [b] -> [(a,b)]
zip [] ys       = []
zip xs []       = []
zip (x:xs) (y:ys) = (x,y) : zip xs ys
```

Two base cases are required because of the simultaneous recursion. The definition might look a little prettier with an extra equation for zip [] [], but that case is covered by the first equation.

Notice what happens if the lengths of the two lists don't match: zip will truncate the longer one, disregarding all of the extra elements. That fact turns out to be very convenient; it allows us to do things like this:

```
> zip [0..] "word"
[(0,'w'),(1,'o'),(2,'r'),(3,'d')]
```

which pairs characters in a string with their positions, counting from 0. It treats [0..] as if it were [0..(length "word" - 1)], without requiring us to specify that list explicitly. Let's see how it works, on a slightly smaller example, recalling that [0..] is shorthand for enumFrom 0:

```
zip [0..] "ab"
       Expanding [0..] and string notation
= zip (enumFrom 0) ('a' : 'b' : [])
       Applying equation for enumFrom, with m = 0
= zip (0 : enumFrom 1) ('a' : 'b' : [])
       Applying third equation for zip, with x = 0, xs = enumFrom 1,
       y = 'a' and ys = 'b' : []
= (0,'a') : zip (enumFrom 1) ('b' : [])
       Applying equation for enumFrom, with m = 1
= (0,'a') : zip (1 : enumFrom 2) ('b' : [])
       Applying third equation for zip, with x = 1, xs = enumFrom 2,
       y = 'b' and ys = []
```

The command
:set -Wincomplete-patterns
asks Haskell to warn you about missing cases when you use pattern matching. Use :set -W to include more warnings about potential mistakes in your code. Notice the upper case W!

If you feel that you understand recursion, these step-by-step computations are probably getting a bit tedious by now. If so, feel free to skip them. Or visit ▶ https://chrisuehlinger.com/LambdaBubblePop/ and play with the fun animation there. In any case, make sure that you really do understand recursion, meaning that you can write answers to the exercises!

```
= (0,'a') : (1,'b') : zip (enumFrom 2) []
```
        Applying equation for enumFrom, with $m = 2$
```
= (0,'a') : (1,'b') : zip (2 : enumFrom 3) []
```
        Applying second equation for zip, with $xs = 2$: enumFrom 3
```
= (0,'a') : (1,'b') : []
```
        Using list notation
```
= [(0,'a'),(1,'b')]
```

Here's another useful way of using zip:

```
> zip "word" (tail "word")
[('w','o'),('o','r'),('r','d')]
```

This is handy when you want to relate successive elements of a list. For example, here's a function that counts the number of doubled letters in a string:

```
countDoubled :: String -> Int
countDoubled [] = 0
countDoubled xs = length [ x | (x,y) <- zip xs (tail xs), x==y ]
```

Now consider the problem of searching a string for occurrences of a character. We want a list of all of the positions where it occurs, counting from 0. We can do this easily with a list comprehension and zip, using the idea of first pairing each character with its position in the string:

```
search :: String -> Char -> [Int]
search xs y = [ i | (i,x) <- zip [0..] xs, x==y ]
```

Here is a recursive version using a helper function:

```
searchRec :: String -> Char -> [Int]
searchRec xs y = srch xs y 0
  where
    -- i is the index of the start of the substring
    srch :: String -> Char -> Int -> [Int]
    srch [] y i     = []
    srch (x:xs) y i
      | x == y    = i : srch xs y (i+1)
      | otherwise = srch xs y (i+1)
```

Let's see how this works:

```
searchRec "book" 'o'
```
        Applying equation for searchRec, with $xs = $ "book" and $y = $ 'o'
```
= srch "book" 'o' 0
```
        Expanding string notation
```
= srch ('b' : 'o' : 'o' : 'k' : []) 'o' 0
```
        Applying third equation for srch, with $x = $ 'b',
        $xs = $ 'o':'o':'k':[], $y = $ 'o' and $i = 0$,
        since $x == y = $ **False**
```
= srch ('o' : 'o' : 'k' : []) 'o' 1
```
        Applying second equation for srch, with $x = $ 'o',
        $xs = $ 'o':'k':[], $y = $ 'o' and $i = 1$,
        since $x == y = $ **True**
```
= 1 : srch ('o' : 'k' : []) 'o' 2
```
        Applying second equation for srch, with $x = $ 'o',
        $xs = $ 'k':[], $y = $ 'o' and $i = 2$,
        since $x == y = $ **True**
```
= 1 : 2 : srch ('k' : []) 'o' 3
```
        Applying third equation for srch, with $x = $ 'k',
        $xs = $ [], $y = $ 'o' and $i = 3$,
        since $x == y = $ **False**

$$= 1 : 2 : \underline{\text{srch [] 'o' 4}}$$
  Applying first equation for srch, with y = 'o' and i = 4
$$= 1 : 2 : \underline{\text{[]}}$$
  Using list notation
$$= [1,2]$$

Using polymorphism, we can search in lists of any type, not just strings. Here's the version using list comprehension again, but with its most general type:

```
search :: Eq a => [a] -> a -> [Int]
search xs y = [ i | (i,x) <- zip [0..] xs, x==y ]
```

Recall that the requirement Eq a => means that any type a of list elements is okay, provided equality testing (==) works on values of type a. So, for example, search will work on lists of strings but not on lists of functions:

```
> search [square,abs] factorial
<interactive>:1:1: error:
    • No instance for (Eq (Int -> Int)) arising from a use of 'search'
        (maybe you haven't applied a function to enough arguments?)
    • In the expression: search [square, abs] factorial
      In an equation for 'it': it = search [square, abs] factorial
```

Select, Take and Drop

We will now look at three related functions from the Prelude that have integer parameters.

- *xs* !! *n* returns the element in the *n*th position of *xs*, starting from 0.
- take *n xs* returns the first *n* elements of *xs*.
- drop *n xs* returns all of the elements of *xs* after the first *n*.

!! is pronounced "select". If you are familiar with programming languages that use arrays, you might expect that !! is a very commonly-used function in Haskell. You would be wrong.

```
(!!) :: [a] -> Int -> a
(x:xs) !! 0 = x
[] !! i      = error "index too large"
(x:xs) !! i = xs !! (i-1)

take :: Int -> [a] -> [a]
take 0 xs     = []
take i []     = []
take i (x:xs) = x : take (i-1) xs

drop :: Int -> [a] -> [a]
drop 0 xs     = xs
drop i []     = []
drop i (x:xs) = drop (i-1) xs
```

These definitions do simultaneous recursion on i and xs. That's why they have two base cases: one for when i is 0, and one for when xs is empty.

Natural Numbers

Let's now take a closer look at functions like !!, take and drop that we defined by recursion on integers.

All three of these functions only make sense for actual parameters that are *natural numbers* (non-negative integers). The recursion counts down, with 0 as a base case.

Recall how recursive definitions of functions on lists was explained by reference to the (recursive) definition of lists:

Definition. A **list** of type [*t*] is either

1. *empty*, written [], or
2. *constructed*, written *x*:*xs*, with *head x* (an element of type *t*) and *tail xs* (a list of type [*t*]).

We can define natural numbers by recursion in the same style:

Definition. A **natural number** is either

1. *zero*, written 0, or
2. the *successor*, written *n*+1, of its *predecessor n* (a natural number).

For lists, we use [] and : for pattern matching, and recursive definitions of functions on lists typically have the same structure as the definition of lists. For natural numbers, we could regard 0 and +1 as constructors, and use them for pattern matching and recursion. Instead, we use *n* and *n*−1 (the predecessor of *n*), once we have dealt with 0, but the idea is exactly the same.

In fact, Haskell once did allow such patterns, but they were removed in Haskell 2010.

Here are recursive definitions of addition, multiplication, and exponentiation in this style:

```
plus :: Int -> Int -> Int
plus m 0 = m
plus m n = (plus m (n-1)) + 1

times :: Int -> Int -> Int
times m 0 = 0
times m n = plus (times m (n-1)) m

power :: Int -> Int -> Int
power m 0 = 1
power m n = times (power m (n-1)) m
```

This is called **Peano arithmetic**, after Giuseppe Peano (1858–1932), an Italian mathematician and linguist who was responsible for the definition of the natural numbers and the modern treatment of proof by induction, see ▶ https://en. wikipedia.org/wiki/Giuseppe_Peano

Recursion and Induction

You are probably familiar with the following method of **proof by induction**.

Proof method (Induction). To prove that a property P holds for all natural numbers:

Base case: Show that P holds for 0; and

Induction step: Show that if P holds for a given natural number n (the **induction hypothesis**), then it also holds for $n + 1$.

Here is an example of a proof by induction that

$$0 + 1 + \cdots + n = \frac{n(n + 1)}{2}$$

for all natural numbers n.

Base case: $0 = \frac{0(0+1)}{2}$

Induction step: Suppose that the property holds for a given natural number n:

$$0 + 1 + \cdots + n = \frac{n(n + 1)}{2}.$$

Then we show that it holds for $n + 1$:

$$0 + 1 + \cdots + n + (n+1) = \frac{n(n+1)}{2} + (n+1)$$

$$= \frac{n(n+1) + 2(n+1)}{2}$$

$$= \frac{(n+1)((n+1)+1)}{2}$$

Why does this work? The justification is given by the recursive definition of natural numbers. First, consider the explanation of why 3 is a natural number:

1. 0 is a natural number, by case (1);
2. and so 1 is a natural number, with predecessor 0, by case (2);
3. and so 2 is a natural number, with predecessor 1, by case (2);
4. and so 3 is a natural number, with predecessor 2, by case (2).

Now, look at what happens when we use the parts of the induction proof in place of the parts of the definition of natural numbers:

1. $0 + 1 + \cdots + n = \frac{n(n+1)}{2}$ for $n = 0$, by the base case
2. and so $0 + 1 + \cdots + n = \frac{n(n+1)}{2}$ for $n = 1$, by the induction step applied to its predecessor 0;
3. and so $0 + 1 + \cdots + n = \frac{n(n+1)}{2}$ for $n = 2$, by the induction step applied to its predecessor 1;
4. and so $0 + 1 + \cdots + n = \frac{n(n+1)}{2}$ for $n = 3$, by the induction step applied to its predecessor 2.

The same explanation works for every natural number, so the proof shows that the property holds for all natural numbers.

By analogy, we can use the definition of lists as justification of an induction method for lists.

Proof method (Structural Induction). To prove that a property P holds for all finite lists of type $[t]$:

Base case: Show that P holds for $[]$; and

Induction step: Show that if P holds for a given list xs of type $[t]$ (the **induction hypothesis**), then it also holds for $x:xs$ for any value x of type t.

Rod Burstall (1934–), a British computer scientist and Professor Emeritus at the University of Edinburgh, was the first to recognise the role of structural induction in proving properties of programs, see ▶ https://en.wikipedia.org/wiki/Rod_Burstall.

Let's prove by structural induction that ++ (the function that appends two lists) is associative. First, here is the function definition:

```
(++) :: [a] -> [a] -> [a]
[] ++ ys     = ys
(x:xs) ++ ys = x : (xs ++ ys)
```

We'll prove that

$$(xs \mathbin{++} ys) \mathbin{++} zs = xs \mathbin{++} (ys \mathbin{++} zs)$$

When there's more than one variable, there is a choice of which one to use in an induction proof.

by induction on xs. We use the technique of proving that an equation holds by showing that both sides are equal to the same thing.

Base case:
$$([] \mathbin{++} ys) \mathbin{++} zs = ys \mathbin{++} zs$$
$$[] \mathbin{++} (ys \mathbin{++} zs) = ys \mathbin{++} zs$$

Induction step: Suppose that $(xs \mathbin{++} ys) \mathbin{++} zs = xs \mathbin{++} (ys \mathbin{++} zs)$ for a given list xs. Then we show that it holds for $x\!:\!xs$.

$$
\begin{aligned}
&((x \,:\, xs) \mathbin{++} ys) \mathbin{++} zs \\
&\quad = (x \,:\, (xs \mathbin{++} ys)) \mathbin{++} zs \\
&\quad = x \,:\, ((xs \mathbin{++} ys) \mathbin{++} zs) \\
&\quad = x \,:\, (xs \mathbin{++} (ys \mathbin{++} zs)) \text{ (applying the induction hypothesis)} \\
&(x \,:\, xs) \mathbin{++} (ys \mathbin{++} zs) \\
&\quad = x \,:\, (xs \mathbin{++} (ys \mathbin{++} zs))
\end{aligned}
$$

Exercises

1. The definition of `enumFromTo` "counts up": `enumFromTo m n` is defined in terms of `enumFromTo (m+1) n`. Write a definition of `enumFromTo` that counts down. Test that the two functions produce the same result.

2. Write a recursive definition of factorial without using `prodFromTo`. Test that it produces the same result as the definition on page 90.

3. Using list comprehension and `zip`, write a version of the function `angleVectors` from Chap. 3 that represents n-dimensional vectors using the type `[Float]`. Test that when $n = 2$ it produces the same result as the earlier version.

4. Use list comprehension and `zip` to write versions of `!!`, `take` and `drop` (**Hint:** for inspiration, look at the definition of `search` on page 93) and test that they produce the same results as the versions above. Investigate the behaviour of both versions on infinite lists, and explain the differences that you observe.

5. Give a definition of `zip` that requires the lengths of the lists to match.

6. Give recursive definitions of subtraction and division in the style of the definitions of `plus`, `times`, and `power` above. Your definition of subtraction should produce an error when the result would otherwise be negative.

7. A recursive definition is **tail recursive** if every recursive function application it contains is the "last action" in that case of the definition. For example, our definition of `plus`

```
plus :: Int -> Int -> Int
plus m 0 = m
plus m n = (plus m (n-1)) + 1
```

is *not* tail recursive, because the recursive application of `plus` is followed by an addition. This would also be the case it we had written the second equation as

```
plus m n = 1 + (plus m (n-1))
```

But the following definition of the same function

```
plus' :: Int -> Int -> Int
plus' m 0 = m
plus' m n = plus' (m+1) (n-1)
```

is tail recursive.
Write a version of Haskell's `reverse` function

```
reverse :: [a] -> [a]
reverse []     = []
reverse (x:xs) = reverse xs ++ [x]
```

that is tail recursive, by completing the following skeleton:

Tail recursive definitions can be implemented very efficiently on conventional hardware, since the recursive function applications can be implemented as jumps rather than as function calls. See ▶ https://en. wikipedia.org/wiki/ Recursion_(computer_science)# Tail-recursive_functions.

```
reverse' :: [a] -> [a]
reverse' xs = rev xs []
  where rev :: [a] -> [a] -> [a]
        rev [] ys     = ys
        rev (x:xs) ys = rev (...) (...)
```

Test that `reverse'` produces the same result as `reverse`.

8. Use structural induction and the associativity of ++ to prove that

$$\text{reverse } (xs \mathbin{+\!+} ys) = \text{reverse } ys \mathbin{+\!+} \text{reverse } xs$$

You can assume that $xs \mathbin{+\!+} [] = xs$, or else prove it as well by structural induction.

11

Higher-Order Functions

Contents

© The Author(s), under exclusive license to Springer Nature Switzerland AG 2021
D. Sannella et al., *Introduction to Computation*, Undergraduate Topics
in Computer Science, https://doi.org/10.1007/978-3-030-76908-6_12

12

A "first-order" function takes ordinary values (integers, etc.) as parameters, and returns such a value as its result. A "second-order" function takes first-order functions, and possibly also ordinary values, as parameters, and returns such a value as its result. And so on. A "higher-order function" is a function of order higher than 1.

This situation is captured by the motto "functions are first-class citizens", see ▶ https://en.wikipedia. org/wiki/First-class_citizen.

Patterns of Computation

The same patterns of computation keep coming up in function definitions. An example from the beginning of Chap. 11 was in the definitions of `enumFromTo` and `prodFromTo`:

```
enumFromTo :: Int -> Int -> [Int]
enumFromTo m n | m > n  = []
               | m <= n = m : enumFromTo (m+1) n

prodFromTo :: Int -> Int -> Int
prodFromTo m n | m > n  = 1
               | m <= n = m * prodFromTo (m+1) n
```

The only difference between these two definitions is that : and [] in `enumFromTo` are replaced by * and 1 in `prodFromTo`. The relationship between the two function definitions is also apparent in the results of the function applications: `enumFromTo` m n yields

$$m : (m+1) : \cdots : n : []$$

and `prodFromTo` m n yields

$$m * (m+1) * \cdots * n * 1$$

In Haskell, a pattern of computation—like the one that appears in both `enumFromTo` and `prodFromTo`—can be captured as a function. This function can then be instantiated in different ways to define the particular functions that exhibit that pattern of computation. This allows us to replace the definitions of `enumFromTo` and `prodFromTo` by something like

```
enumFromTo = enumPattern (:) []
prodFromTo = enumPattern (*) 1
```

where `enumPattern` is a function that expresses the pattern of computation in `enumFromTo` and `prodFromTo`.

So far so good, but notice that this involves the use of *functions*—the function : for `enumFromTo` and the function * for `prodFromTo`—as actual parameters of `enumPattern`. So `enumPattern` is different from all of the functions that have appeared up to now: it's a **higher-order function**, because it takes another function as a parameter. Here is the definition of the function `enumPattern`:

```
enumPattern :: (Int -> t -> t) -> t -> Int -> Int -> t
enumPattern f e m n | m > n  = e
                    | m <= n = f m (enumPattern f e (m+1) n)
```

According to the type of `enumPattern`, its first parameter is indeed a function. You'll be able to fully understand this definition, and its type, once you've seen some more examples. But what is clear already is that Haskell is able to treat functions as ordinary data values. As you'll see, this gives surprising power.

Although our first examples of higher-order functions appear in this chapter, there is actually nothing new here. Functions take values as parameters and return values as results. Therefore, functions as parameters and results of other functions are supported simply because functions are values.

Map

An extremely common pattern of computation appears in the following function definitions using list comprehensions:

Map

101   **12**

```
squares :: [Int] -> [Int]
squares ns = [ n*n | n <- ns ]
```

```
ords :: [Char] -> [Int]
ords xs = [ ord x | x <- xs ]
```

You've seen the function squares before. The function ords computes the numeric code of each character in a list, using the function ord :: Char -> Int from the **Data.Char** library module.

```
> ords "cat"
[99,97,116]
```

Both squares and ords apply a function—the function that squares an integer, in squares, or ord, in ords—to each element of a list, returning a list of the results. The general pattern, which takes the function to be applied to each list element as a parameter, is the Prelude function map.

$$
\begin{aligned}
\text{squares } [a_1, a_2, \ldots, a_n] &= [a_1*a_1, \quad a_2*a_2, \quad \ldots, \quad a_n*a_n] \\
\text{ords } \quad [a_1, a_2, \ldots, a_n] &= [\text{ord } a_1, \quad \text{ord } a_2, \quad \ldots, \quad \text{ord } a_n] \\
\text{map f } \quad [a_1, a_2, \ldots, a_n] &= [\text{f } a_1, \quad \text{f } a_2, \quad \ldots, \quad \text{f } a_n]
\end{aligned}
$$

Here is the definition of map:

```
map :: (a -> b) -> [a] -> [b]
map f xs = [ f x | x <- xs ]
```

To instantiate the pattern of computation that map captures, we simply apply it to the function that we want to apply to each list element. This allows the functions square and ords to be defined in terms of map as follows:

```
squares :: [Int] -> [Int]
squares = map sqr
          where sqr x = x*x
```

```
ords :: [Char] -> [Int]
ords = map ord
```

To see how this works, let's work through an example.

squares [1,2,3]
   Applying equation for squares
= map sqr [1,2,3]
   Applying equation for map, with f = sqr and xs = [1,2,3]
= [sqr x | x <- [1,2,3]]
   Expanding list comprehension
= [sqr 1, sqr 2, sqr 3]
   Applying sqr and doing the multiplications
= [1,4,9]

This shows that using functions as parameters is not really any different from using ordinary values as parameters. What *is* a little different from what you have seen so far is the way that we have defined squares and ords as the application of map to just one of its two parameters. The type of map shows why this works:

```
map :: (a -> b) -> [a] -> [b] and sqr :: Int -> Int
so map sqr :: [Int] -> [Int]
```

```
map :: (a -> b) -> [a] -> [b] and ord :: Char -> Int
so map ord :: [Char] -> [Int]
```

The function chr :: Int -> Char from Data.Char converts an integer to the corresponding character. Character codes for the Latin alphabet are arranged in such a way that for each lower case letter c, the corresponding upper case letter is given by chr (ord c – 32). But it's better to use the function toUpper from Data.Char since it also works for non-Latin letters.

What we have done in the definition of squares and ords, called **partial application**, is an important technique in functional programming. There's more about it coming later in this chapter.

The history of the map function shows how ideas from functional programming eventually make their way into other programming languages. It first appeared in 1959 in the first functional programming language, LISP. Now, more than 60 years later, map is included in some form in most other programming languages, but for instance, it first became available in Java in 2014.

Exactly the same pattern of computation that is captured by map appears in a different form in the recursive versions of squares and ords:

```
squaresRec :: [Int] -> [Int]
squaresRec []     = []
squaresRec (x:xs) = x*x : squaresRec xs
```

```
ordsRec :: [Char] -> [Int]
ordsRec []     = []
ordsRec (x:xs) = ord x : ordsRec xs
```

Again, the only difference between these two definitions is the function that is applied to x in the second equation of each definition. This pattern is captured by the following recursive definition of map:

```
mapRec :: (a -> b) -> [a] -> [b]
mapRec f []     = []
mapRec f (x:xs) = f x : mapRec f xs
```

As before, we can instantiate the pattern by applying mapRec to a function, for example,

```
squaresRec :: [Int] -> [Int]
squaresRec = mapRec sqr
                 where sqr x = x*x
```

And here's the same example as before, using the recursive version of mapRec:

squaresRec [1,2,3]
        Applying equation for squaresRec
= mapRec sqr [1,2,3]
        Expanding list notation
= mapRec sqr (1 : (2 : (3 : [])))
        Applying second equation, with x = 1 and xs = 2 : (3 : [])
= sqr 1 : (mapRec sqr (2 : (3 : [])))
        Applying second equation , with x = 2 and xs = 3 : []
= sqr 1 : (sqr 2 : (mapRec sqr (3 : [])))
        Applying second equation , with x = 3 and xs = []
= sqr 1 : (sqr 2 : (sqr 3 : (mapRec sqr [])))
        Applying first equation
= sqr 1 : (sqr 2 : (sqr 3 : []))
        Applying sqr and doing the multiplications
= 1 : (4 : (9 : []))
        Using list notation
= [1,4,9]

Filter

Another very common pattern is extracting all of the elements of a list that have some property. Here are two examples, written using both list comprehension and recursion:

```
odds :: [Int] -> [Int]
odds xs = [ x | x <- xs, odd x ]
```

```
oddsRec :: [Int] -> [Int]
```

12

```
oddsRec []                    = []
oddsRec (x:xs) | odd x        = x : oddsRec xs
               | otherwise = oddsRec xs

digits :: [Char] -> [Char]
digits xs = [ x | x <- xs, isDigit x ]

digitsRec :: [Char] -> [Char]
digitsRec []                  = []
digitsRec (x:xs) | isDigit x = x : digitsRec xs
                 | otherwise = digitsRec xs
```

You've seen odds and oddsRec before. The function digits returns all of the characters in a list that are digits, '0'..'9', using **Data.Char.isDigit**.

The difference between odds and digits, and between oddsRec and digitsRec, is the function that is applied in the guard: odd versus isDigit. Only a function that produces a result of type Bool—a **predicate**, see Chap. 6—would make sense because of its use in the guard.

This pattern is captured by the Prelude function filter. Here are definitions of filter using list comprehension and recursion.

```
filter :: (a -> Bool) -> [a] -> [a]
filter p xs = [ x | x <- xs, p x ]

filterRec :: (a -> Bool) -> [a] -> [a]
filterRec p []                 = []
filterRec p (x:xs) | p x        = x : filterRec p xs
                   | otherwise = filterRec p xs
```

Since we have defined
type Predicate u = u -> Bool,
we can also say
filter :: Predicate a -> [a] -> [a]
and ditto for filterRec.

To instantiate filter or filterRec, we apply it to an appropriate function.

```
odds :: [Int] -> [Int]
odds = filter odd

digits :: [Char] -> [Char]
digits = filter isDigit
```

All of the functions that we have seen so far being used as parameters to map and filter have been very simple. This is not required: functional parameters can be as complicated as desired. The only restriction is that they need to have the required type.

Fold

The next pattern of computation that we are going to consider is also very common, but is a little more complicated than map and filter. It is demonstrated by the definitions of the functions sum, product and and, which you have seen before, as well as the Prelude function concat, which appends ("concatenates") all of the elements in a list of lists.

```
sum :: [Int] -> Int
sum []     = 0
sum (x:xs) = x + sum xs

product :: [Int] -> Int
product []     = 1
product (x:xs) = x * product xs
```

```
and :: [Bool] -> Bool
and []     = True
and (x:xs) = x && and xs
```

```
concat :: [[a]] -> [a]
concat []       = []
concat (xs:xss) = xs ++ concat xss
```

Here are some examples of concat in action:

```
> concat [[1,2,3],[4,5]]
[1,2,3,4,5]
> concat ["con","cat","en","ate"]
"concatenate"
```

This time there are *two* differences between these definitions, not just one:

1. the value that is returned in the base case: 0 for sum, 1 for product, True for and, [] for concat; and
2. the function that is used to combine the head of the list with the result of the recursive function application to the tail of the list: + for sum, * for product, && for and, ++ for concat.

In each of these examples, the base case value is the **identity element** of the "combining" function: 0 for +, 1 for *, True for &&, [] for ++. That is, $0 + x = x = x + 0$, etc. This relationship is not required, but it is a common situation.

The pattern that is used in all of these examples is captured by the Prelude function foldr:

foldr is pronounced "fold-R", or sometimes "fold right".

```
foldr :: (a -> a -> a) -> a -> [a] -> a
foldr f v []     = v
foldr f v (x:xs) = f x (foldr f v xs)
```

This definition is a little easier to understand when the application of the function f is written using infix notation:

```
foldr :: (a -> a -> a) -> a -> [a] -> a
foldr f v []     = v
foldr f v (x:xs) = x `f` (foldr f v xs)
```

foldr is called "reduce" in some other languages. Google's MapReduce and Apache's Hadoop frameworks for distributed computing are based on a combination of map and foldr, see ▶ https://en.wikipedia.org/wiki/MapReduce.

Then, writing out the computations makes it easy to see what's going on:

$$
\begin{aligned}
\text{sum} && [a_1,\ldots,a_n] &= a_1 & + & (\cdots & + & (a_n & + & 0)\cdots) \\
\text{product} && [a_1,\ldots,a_n] &= a_1 & * & (\cdots & * & (a_n & * & 1)\cdots) \\
\text{and} && [b_1,\ldots,b_n] &= b_1 & \&\& & (\cdots & \&\& & (b_n & \&\& & \text{True})\cdots) \\
\text{concat} && [xs_1,\ldots,xs_n] &= xs_1 & ++ & (\cdots & ++ & (xs_n & ++ & [])\cdots) \\
\text{foldr } f\ v && [x_1,\ldots,x_n] &= x_1 & `f` & (\cdots & `f` & (x_n & `f` & v)\cdots)
\end{aligned}
$$

In foldr $f\ v\ xs$, the result is computed by combining the elements of the list xs using the function f, with the value v as the result when we get to the end. Or, reading from right to left, v is the *starting value* for combining the elements of xs using f.

The following diagram gives another way of seeing what's happening.

This way of drawing trees upside down, with the "root" at the top and the "leaves" at the bottom, is standard in Informatics and Linguistics.

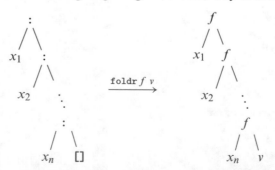

The tree on the left represents the expression

$$x_1 : (x_2 : (\cdots : (x_n : [])))$$

with the tree structure used in place of parentheses to represent nesting. Using foldr f v to go from there to the tree on the right replaces [] and all occurrences of : with v and f, respectively, without changing the structure of the expression.

We can now define sum, etc., by applying foldr to appropriate parameters:

```
sum :: [Int] -> Int
sum = foldr (+) 0

product :: [Int] -> Int
product = foldr (*) 1

and :: [Bool] -> Bool
and = foldr (&&) True

concat :: [[a]] -> [a]
concat = foldr (++) []
```

foldr and foldl

The following question may be nagging you at this point: what does the "r" in foldr refer to? It seems that it has something to do with "right", but what? And what's the equivalent thing for "left"?

Good question! The "r" in foldr refers to the fact that the applications of f are nested "to the right", as can be seen from the structure of the expressions and trees above. The relationship between the structure of the parameter list and the result of foldr f v makes that the most natural option.

Nesting "to the left" is also possible, and that's what the Prelude function foldl does:

$$\text{foldl } f \ v \ [x_1,\ldots,x_n] \ = \ (\cdots (v \ f^{\backprime} \ x_1) \ f^{\backprime} \cdots \ ) \ f^{\backprime} \ x_n$$

Here's what this looks like as a tree:

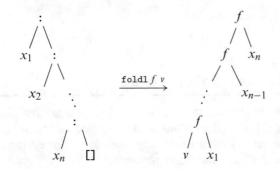

If f is associative and commutative then the results of foldr and foldl will be the same. This is the case for all of the uses of foldr above, so we could instead define

```
sum :: [Int] -> Int
sum = foldl (+) 0

product :: [Int] -> Int
product = foldl (*) 1
```

```
and :: [Bool] -> Bool
and = foldl (&&) True

concat :: [[a]] -> [a]
concat = foldl (++) []
```

Here's an example where they are different:

```
cumulativeDivide :: Int -> [Int] ->Int
cumulativeDivide i = foldl div i
```

`cumulativeDivide`, which produces the quotient of its first parameter by all of the integers in its second parameter, is a possible component of a function for factorising integers. Using `foldr` in place of `foldl` gives a completely different function.

If we define a version of : with the parameters in the opposite order

```
(<:) :: [a] -> a -> [a]
xs <: x = x : xs
```

then we can use `foldl` to reverse a list

```
reverse :: [a] -> [a]
reverse = foldl (<:) []
```

while `foldr (<:) []` causes a type error, and `foldr (:) []` is just the identity function on lists. (This requires a more general type for `foldr`, see Exercise 2.)

`foldr1` and `foldl1` are variants of `foldr` and `foldl` for use on lists that are guaranteed to be non-empty. In that case, no "starting value" is required: the last element of the list (for `foldr1`) or first element (for `foldl1`) can be used instead. Here is the definition of `foldr1`:

```
foldr1 :: (a -> a -> a) -> [a] -> a
foldr1 f []    = error "empty list"
foldr1 f [x]   = x
foldr1 f (x:xs) = x `f` (foldr1 f xs)
```

We can define the Prelude function `maximum`, which doesn't make sense for an empty list, using `foldr1` (or `foldl1`) and `max`:

```
maximum :: [Int] -> Int
maximum = foldr1 max
```

Combining `map`, `filter` and `foldr`/`foldl`

The patterns of computation that are captured by `map` and `filter` are familiar from earlier examples using list comprehensions, and these two functions can be combined to give the same result as many list comprehensions. For example, consider a function that doubles all of the prime numbers in a list:

```
dblPrimes :: [Int] -> [Int]
dblPrimes ns = [ 2*n | n <- ns, isPrime n ]

dblPrimes' :: [Int] -> [Int]
dblPrimes' ns = map dbl (filter isPrime ns)
                where dbl x = 2*x
```

In general,

map f (filter p xs) = [f x | x <- xs, p x]

and for multiple guards

map f (filter p (filter q xs)) = [f x | x <- xs, p x, q x]

Adding an application of `foldr` or `foldl` is required for examples that involve the use of an accumulator. Looking at the example of the sum of the squares of the odd numbers in a list:

```
sumSqOdds :: [Int] -> Int
sumSqOdds ns = sum [ n*n | n <- ns, odd n ]

sumSqOdds' :: [Int] -> Int
sumSqOdds' ns = foldr (+) 0 (map sqr (filter odd ns))
                where sqr x = x*x
```

Curried Types and Partial Application

In Haskell, functions with two parameters are usually defined to have a type of the form s -> t -> u, and analogously for larger values of two. For example, the type of `div` is `Int -> Int -> Int`, and the type of `max3` is `Int -> Int -> Int -> Int`. An alternative would be to combine the parameters into a tuple, which for `div` would give the type `(Int,Int) -> Int`.

One reason for not combining the parameters into a tuple is to allow **partial application**. We've seen this when defining functions like `sum` by instantiating the pattern of computation that is captured in `foldr`:

```
sum :: [Int] -> Int
sum = foldr (+) 0
```

The type of `foldr` is (a -> a -> a) -> a -> [a] -> a. Applying it to + gives a function of type `Int -> [Int] -> Int`—this is the function that adds all of the integers in a list (the second parameter) to an integer (the first parameter)—and then applying this function to 0 gives `sum :: [Int] -> Int`.

Simpler examples of partial application are the application of + to 1, to get a function for increasing a number by 1; the application of <= to 0, to get a function for testing if an integer is positive; and the application of ˆ to 2, for computing powers of 2.

```
increment :: Int -> Int
increment = (+) 1

isPositive :: Int -> Bool
isPositive = (<=) 0

pow2 :: Int -> Int
pow2 = (^) 2
```

This only works when we need to partially apply a function to its first parameter(s). Haskell provides a handy notation, called **sections**, for writing the partial application of an infix function to either of its two parameters. A section is written by surrounding an infix function with parentheses and supplying one of its parameters, for example:

```
pow2 = (2 ^)
isVowel = (`elem` "aeiouAEIOU")
squares = map (^ 2)
```

A section yields a function from the missing parameter to the result.

Let's take a closer look at the type `Int -> Int -> Int`. All infix operators, including ->, come with a **precedence**—for instance, * and / have higher precedence than + and -, as in arithmetic—and an **associativity**. These attributes determine how Haskell understands expressions in which the operator appears.

A type like `Int -> Int -> Int` is called a "curried" type, with the alternative `(Int,Int) -> Int` being called an "uncurried" type. The name comes from American mathematician and logician Haskell B. Curry (1900–1982), see ▶ https://en.wikipedia.org/wiki/Haskell_Curry. Curry developed combinatory logic, used in Informatics as a simplified model of computation. A well-known programming language is also named after him.

Unfortunately, sections don't work for partially applying subtraction to its second parameter: (- 1) is the negative number −1.

You might think that right/left associativity of + doesn't matter, since addition is associative, $(a + b) + c = a + (b + c)$. Except that this doesn't actually hold for computer arithmetic: compare `(1e100 + (-1e100)) + 1` and `1e100 + ((-1e100) + 1)` in Haskell.

Many operators are left associative, for instance +, so $a + b + c$ means $(a + b) + c$. But the function-type operator `->` is right associative, so `Int -> Int -> Int` means `Int -> (Int -> Int)`. (Likewise for `:`, as mentioned earlier, so `1:2:3:[]` means `1:(2:(3:[]))`.)

On the other hand, the function application operation that is being used in expressions like $f\ x$ is *left* associative, so `(+) 1 2` means `((+) 1) 2`. But all of this fits perfectly together:

```
(+) :: Int -> Int -> Int means (+) :: Int -> (Int -> Int)
                         so    ((+) 1) :: Int -> Int
                         so    ((+) 1) 2 :: Int
i.e. (+) 1 2 :: Int
```

The ability to use partial application is what makes it possible to write function definitions like

```
sum :: [Int] -> Int
sum = foldr (+) 0
```

in place of the slightly longer winded

```
sum :: [Int] -> Int
sum ns = foldr (+) 0 ns
```

where the list parameter has been made explicit. These are equivalent; which of them you find clearer is a matter of taste.

The first definition of `sum` is written in so-called **point-free** style, while the second definition is written in **pointed** style, referring to "points" in the parameter space. See ▶ https://wiki.haskell.org/Pointfree. People who find point-free style hard to understand call it "pointless" style.

Exercises

1. The Prelude function `all :: (a -> Bool) -> [a] -> Bool` returns `True` if all of the elements of its second parameter (a list) satisfy the property given in its first parameter (a predicate). Define `all` in terms of `map`, `filter` and/or `foldr`/`foldl`. Use `all` to define a function `allPosDiv3 :: [Int] -> Bool` which returns `True` if all of the elements of the parameter list are both positive and divisible by 3.

2. The type of `foldr` is more general than `(a -> a -> a) -> a -> [a] -> a`. Work out its most general type. **Hint:** In the definition of `foldr`, suppose `v :: b` and `xs :: [a]`. What type does `f` need to have?

3. Using `filter`, write a function `rmChar :: Char -> String -> String` that removes all occurrences of a character from a string. Using `foldr` or `foldl`, and `rmChar`, write a function `rmChars :: String -> String -> String` that removes all characters in the first string from the second string.

4. Write a function `halveEvensHO :: [Int] -> [Int]` that returns half of each even number in a list. Use `map`, `filter` and/or `foldr`/`foldl`, not recursion or list comprehension.
   Use `QuickCheck` to test that `halveEvensHO` returns the same result as `halveEvens` in Exercise 5.3 and `halveEvensRec` in Exercise 10.2.

5. Write a function `countPositivesHO` to count the number of positive numbers in a list. Use `map`, `filter` and/or `foldr`/`foldl`, not recursion or list comprehension.
   Use `QuickCheck` to test that `countPositivesHO` returns the same result as `countPositives` in Exercise 5.5 and `countPositivesRec` in Exercise 10.4.

6. Write a function `multDigitsRecHO :: String -> Int` that returns the product of all the digits in the input string. If there are no digits, your function should return 1. Use `map`, `filter` and/or `foldr`/`foldl`, not recursion or list comprehension.

   Use `QuickCheck` to test that `multDigitsHO` returns the same result as `multDigits` in Exercise 5.6 and `multDigitsRec` in Exercise 10.5.

7. Define the function `foldl`. What is its most general type?

8. Using recursion, define a function `foldr'`, with the same type as `foldr`, such that

   $$\texttt{foldr'}\ f\ v\ [x_1,\ldots,x_n]\ =\ x_n\ f\texttt{`}\ (\cdots\ f\texttt{`}\ (x_1\ f\texttt{`}\ v))$$

   Define `reverse :: [a] -> [a]` as the application of `foldr'` to appropriate parameters.

   Now define `foldr'` again, this time in terms of `foldr` and `reverse`.

9. Define `map` f and `filter` p as the applications of `foldr` to appropriate parameters.

10. The sentence

    » The patterns of computation that are captured by `map` and `filter` are familiar from earlier examples using list comprehensions, and these two functions can be combined to give the same result as *many list comprehensions*.

    suggests that there are some list comprehensions that cannot be captured using `map` and `filter`. Give an example of such a list comprehension.

Higher and Higher

Contents

© The Author(s), under exclusive license to Springer Nature Switzerland AG 2021
D. Sannella et al., *Introduction to Computation*, Undergraduate Topics
in Computer Science, https://doi.org/10.1007/978-3-030-76908-6_13

13

Lambda Expressions

Sections are convenient for supplying functional arguments to higher-order functions. For instance, we can replace

```
f :: [Int] -> Int
f ns = foldr (+) 0 (map sqr (filter pos ns))
            where sqr x = x * x
                  pos x = x >= 0
```

which involves two helper functions, with

```
f :: [Int] -> Int
f ns = foldr (+) 0 (map (^ 2) (filter (>= 0) ns))
```

But sections are a very specific trick: they only work when the functional argument is a single partially applied infix operation.

Let's try an experiment: simply putting the definitions of the helper functions in place of their names:

```
f :: [Int] -> Int
f ns = foldr (+) 0 (map (x * x) (filter (x >= 0) ns))
```

Unfortunately, the result is a sequence of error messages:

```
f.hs:2:26: error:
    Variable not in scope: x :: Int -> Int

f.hs:2:30: error:
    Variable not in scope: x :: Int -> Int

f.hs:2:42: error:
    Variable not in scope: x :: Integer
```

Haskell is complaining that none of the occurrences of x makes sense. Each of the helper functions introduces x as a formal parameter—the same one in both helper functions, but that's not important—and the scope of each of these formal parameters is the body of the helper function. Using the bodies of the helper functions without their left-hand sides removes the uses of x from the scopes of their bindings.

So is there a way to do something similar without the need to define boring single-use helper functions?

Yes! The key is **lambda expressions**: expressions denoting nameless functions. Here is the example above again, this time done properly using lambda expressions:

```
f :: [Int] -> Int
f ns = foldr (+) 0
         (map (\x -> x * x)
           (filter (\x -> x >= 0) ns))
```

In a lambda expression, the formal parameter—a pattern, possibly containing more than one variable—follows the \, and then the function body is given after ->. So \x -> x * x is the helper function sqr, but without that name.

Lambda expressions can be used anywhere that a function is required. To evaluate a lambda expression applied to an actual parameter, Haskell simply substitutes the actual parameter for the formal parameter. For instance:

$$\frac{(\backslash x -> (\backslash y -> x + y + 1))\ 3\ 4}{= (\backslash y -> 3 + y + 1)\ 4}$$
$$= 3 + 4 + 1$$
$$= 8$$

\x -> x * x is pronounced "lambda ex, ex times ex".

Lambda expressions—also called **lambda abstractions**—were introduced before the invention of computers by Alonzo Church (1903–1995), see ▶ https://en.wikipedia.org/wiki/Alonzo_Church. In the **lambda calculus**, the Greek letter lambda (λ) starts a lambda expression. Haskell uses a backslash instead since it is the closest symbol to a lambda on the keyboard.

This example shows in detail what happens when a function with a curried type—in this case, `\x -> (\y -> x + y + 1)` has type `Int -> Int -> Int`—is applied.

However, "Haskell simply substitutes the actual parameter for the formal parameter" hides an important point: substitution is only done *within the scope of* the formal parameter. Here's an example which shows why that qualification is necessary:

$$
\begin{aligned}
&(\texttt{\textbackslash x -> (\textbackslash y -> (\textbackslash x -> x + y) (x + 7))) 3 4} \\
=\ &(\texttt{\textbackslash y -> (\textbackslash x -> x + y) (3 + 7)) 4} \\
=\ &(\texttt{\textbackslash x -> x + 4) (3 + 7)} \\
=\ &(\texttt{\textbackslash x -> x + 4) 10} \\
=\ &\texttt{10 + 4} \\
=\ &\texttt{14}
\end{aligned}
$$

In the first step, the actual parameter 3 is substituted for the formal parameter x but *not* within the inner lambda expression `\x -> x + y`. The reuse of the formal parameter x here, whose scope is the expression x + y, makes a "hole" in the scope of the outer binding of x.

Function Composition

Recall function composition from Mathematics: the composition of two functions $f : T \rightarrow U$ and $g : S \rightarrow T$ is the function $f \circ g : S \rightarrow U$ defined by $(f \circ g)(x) = f(g(x))$.

Function composition is the Prelude function . (dot) in Haskell, with the same definition:

```haskell
(.) :: (b -> c) -> (a -> b) -> a -> c
(f . g) x = f (g x)
```

f . g is pronounced "f after g", taking account of the order in which f and g are applied.

This is another higher-order function, taking two functions as parameters and producing a function as its result.

Function composition can be used as "plumbing" to join together the stages of a computation. For instance, consider the earlier example of defining the sum of the squares of the odd numbers in a list using `map`, `filter` and `foldr`:

```haskell
sumSqOdds' :: [Int] -> Int
sumSqOdds' ns = foldr (+) 0 (map (^ 2) (filter odd ns))
```

The same definition can be written as follows, making the three stages of the computation explicit as functions and joining them together in a pipeline:

```haskell
sumSqOdds'' :: [Int] -> Int
sumSqOdds'' = foldr (+) 0 . map (^ 2) . filter odd
```

Function composition is associative, so it doesn't matter whether we put in parentheses or not.

Another use for function composition is to build function parameters for higher-order functions. To compute the squares of the non-prime numbers in a list, we can write

The identity element of function composition is the identity function `id :: a -> a`, defined by `id x = x`.

```haskell
squareNonprimes :: [Int] -> [Int]
squareNonprimes = map (^ 2) . filter (not . isPrime)
```

using the negation of the `isPrime` predicate as the actual parameter of `filter`.

The Function Application Operator $

Here's the definition of another higher-order Prelude function.

```
($) :: (a -> b) -> a -> b
f $ x = f x
```

But this is simply function application. Why in the world would we want to write *f* $ *x* instead of *f* *x*?

Here's why. Expressions in Haskell sometimes become a little complicated, with masses of nested parentheses, and that makes them hard to read. Visually, it's sometimes difficult to find the right parenthesis that matches a particular left parenthesis. The $ operator gives you a way of avoiding parentheses in expressions like *f* (*g* (*h* (*j* *a*))).

Normal function application—juxtaposition of expressions, as in *j* *a*—has high precedence, and is left associative, so *f* *g* *h* *j* *a* amounts to (((*f* *g*) *h*) *j*) *a*. If you instead want *f* (*g* (*h* (*j* *a*))), you need to add the parentheses. The function application operator $, in contrast, has very *low* precedence, and is *right* associative. Thus, *f* $ *g* $ *h* $ *j* $ *a*, with no parentheses, amounts to *f* $ (*g* $ (*h* $ (*j* $ *a*))), and that has the same meaning as *f* (*g* (*h* (*j* *a*))). Voilà!

We can apply this little trick to get rid of some of the parentheses in the definition of sumSqOdds' above:

```
sumSqOdds' :: [Int] -> Int
sumSqOdds' ns = foldr (+) 0 $ map (^ 2) $ filter odd ns
```

Currying and Uncurrying Functions

Functions with curried function types like `Int -> Int -> Int` are convenient because they enable partial application, and most Haskell functions have curried types. But functions with uncurried types like `(Int,Int) -> Int` are sometimes required, especially in combination with the `zip` function. Here's an example which compares the alphabetical order of the English and German names of numbers, using an uncurried version of the < function:

```
> filter (\(x,y) -> x < y)
         (zip ["one", "two", "three", "four"]
              ["eins", "zwei", "drei", "vier"])
[("two","zwei"),("four","vier")]
```

The Prelude function `curry` "curries" an uncurried function:

```
curry :: ((a,b) -> c) -> a -> b -> c
curry f x y = f (x,y)
```

This example uses the fact that the functions <, >, etc. work on **String**—and in general, on lists over types like **Char** that are ordered, see Chap. 24.

The idea of currying first appeared in the work of Gottlob Frege. It was further developed by Moses Schönfinkel, and only later by Curry. So alternative names would be "fregeing" or "schönfinkelisation".

Because `curry` takes a function to a function, the following equivalent definition might be easier to understand:

```
curry :: ((a,b) -> c) -> a -> b -> c
curry f = \x -> \y -> f (x,y)
```

The opposite conversion is also a Prelude function:

```
uncurry :: (a -> b -> c) -> (a,b) -> c
uncurry f (x,y) = f x y
```

Or, using a lambda expression:

```
uncurry :: (a -> b -> c) -> (a,b) -> c
uncurry f = \(x,y) -> f x y
```

Here's the example above with uncurry:

```
> filter (uncurry (<))
        (zip ["one", "two", "three", "four"]
             ["eins", "zwei", "drei", "vier"])
[("two","zwei"),("four","vier")]
```

The definitions of curry and uncurry are short and simple. Perhaps less obvious is that they practically write themselves, with the help of the type system. This is actually pretty common with higher-order functions.

For example, consider curry. Let's write down the required type, and then write curry applied to all of the formal parameters that the type requires:

```
curry :: ((a,b) -> c) -> a -> b -> c
curry f x y = ...
```

Now, what can you write on the right-hand side of the equation that has type c, using f :: (a,b) -> c, x :: a and y :: b? The only thing that has the right type, f (x,y), is the correct answer!

> This point is demonstrated by Djinn (▶ https://hackage.haskell.org/ package/djinn) which takes a Haskell type and uses a theorem prover to produce the definition of a function of that type if one exists.

Bindings and Lambda Expressions

A Haskell program consists of a sequence of definitions of functions and variables. Individual function/variable definitions can include nested **where** clauses which define additional variables and functions. All such bindings can be eliminated using lambda expressions. Doing this will make your programs much harder to understand, so it's not recommended! The point is just to show that "bare" lambda expressions are enough to express everything in Haskell.

Here's the key transformation, where a binding is replaced by an extra lambda expression and an application:

$$exp \ \textbf{where} \ var = exp' \quad \longrightarrow \quad (\backslash var \ \text{->} \ exp) \ exp'$$

Let's look at how this works on a simple expression with nested **where** clauses:

> Everything in Haskell can be explained in terms of lambda expressions. You could view them as the "machine code" of functional programming. Even lower level than lambda expressions are **combinators**: all lambda expressions can be boiled down to expressions composed of S and K, where S x y z = (x z) (y z) and K x y = x, without variables or lambdas! Combinators are like the quarks of computing. See ▶ https:// en.wikipedia.org/wiki/ Combinatory_logic.

```
f 2
  where f x = x + y * y
                where y = x + 1
-->
f 2
  where f = \x -> (x + y * y where y = x + 1)
-->
f 2
  where f = \x -> ((\y -> x + y * y) (x + 1))
-->
(\f -> f 2) (\x -> ((\y -> x + y * y) (x + 1)))
```

Evaluating this produces the same result as the original expression:

$$\frac{(\backslash f \rightarrow f\ 2)\ (\backslash x \rightarrow ((\backslash y \rightarrow x + y * y)\ (x + 1)))}{= (\backslash x \rightarrow ((\backslash y \rightarrow x + y * y)\ (x + 1)))\ 2}$$
$$= (\backslash y \rightarrow 2 + y * y)\ (2 + 1)$$
$$= (\backslash y \rightarrow 2 + y * y)\ 3$$
$$= 2 + 3 * 3$$
$$= 11$$

Here's a more complicated example, with two function definitions followed by an expression:

```
f x = w ^ 2
      where w = x + 2
g y = f (y * z)
      where z = y + 1
g b
  where b = f 3
```

Before translating bindings to application of lambda expressions, we need to reorganise this into a single expression with nested **where** clauses:

```
g b
  where b = f 3
           where f x = w ^ 2
                      where w = x + 2
        g y = f (y * z)
            where z = y + 1
                  f x = w ^ 2
                      where w = x + 2
```

Removing bindings gives the result:

```
(\b -> \g -> g b)
  (((\f -> f 3) (\x -> (\w -> w ^ 2) (x + 2)))
  (\y -> (\z -> \f -> f (y * z))
         (y + 1)
         (\x -> (\w -> w ^ 2) (x + 2)))
```

13

This isn't the whole story: properly dealing with recursive definitions requires the use of a **fixpoint combinator**, see ▶ https://en.wikipedia.org/wiki/Fixed-point_combinator.

Exercises

1. Give two definitions of the function

    ```
    iter :: Int -> (a -> a) -> (a -> a)
    ```

    which composes a function with itself the given number of times. One definition should use recursion on natural numbers. The other should use the Prelude function `replicate :: Int -> a -> [a]` to create a list of copies of the function and `foldr`/`foldl` to compose them.

2. Define the Prelude function

    ```
    flip :: (a -> b -> c) -> b -> a -> c
    ```

    which reverses the order of the parameters of a function. `flip` is sometimes useful if you want to partially apply a function to its second parameter rather than its first parameter.

3. Define the Prelude functions

    ```
    takeWhile :: (a -> Bool) -> [a] -> [a]
    dropWhile :: (a -> Bool) -> [a] -> [a]
    ```

which take/drop elements as long as the given predicate holds. For example, `takeWhile isPrime [2..] == [2,3]`.

Use `takeWhile` and `dropWhile` to define the Prelude function `words :: String -> [String]` which splits a string into words, using as separators those characters that satisfy the predicate `isSpace` from the `Data.Char` library module. Test that your version of `words` gives the same result as Haskell's version.

4. The Prelude function

   ```
   zipWith :: (a -> b -> c) -> [a] -> [b] -> [c]
   ```

   applies the given function to the corresponding elements of two lists. For example, `zipWith (+) [1,2,3] [10,20,30] == [11,22,33]`. Give a definition of `zipWith` using recursion. Give another definition using `zip` and `map`.

5. Work out the types of the following expressions, without using `:t` in GHCi.

   - `map . map`
   - `uncurry curry`
   - `zipWith . zipWith`
   - `(.).(.)`

   Do any of them do anything useful?

6. Consider the Prelude function `unzip`:

   ```
   unzip :: [(a,b)] -> ([a],[b])
   unzip [] = ([],[])
   unzip ((x,y):xys) = (x:xs,y:ys)
                    where (xs,ys) = unzip xys
   ```

   What is its inverse? It's not `zip :: [a] -> [b] -> [(a,b)]`, because the types don't match.

7. Use structural induction to show that map $(f \cdot g)$ $xs =$ map f (map g xs) for all finite lists xs.

8. **Church numerals** are a way of representing the natural numbers using higher-order functions: n is represented by the function of type `(a -> a) -> a -> a` which maps f :: `a -> a` to its n-fold composition, see Exercise 1:

   For more on Church numerals and similar encodings of other types of data, see ▶ https://en.wikipedia.org/wiki/Church_encoding.

   ```
   type Church a = (a -> a) -> a -> a

   church :: Int -> Church a
   church n = iter n

   succ :: Church a -> Church a
   succ cm = \f -> \x -> f (cm f x)

   plus :: Church a -> Church a -> Church a
   plus cm cn = \f -> \x -> cm f (cn f x)
   ```

   Play around with these definitions, using the following function to see the results:

```
unchurch :: Church Int -> Int
unchurch cn = cn (+ 1) 0
```

Once you've understood the definitions of succ and plus, define a function times :: Church a -> Church a -> Church a for multiplying Church numerals.

13

Sequent Calculus

Contents

Combining Predicates

Symbolic logic began with the work of George Boole (1815–1864) who invented what is now called **Boolean algebra**, where variables stand for truth values instead of numbers and operations are conjunction, disjunction, and negation instead of addition, multiplication, etc. See ► https://en.wikipedia.org/wiki/Boolean_algebra.

Chapters 8 and 9 covered 2000 years of logic, up to the mid-nineteenth century, but using modern notation which makes things much simpler. We're now going to study modern symbolic logic, keeping the same notation and maintaining consistency with what we've learned up to now but going well beyond Aristotle's syllogisms. The main difference is the addition of logical connectives—not just negation—for combining predicates. This makes it possible to build and analyse logical arguments that involve much more complicated statements than we've seen up to now.

You've seen already in Chap. 8 that **negation** of a predicate

$$(\neg a)\, x = \neg(a\, x)$$

corresponds to complement of the set of things that satisfy the predicate, with the Haskell definition (Exercise 8.6)

```
neg :: Predicate u -> Predicate u
(neg a) x = not (a x)
```

We can give a similar definition of the **conjunction** $a \wedge b$ of two predicates a and b—the predicate that's satisfied when both a and b are true:

$$(a \wedge b)\, x = a\, x \wedge b\, x$$

As we've seen already, conjunction of predicates corresponds to intersection of sets.

What does this say about a sequent involving a conjunction of predicates, and its relationship to the sequents involving the individual predicates? The situation is represented by the following Euler diagram:

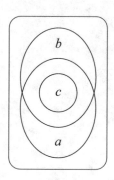

This diagram shows that $c \vDash a$ and $c \vDash b$ is exactly the same as $c \vDash a \wedge b$, since $a \wedge b$ is the intersection of a and b. In symbols:

$$c \vDash a \text{ and } c \vDash b \ \text{ iff } \ c \subseteq a \text{ and } c \subseteq b \ \text{ iff } \ c \subseteq a \cap b \ \text{ iff } \ c \vDash a \wedge b$$

That justifies the following rule, which we write using a double bar since it's an equivalence:

$$\frac{c \vDash a \qquad c \vDash b}{c \vDash a \wedge b} \wedge$$

And we can define conjunction on predicates in Haskell (Exercise 8.6):

It's important for you to understand that we can't just use `&&` for conjunction of predicates because it has the wrong type:
`(&&) :: Bool -> Bool -> Bool`. The same applies to `not` vs. `neg` above and `||` vs. `|:|` below.

```
(&:&) :: Predicate u -> Predicate u -> Predicate u
(a &:& b) x = a x && b x
```

We can do the same thing for **disjunction**. Given predicates a and b, their disjunction $a \vee b$ is the predicate that's satisfied when *either a or b* is true:

$$(a \vee b) \, x = a \, x \vee b \, x$$

Disjunction of predicates corresponds to union of sets.

Considering sequents involving a disjunction of predicates, the following Euler diagram is the counterpart of the one above for conjunction:

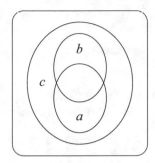

This diagram shows that $a \vDash c$ and $b \vDash c$ is exactly the same as $a \vee b \vDash c$, since $a \vee b$ is the union of a and b. In symbols:

$$a \vDash c \text{ and } b \vDash c \text{ iff } a \subseteq c \text{ and } b \subseteq c \text{ iff } a \cup b \subseteq c \text{ iff } a \vee b \vDash c$$

But wait a minute. The corresponding explanation for conjunction involved "and", and that seemed natural because conjunction is expressed in English using "and". So why do we now have something for disjunction that also involves "and", rather than "or"? The reason is that now the disjunction is in the *antecedent* of the sequent (because a and b are subsets of c) rather than the *succedent* (which we had before, because c was a subset of a and b). We'll come to conjunction in the antecedent and disjunction in the succedent soon.

Anyway, this justifies the rule

$$\frac{a \vDash c \qquad b \vDash c}{a \vee b \vDash c} \vee$$

which is intuitively correct: to be sure that $a \vee b \vDash c$ is valid, we need to know that *both* $a \vDash c$ is valid (in case b is false) *and* that $b \vDash c$ is valid (in case a is false).

And we can define disjunction on predicates in Haskell (Exercise 8.6):

```
(|:|) :: Predicate u -> Predicate u -> Predicate u
(a |:| b) x = a x || b x
```

The "Immediate" Rule

We'll soon need the following rule. It's so obvious that it almost doesn't deserve to be called a rule, so we won't spend much time on it:

$$\frac{}{a \vDash a} \text{ immediate}$$

This says that the sequent $a \vDash a$ follows from no assumptions. An English example of this is: every triangle is a triangle.

We will often need this rule as a way of finishing off a proof.

De Morgan's Laws

In Exercise 1.7, we looked at two important relationships in set theory between union, intersection, and complement, called De Morgan's laws:

$$\overline{A \cup B} = \bar{A} \cap \bar{B} \qquad\qquad \overline{A \cap B} = \bar{A} \cup \bar{B}$$

Augustus De Morgan (1806–1871) was a British mathematician and logician, see ▶ https://en.wikipedia.org/wiki/Augustus_De_Morgan

There was a hint of this back in Exercise 4.5.

These correspond exactly to laws in logic that involve conjunction, disjunction, and negation, and that turn out to follow directly from the rules above.

We start with the following proofs built using those rules, both starting with premises $c \vDash \neg a$ and $c \vDash \neg b$:

$$\dfrac{\dfrac{\dfrac{c \vDash \neg a}{a \vDash \neg c}\text{ contra-}\atop\text{position}\quad\dfrac{c \vDash \neg b}{b \vDash \neg c}\text{ contra-}\atop\text{position}}{a \vee b \vDash \neg c}\vee}{c \vDash \neg(a \vee b)}\text{contraposition} \qquad \dfrac{c \vDash \neg a \quad c \vDash \neg b}{c \vDash \neg a \wedge \neg b}\wedge$$

Both of these proofs involve only equivalences, and both of them start from the same two premises. It follows that they can be combined to give the equivalence

$$\dfrac{c \vDash \neg(a \vee b)}{c \vDash \neg a \wedge \neg b}$$

Now, we can use this equivalence in the following proofs:

$$\dfrac{\dfrac{}{\neg a \wedge \neg b \vDash \neg a \wedge \neg b}\text{ immediate}}{\neg a \wedge \neg b \vDash \neg(a \vee b)} \qquad \dfrac{\dfrac{}{\neg(a \vee b) \vDash \neg(a \vee b)}\text{ immediate}}{\neg(a \vee b) \vDash \neg a \wedge \neg b}$$

Interpreting $\vDash$ as set inclusion, this means that $\neg a \wedge \neg b \subseteq \neg(a \vee b)$ and $\neg(a \vee b) \subseteq \neg a \wedge \neg b$, that is,

$$\neg(a \vee b) = \neg a \wedge \neg b$$

which is the first of De Morgan's laws.

The second of De Morgan's laws arises in a similar way, see Exercise 2.

Sequents Again

A sequent with two antecedents is valid if everything in the universe of discourse that satisfies both of the antecedents also satisfies the succedent. This is equivalent to a sequent with a single antecedent that is the conjunction of those two, since the conjunction is satisfied whenever both of the conjuncts are satisfied. In terms of our Haskell implementation, the following are all equivalent:

```
  and [ c x | x <- things, a x, b x ]
= and [ c x | x <- things, a x && b x ]
= and [ c x | x <- things, (a &:& b) x ]
```

It follows that a proof that a sequent $a, b \vDash c$ is valid also shows that the sequent $a \wedge b \vDash c$ is valid. That gives us the rule

$$\dfrac{a, b \vDash c}{a \wedge b \vDash c}$$

So now we have two rules for conjunction: one for conjunction on the left of a sequent, and one for conjunction on the right of a sequent. Let's give them names which include that information:

$$\dfrac{a, b \vDash c}{a \wedge b \vDash c}\wedge L \qquad\qquad \dfrac{c \vDash a \quad c \vDash b}{c \vDash a \wedge b}\wedge R$$

What about disjunction? So far, we have a rule for disjunction on the left:

$$\frac{a \vDash c \qquad b \vDash c}{a \vee b \vDash c} \vee L$$

but what do we do about disjunction on the right?

Let's see what happens when we use the other rules to work with a sequent having a disjunction on the right.

$$\frac{\dfrac{\dfrac{\dfrac{c \vDash a \vee b}{\neg(a \vee b) \vDash \neg c} \text{ contraposition}}{\neg a \wedge \neg b \vDash \neg c} \text{ De Morgan}}{\neg a, \neg b \vDash \neg c} \wedge L}{c \vDash a, b} \text{ contraposition??}$$

The final step is like contraposition, except that we have moved *both* antecedents to the other side of $\vDash$. But this gives a sequent with two predicates on the right! What does that mean?

Looking at the whole thing, we have deduced—provided that the final step is correct—that $c \vDash a, b$ is equivalent to $c \vDash a \vee b$. So, while multiple antecedents (predicates on the left) corresponds to *conjunction* of conditions, multiple succedents (predicates on the right) corresponds to *disjunction*. That is, a sequent

$$a_0, a_1, \ldots, a_{m-1} \vDash s_0, s_1, \ldots, s_{n-1}$$

is valid if everything in the universe of discourse that satisfies *every* antecedent predicate $a_0, a_1, \ldots, a_{m-1}$ satisfies *at least one* of the succedent predicates $s_0, s_1, \ldots, s_{n-1}$. In symbols:

$$\Gamma \vDash \Delta \text{ if } \bigwedge \Gamma \subseteq \bigvee \Delta$$

(The upper case Greek letters Γ and Δ are traditionally used for the antecedents and succedents of sequents.)

Γ is pronounced "Gamma" and Δ is pronounced "Delta".

And with this interpretation, the following rule deals with disjunction on the right:

$$\frac{c \vDash a, b}{c \vDash a \vee b} \vee R$$

Up until now we've been working with the special case of sequents where there was only one succedent, s_0. We're going to consider the general case from now on. Disjunctive succedents are a little harder to understand than the conjunctive antecedents we've been using up to now, but we need them to give a good set of rules.

Adding Antecedents and Succedents

We've seen that $a \vDash b$ (every a is b) can be drawn as a Venn diagram with a region that's designated as empty:

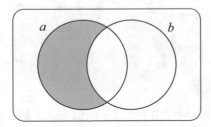

What does it mean to add an additional antecedent g to this sequent to give $g, a \vDash b$?

Instead of adding a circle to the Venn diagram for g, let's draw a line to separate everything for which g is true (the light grey top part of the diagram) from everything for which g is false (the white bottom part):

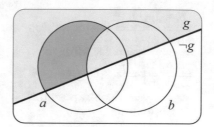

We know that $g, a \vDash b$ is valid if all of the things in the universe that satisfy both g and a also satisfy b. But looking at it in another way, g carves out a subset of the universe, the part where g is true, $\{x \in Universe \mid g\ x\}$. Then, $g, a \vDash b$ is valid provided $a \vDash b$ is valid in that subset of the universe.

For this to be true, the region of a inside g but outside b needs to be empty. It doesn't matter what happens in the other part of the diagram, where g is false. So adding antecedents can be seen as focusing attention on a smaller universe.

The same reasoning holds if we add more than one antecedent, with essentially the same diagram for $\Gamma, a \vDash b$:

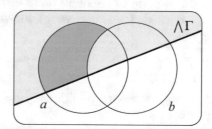

That is, $\Gamma, a \vDash b$ is valid if $a \vDash b$ is valid in $\{x \in Universe \mid \bigwedge \Gamma\ x\}$. Here, we're focusing attention on the part of the universe in which everything in Γ is true.

Now let's look at what it means to add an additional *succedent* d to a sequent $a \vDash b$ to give $a \vDash b, d$. To explain this, we need the following rule for negation on the left of a sequent, which can be viewed as half of contraposition:

$$\frac{a \vDash b, d}{\neg d, a \vDash b}\ \neg L$$

Proving soundness of $\neg L$ is part of Exercise 4.

The sequent on the top of the rule is the one we're interested in, but the rule says that it's equivalent to the sequent on the bottom of the rule, so here's a diagram corresponding to that one:

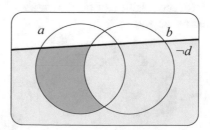

Taking the rule into account, $a \vDash b, d$ is valid if $a \vDash b$ is valid in the subset of the universe where d is *false*, namely $\{x \in Universe \mid \neg d\ x\}$.

If we add more than one succedent, we have the rule

$$\frac{a \vDash b, \Delta}{\neg \bigvee \Delta, a \vDash b}$$

where $\neg \bigvee \Delta$ is the result of taking the disjunction of all the predicates in Δ to create a single predicate which is then moved to the other side of the sequent and negated, using the $\neg L$ rule above.

We could instead move each of the predicates in Δ to the other side of the sequent using the $\neg L$ rule, one at a time, to give

$$\frac{a \vDash b, d_0, d_1, \ldots, d_{m-1}}{\neg d_0, \neg d_1, \ldots, \neg d_{m-1}, a \vDash b}$$

The result is the same, since

$$\neg(d_0 \vee d_1 \vee \cdots \vee d_{m-1}) = \neg d_0 \wedge \neg d_1 \wedge \cdots \wedge \neg d_{m-1}$$

by De Morgan's laws.

The diagram is essentially the same as the one before:

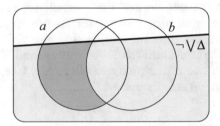

So $a \vDash b, \Delta$ is valid if $a \vDash b$ is valid in the subset of the universe where all of the predicates in Δ are false, namely $\{x \in Universe \mid \neg \bigvee \Delta\, x\}$.

Putting these together gives the following diagram:

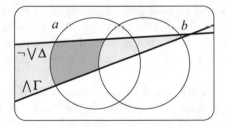

which says that $\Gamma, a \vDash b, \Delta$ is valid if $a \vDash b$ is valid in the subset of the universe $\{x \in Universe \mid \bigwedge \Gamma\, x \wedge \neg \bigvee \Delta\, x\}$, where all of the predicates in Γ are true and all of the predicates in Δ are false.

The upshot is that, since the following rules apply in any universe:

$$\frac{a, b \vDash c}{a \wedge b \vDash c} \wedge L \qquad\qquad \frac{c \vDash a \quad c \vDash b}{c \vDash a \wedge b} \wedge R$$

$$\frac{a \vDash c \quad b \vDash c}{a \vee b \vDash c} \vee L \qquad\qquad \frac{c \vDash a, b}{c \vDash a \vee b} \vee R$$

the following rules are sound, with arbitrary sets of additional antecedents and succedents—the same ones added to all of the premises and to the conclusion of the rule:

$$\frac{\Gamma, a, b \vDash c, \Delta}{\Gamma, a \wedge b \vDash c, \Delta} \wedge L \qquad \frac{\Gamma, c \vDash a, \Delta \qquad \Gamma, c \vDash b, \Delta}{\Gamma, c \vDash a \wedge b, \Delta} \wedge R$$

$$\frac{\Gamma, a \vDash c, \Delta \qquad \Gamma, b \vDash c, \Delta}{\Gamma, a \vee b \vDash c, \Delta} \vee L \qquad \frac{\Gamma, c \vDash a, b, \Delta}{\Gamma, c \vDash a \vee b, \Delta} \vee R$$

since the additional antecedents and succedents simply restrict to the subset of the universe in which all of the predicates in Γ are true and all of the predicates in Δ are false.

Finally, renaming c, Δ as Δ in $\wedge L$ and $\vee L$ and Γ, c as Γ in $\wedge R$ and $\vee R$ gives the final versions of these rules:

$$\frac{\Gamma, a, b \vDash \Delta}{\Gamma, a \wedge b \vDash \Delta} \wedge L \qquad \frac{\Gamma \vDash a, \Delta \qquad \Gamma \vDash b, \Delta}{\Gamma \vDash a \wedge b, \Delta} \wedge R$$

$$\frac{\Gamma, a \vDash \Delta \qquad \Gamma, b \vDash \Delta}{\Gamma, a \vee b \vDash \Delta} \vee L \qquad \frac{\Gamma \vDash a, b, \Delta}{\Gamma \vDash a \vee b, \Delta} \vee R$$

Sequent Calculus

To complete the set of rules, we add the "immediate" rule from above, with additional antecedents and succedents, and rules for negation on the left and right of a sequent, to the rules for conjunction and disjunction above. This generalised form of the immediate rule is obviously sound: if all of the predicates in Γ, a are true, then one of the predicates in a, Δ is true, namely a. For the soundness of $\neg L$ and $\neg R$, see Exercise 4.

$$\frac{}{\Gamma, a \vDash a, \Delta} I$$

$$\frac{\Gamma \vDash a, \Delta}{\Gamma, \neg a \vDash \Delta} \neg L \qquad \frac{\Gamma, a \vDash \Delta}{\Gamma \vDash \neg a, \Delta} \neg R$$

These are the rules of the **sequent calculus**.

We will generally be building proofs from these rules "bottom-up", starting from a desired conclusion and working upwards to discover from which premises that conclusion follows. At each stage, one connective ($\wedge, \vee, \neg$) is eliminated. The names of the rules therefore refer to the connective in the sequent *below* the line that the rule eliminates, and where in the sequent—left (L) or right (R)—it appears.

We've been writing the antecedents and succedents of sequents as lists, with the notation a, Δ suggesting that a is added to the front of Δ, but they're actually (finite) sets. So order doesn't matter—the predicate containing a connective that a rule eliminates can occur anywhere in the antecedents/succedents of the sequent. It's not required to be the last antecedent or the first succedent, which the format of the rules appears to suggest. We could use a notation like $\{a\} \cup \Delta$ instead, but using commas is less clumsy.

14

The idea of sequent and the elegant and symmetric rules of the sequent calculus are due to the German logician Gerhard Gentzen (1909–1945), see ▶ https://en. wikipedia.org/wiki/ Gerhard_Gentzen.

$$\frac{}{\Gamma, a \vDash a, \Delta} I$$

$$\frac{\Gamma \vDash a, \Delta}{\Gamma, \neg a \vDash \Delta} \neg L \qquad \frac{\Gamma, a \vDash \Delta}{\Gamma \vDash \neg a, \Delta} \neg R$$

$$\frac{\Gamma, a, b \vDash \Delta}{\Gamma, a \wedge b \vDash \Delta} \wedge L \qquad \frac{\Gamma \vDash a, \Delta \quad \Gamma \vDash b, \Delta}{\Gamma \vDash a \wedge b, \Delta} \wedge R$$

$$\frac{\Gamma, a \vDash \Delta \quad \Gamma, b \vDash \Delta}{\Gamma, a \vee b \vDash \Delta} \vee L \qquad \frac{\Gamma \vDash a, b, \Delta}{\Gamma \vDash a \vee b, \Delta} \vee R$$

The sequent calculus

Proofs in Sequent Calculus

Let's do a proof with the rules of the sequent calculus.

We'll start with the conclusion $\vDash ((\neg p \vee q) \wedge \neg p) \vee p$ and work upwards, using the rules to eliminate connectives until we have a set of premises—which will be so-called **simple sequents** involving only "bare" predicates, not containing any connectives—from which the conclusion follows. At each step, we'll apply a rule to the "main" connective on the left or on the right.

Given the sequent $\vDash ((\neg p \vee q) \wedge \neg p) \vee p$, we can only apply rules that operate on the right since there's nothing on the left. The main connective is $\vee$, which combines $(\neg p \vee q) \wedge \neg p$ and p to form $((\neg p \vee q) \wedge \neg p) \vee p$, so we need to use the $\vee R$ rule. We can't use $\wedge R$ or $\neg R$ to eliminate the other connectives in $((\neg p \vee q) \wedge \neg p) \vee p$, and we can't apply $\vee R$ to eliminate the other instance of $\vee$, since they're nested inside the "main" application of $\vee$. Applying the $\vee R$ rule—where Γ and Δ are both the empty set of predicates, a is $(\neg p \vee q) \wedge \neg p$ and b is p—gives

$$\frac{\vDash (\neg p \vee q) \wedge \neg p, p}{\vDash ((\neg p \vee q) \wedge \neg p) \vee p} \vee R$$

We now have two predicates on the right. The second one, p, can't be reduced. So, we apply the $\wedge R$ rule, which eliminates the main connective, $\wedge$, of the first one. In this rule application, Γ is empty, $\Delta = \{p\}$, a is $\neg p \vee q$ and b is $\neg p$. That gives:

$$\frac{\dfrac{\vDash \neg p \vee q, p \quad \vDash \neg p, p}{\vDash (\neg p \vee q) \wedge \neg p, p} \wedge R}{\vDash ((\neg p \vee q) \wedge \neg p) \vee p} \vee R$$

We now have two premises, which we have to consider separately. Starting with the first one, we can apply $\vee R$ again to get

$$\frac{\dfrac{\dfrac{\vDash \neg p, q, p}{\vDash \neg p \vee q, p} \vee R \quad \vDash \neg p, p}{\vDash (\neg p \vee q) \wedge \neg p, p} \wedge R}{\vDash ((\neg p \vee q) \wedge \neg p) \vee p} \vee R$$

Continuing, we eventually get to the following proof, which shows that $\vDash ((\neg p \vee q) \wedge \neg p) \vee p$ follows from the empty set of premises:

$$\frac{\dfrac{\dfrac{\dfrac{\boxed{p} \vDash q, \boxed{p}}{\vDash \boxed{\neg} p, q, p} \neg R}{\vDash \neg p \boxed{\vee} q, p} \vee R \quad \dfrac{\dfrac{\boxed{p} \vDash \boxed{p}}{\vDash \boxed{\neg} p, p} \neg R}{} }{\vDash (\neg p \vee q) \boxed{\wedge} \neg p, p} \wedge R}{\vDash ((\neg p \vee q) \wedge \neg p) \boxed{\vee} p} \vee R$$

(In each step, the main connective—or the matching predicates, in the immediate rule—is indicated with a box.) Because there's an empty set of premises, this proof shows that the sequent $\vDash ((\neg p \vee q) \wedge \neg p) \vee p$ is valid in every universe. Such a sequent is called **universally valid**.

Let's look at a slightly more complicated example, and try to prove the conclusion $\vDash \neg ((\neg a \vee b) \wedge (\neg c \vee b)) \vee (\neg a \vee c)$. Applying the same procedure, which stops once no further rule can be applied, we obtain the following proof:

$$\frac{}{\Gamma, a \vDash a, \Delta} I$$

$$\frac{\Gamma \vDash a, \Delta}{\Gamma, \neg a \vDash \Delta} \neg L \qquad \frac{\Gamma, a \vDash \Delta}{\Gamma \vDash \neg a, \Delta} \neg R$$

$$\frac{\Gamma, a, b \vDash \Delta}{\Gamma, a \wedge b \vDash \Delta} \wedge L \qquad \frac{\Gamma \vDash a, \Delta \quad \Gamma \vDash b, \Delta}{\Gamma \vDash a \wedge b, \Delta} \wedge R$$

$$\frac{\Gamma, a \vDash \Delta \quad \Gamma, b \vDash \Delta}{\Gamma, a \vee b \vDash \Delta} \vee L \qquad \frac{\Gamma \vDash a, b, \Delta}{\Gamma \vDash a \vee b, \Delta} \vee R$$

The sequent calculus

Make sure that you understand every step in this proof!

$$\overline{\Gamma, a \vDash a, \Delta}\ I$$

$$\frac{\Gamma \vDash a, \Delta}{\Gamma, \neg a \vDash \Delta}\ \neg L \qquad \frac{\Gamma, a \vDash \Delta}{\Gamma \vDash \neg a, \Delta}\ \neg R$$

$$\frac{\Gamma, a, b \vDash \Delta}{\Gamma, a \wedge b \vDash \Delta}\ \wedge L \qquad \frac{\Gamma \vDash a, \Delta \quad \Gamma \vDash b, \Delta}{\Gamma \vDash a \wedge b, \Delta}\ \wedge R$$

$$\frac{\Gamma, a \vDash \Delta \quad \Gamma, b \vDash \Delta}{\Gamma, a \vee b \vDash \Delta}\ \vee L \qquad \frac{\Gamma \vDash a, b, \Delta}{\Gamma \vDash a \vee b, \Delta}\ \vee R$$

The sequent calculus

$$
\cfrac{
\cfrac{
\cfrac{
I \quad
\cfrac{
\cfrac{\cfrac{a, b \vDash c}{b \vDash \boxed{\neg a}, c}\ \neg R}{b, \boxed{\neg c} \vDash \neg a, c}\ \neg L \quad
\cfrac{a, b \vDash c}{b \vDash \boxed{\neg a}, c}\ \neg R
}{b, \neg c \boxed{\vee} b \vDash \neg a, c}\ \vee L
}{
\cfrac{
\boxed{\neg a}, \neg c \vee b \vDash \boxed{\neg a}, c \qquad b, \neg c \vee b \vDash \neg a, c
}{\neg a \boxed{\vee} b, \neg c \vee b \vDash \neg a, c}\ \vee L
}
}{(\neg a \vee b) \boxed{\wedge} (\neg c \vee b) \vDash \neg a, c}\ \wedge L
}{(\neg a \vee b) \wedge (\neg c \vee b) \vDash \neg a \boxed{\vee} c}\ \vee R
}{\vDash \boxed{\neg}((\neg a \vee b) \wedge (\neg c \vee b)), \neg a \vee c}\ \neg R
$$

$$\frac{\vDash \boxed{\neg}((\neg a \vee b) \wedge (\neg c \vee b)), \neg a \vee c}{\vDash \neg((\neg a \vee b) \wedge (\neg c \vee b)) \boxed{\vee} (\neg a \vee c)}\ \vee R$$

The same premise appears twice in the proof, but we're interested in the *set* of premises from which the conclusion follows, so using it twice doesn't matter. Remember, the antecedents and succedents of each sequent are also sets, which is why we wrote

$$\frac{b, \neg c \vDash \neg a, c \qquad b \vDash \neg a, c}{b, \neg c \boxed{\vee} b \vDash \neg a, c}\ \vee L$$

rather than

$$\frac{b, \neg c \vDash \neg a, c \qquad b, b \vDash \neg a, c}{b, \neg c \boxed{\vee} b \vDash \neg a, c}\ \vee L$$

at the bottom of the right-hand branch of the proof.

This proof shows that $\vDash \neg((\neg a \vee b) \wedge (\neg c \vee b)) \vee (\neg a \vee c)$ follows from the single premise $a, b \vDash c$, meaning that it's valid whenever that sequent is valid. More than that: since all of the rules are equivalences, it also shows that $a, b \vDash c$ is valid whenever $\vDash \neg((\neg a \vee b) \wedge (\neg c \vee b)) \vee (\neg a \vee c)$ is valid. That is, it shows that the following equivalence is sound:

$$\frac{a, b \vDash c}{\vDash \neg((\neg a \vee b) \wedge (\neg c \vee b)) \vee (\neg a \vee c)}$$

It also tells us that any x in the universe for which $a\,x$ and $b\,x$ are both true and $c\,x$ is false would be a counterexample to the conclusion, since it's a counterexample to the premise. If there's more than one premise, then a counterexample to any one of them gives a counterexample to the conclusion.

Notice that there are other possible proofs of $\vDash \neg((\neg a \vee b) \wedge (\neg c \vee b)) \vee (\neg a \vee c)$, obtained by choosing a different connective to eliminate when there's more than one possibility. For example, in the second step (counting from the bottom), we applied $\neg R$ to the first succedent, but we could have instead applied $\vee R$ to the second succedent. But all proofs lead to the same set of premises.

Using these rules, the procedure of constructing a proof will always terminate. That's easy to see, because each rule eliminates one connective. The starting sequent can only contain a finite number of connectives, and that limits the maximum possible *depth* of the proof.

When there's a choice between a rule having two premises, like $\vee R$, and a rule having one premise, like $\neg R$, it's generally better to apply the rule having one premise. Both rules will need to be applied eventually, and applying the rule with two premises first would lead to the other rule needing to be applied twice, once on each branch.

14

Exercises

1. See Exercise 8.5 for Haskell definitions of validity of sequents with one antecedent and one succedent, and of sequents with a list of antecedents and one succedent. Give a Haskell definition of validity of sequents in the general case, with a list of antecedents *and* a list of succedents. Test that the previous definitions are special cases of this general definition.

   Hint: To do QuickCheck tests involving predicates, you'll need to include the following code, with the first line at the top of your file:

   ```
   {-# LANGUAGE FlexibleInstances #-}
   instance Arbitrary Thing where
     arbitrary = elements [R, S, T, U, V, W, X, Y, Z]

   instance CoArbitrary Thing where
     coarbitrary R = variant 0
     coarbitrary S = variant 1
     coarbitrary T = variant 2
     coarbitrary U = variant 3
     coarbitrary V = variant 4
     coarbitrary W = variant 5
     coarbitrary X = variant 6
     coarbitrary Y = variant 7
     coarbitrary Z = variant 8

   instance Show (u -> Bool) where
     show p = "a predicate"
   ```

   If a test case fails, QuickCheck won't be able to show you the failing predicate because it's a function.

2. Derive the second of De Morgan's laws

   $$\neg(a \wedge b) = \neg a \vee \neg b$$

   using an argument like the one above for the first law.

3. The addition of a succedent d to a sequent $a \vDash b$ to give $a \vDash b, d$ was explained using the diagram

   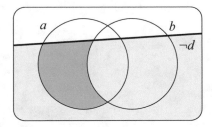

   Use the same circles and lines, with the same identification (a, b, d) as above, to make a diagram to explain the addition of a succedent b to a sequent $a \vDash d$ to give $a \vDash b, d$.

4. Explain why the $\neg L$ and $\neg R$ rules are sound.

$$\frac{}{\Gamma, a \vDash a, \Delta} \, I$$

$$\frac{\Gamma \vDash a, \Delta}{\Gamma, \neg a \vDash \Delta} \, \neg L \qquad \frac{\Gamma, a \vDash \Delta}{\Gamma \vDash \neg a, \Delta} \, \neg R$$

$$\frac{\Gamma, a, b \vDash \Delta}{\Gamma, a \wedge b \vDash \Delta} \, \wedge L \qquad \frac{\Gamma \vDash a, \Delta \quad \Gamma \vDash b, \Delta}{\Gamma \vDash a \wedge b, \Delta} \, \wedge R$$

$$\frac{\Gamma, a \vDash \Delta \quad \Gamma, b \vDash \Delta}{\Gamma, a \vee b \vDash \Delta} \, \vee L \qquad \frac{\Gamma \vDash a, b, \Delta}{\Gamma \vDash a \vee b, \Delta} \, \vee R$$

The sequent calculus

$$\frac{}{\Gamma, a \vDash a, \Delta} \, I$$

$$\frac{\Gamma \vDash a, \Delta}{\Gamma, \neg a \vDash \Delta} \, \neg L \qquad \frac{\Gamma, a \vDash \Delta}{\Gamma \vDash \neg a, \Delta} \, \neg R$$

$$\frac{\Gamma, a, b \vDash \Delta}{\Gamma, a \wedge b \vDash \Delta} \, \wedge L \qquad \frac{\Gamma \vDash a, \Delta \quad \Gamma \vDash b, \Delta}{\Gamma \vDash a \wedge b, \Delta} \, \wedge R$$

$$\frac{\Gamma, a \vDash \Delta \quad \Gamma, b \vDash \Delta}{\Gamma, a \vee b \vDash \Delta} \, \vee L \qquad \frac{\Gamma \vDash a, b, \Delta}{\Gamma \vDash a \vee b, \Delta} \, \vee R$$

The sequent calculus

5. Do a proof that reduces the conclusion $(x \wedge y) \vee (x \wedge z) \vDash x \wedge (y \vee z)$ to premises that can't be reduced further. Is it universally valid? If not, give a counterexample.

6. Do proofs which reduce the conclusions $\neg a \wedge \neg b \vDash \neg(a \wedge b)$ and $\neg(a \wedge b) \vDash \neg a \wedge \neg b$ to premises that can't be reduced further. Is one or both universally valid? If not, give a counterexample. If so, explain how that shows that $\neg a \wedge \neg b = \neg(a \wedge b)$.

7. (a) Consider a rule like

$$\frac{\Gamma, c \vDash a, \Delta \quad \Gamma, c \vDash b, \Delta}{\Gamma, c \vDash a \wedge b, \Delta} \, \wedge R$$

that has two premises. Turning it upside down gives a rule with two conclusions. What interpretation of rules with two conclusions makes this rule sound, as required for $\wedge R$ to be an equivalence?

(b) The only rule in the sequent calculus that has not been written as an equivalence is the immediate rule:

$$\frac{}{\Gamma, a \vDash a, \Delta} \, I$$

Show that this rule is actually also an equivalence. (Hint: first, generalise your answer to (a) to the case of rules with $n \neq 1$ premises.)

8. (a) Show that De Morgan's Laws are universally valid (that is, $\neg(a \vee b) \vDash \neg a \wedge \neg b$ and $\neg a \wedge \neg b \vDash \neg(a \vee b)$, and likewise for the other De Morgan's Law) using sequent calculus.

(b) Show that De Morgan's Laws are sound using truth tables.

14

Algebraic Data Types

Contents

More Types

So far, we've done a lot using the types that come "out of the box" with Haskell. The type of lists has been particularly useful, and higher-order functions have revealed the power of the function type ->. Both of these actually provide an infinite number of types: there is a type [t] for every type t and a type s -> t for every s and t.

But there's more to come, because Haskell provides a powerful mechanism called **algebraic data types** for you to define your own types. In fact, most of the types that you have seen up to now—including lists—could have been left out of Haskell, for you to define yourself. It's useful to have them built in, so that Haskell can provide some special notation for them, and lots of handy pre-defined functions. But the fact that you can define lists yourself, and all of the functions over them, means that having them built in is just a convenience.

These are called *algebraic* data types because new types are created by taking the sum of products of existing types, like polynomials in algebra, see ▶ https://en.wikipedia.org/wiki/Algebraic_data_type.

Haskell's syntax for type definitions is pretty simple but you need to see lots of examples to appreciate what you can do with it. So, we're going to approach this via a sequence of examples, starting with simple ones and then building up.

Booleans

Our first example is the built-in type `Bool` of Boolean values. This is review because we already saw how `Bool` was defined back in Chap. 2:

```
data Bool = False | True
```

This is a very simple example of an algebraic data type. Let's look at the details before going on to more complex examples.

First, "`data`" says that this is an algebraic data type definition. The type being defined, `Bool`, is on the left-hand side of the definition. As with all types, its name begins with an upper case letter. On the right-hand side is a list of all of the ways of forming a value of that type, separated by vertical bars. In this case there are just two values, `False` and `True`, and they are "formed" by simply writing their names. Their names begin with an upper case letter because they are **constructors**. This means that they can be used in patterns, as in the following definitions of Boolean equality and conversion from `Bool` to `String`:

List types, function types, and tuple types don't have names that begin with an upper case letter, because their names are composed of symbols rather than letters.

```
eqBool :: Bool -> Bool -> Bool
eqBool False False = True
eqBool True  True  = True
eqBool _     _     = False

showBool :: Bool -> String
showBool False = "False"
showBool True  = "True"
```

(Because `False` and `True` are literal values, they can be used in patterns even if they weren't constructors. But soon we'll see examples where that isn't the case.)

The algebraic data type definition above is a complete definition of the type `Bool`. Note in particular that there is no need to define a representation of the values `False` and `True` in terms of values of some other type. Conceptually, values of type `Bool` are simply the expressions `False` and `True`. Of course, the computer's internal representation of `Bool` and all other data is in terms of bits, but we don't need to know any of the details of how that works to write programs.

Seasons

Our next example is for the seasons of the year:

```
data Season = Winter | Spring | Summer | Fall
```

This is similar to `Bool` but with four constructors instead of two. `Bool` and `Seasons` are called **enumerated types** because they are defined by simply listing (i.e. enumerating) their values. The types `Weekday` on page 10 and `Thing` on page 46 are other examples.

Here are functions for computing the next season, for equality of seasons, and for converting from `Season` to `String`:

```
next :: Season -> Season
next Winter = Spring
next Spring = Summer
next Summer = Fall
next Fall   = Winter

eqSeason :: Season -> Season -> Bool
eqSeason Winter Winter = True
eqSeason Spring Spring = True
eqSeason Summer Summer = True
eqSeason Fall   Fall   = True
eqSeason _      _      = False

showSeason :: Season -> String
showSeason Winter = "Winter"
showSeason Spring = "Spring"
showSeason Summer = "Summer"
showSeason Fall   = "Fall"
```

It's very tempting to write
`eqSeason s s = True`
to cover the first four cases of this definition, but that doesn't work: remember that repeated variables in patterns aren't allowed!

The definitions of eqSeason and showSeason (and the earlier definitions of eqBool and showBool) are boring, but they're required for computing equality and for displaying values of type **Season**. Luckily, we can get Haskell to work out these definitions automatically for itself by adding a magic incantation to the type definition:

```
data Season = Winter | Spring | Summer | Fall deriving (Eq,Show)
```

This use of the **type classes** Eq and Show, and how type classes work in general, will be explained in Chap. 24.

Not only does this incantation define these functions automatically; it also makes Haskell incorporate them into the built-in == and show functions. The latter is used to display values when working interactively. Observe the difference between the first and second versions of the type definition:

```
> data Season = Winter | Spring | Summer | Fall
> next Winter
<interactive>:2:1: error:
    • No instance for (Show Season) arising from a use of 'print'
    • In a stmt of an interactive GHCi command: print it
> Winter == Spring
<interactive>:3:1: error:
    • No instance for (Eq Season) arising from a use of '=='
    • In the expression: Winter == Spring
      In an equation for 'it': it = Winter == Spring

> data Season = Winter | Spring | Summer | Fall deriving (Eq,Show)
> next Winter
Spring
> Winter == Spring
False
```

The error messages refer to it because that's the name that Haskell gives to the last expression typed during an interactive session. This allows convenient reference to the last thing typed.

If "`deriving (Eq,Show)`" is left out, then Haskell can't display the result of computations producing values of type `Season` or compare them using `==`. Both things work when "`deriving ...`" is added.

Another way of defining the functions `next` and `eqSeason` is in terms of functions that convert back and forth from `Season` to `Int`:

```
toInt :: Season -> Int
toInt Winter = 0
toInt Spring = 1
toInt Summer = 2
toInt Fall   = 3


fromInt :: Int -> Season
fromInt 0 = Winter
fromInt 1 = Spring
fromInt 2 = Summer
fromInt 3 = Fall


next :: Season -> Season
next x = fromInt ((toInt x + 1) `mod` 4)


eqSeason :: Season -> Season -> Bool
eqSeason x y = (toInt x == toInt y)
```

Defining `toInt` and `fromInt` such that `fromInt (toInt s)` = s allows us to give shorter definitions of next and eqSeason.

Shapes

The types `Bool` and `Season` were defined by enumerating their values, represented by constructors. The following example of geometric shapes is different: its values are also formed using constructors, but applied to values of another type.

We'll start by defining a few type synonyms, to help us remember the intended meaning of measurements used in defining geometric shapes:

Remember that these just define different names for `Float`. We have `3.1 :: Float`, `3.1 :: Radius`, etc.

```
type Radius = Float
type Width  = Float
type Height = Float
```

Now we give the type definition. To keep things simple, we'll only use two geometric shapes, circles and rectangles:

```
data Shape = Circle Radius
           | Rect Width Height
  deriving (Eq,Show)
```

The right-hand side of the definition says that there are two kinds of shapes. One kind is formed by applying the constructor `Circle` to a value of type `Radius`. The other kind is formed by applying the constructor `Rect` to two values, one of type `Width` and one of type `Height`. So the constructors are functions, `Circle :: Radius -> Shape` and `Rect :: Width -> Height -> Shape`. The expression `Circle 3.1` represents a circle with radius 3.1, and the expression `Rect 2.0 1.3` represents a rectangle with width 2.0 and height 1.3. The type of value that `Circle` takes as its parameter is listed after the constructor name in the type definition, and the same for `Rect`.

We can now define the area of a shape by cases, using pattern matching with the constructors `Circle` and `Rect` to discriminate between cases and to extract the radius or width/height:

```
area :: Shape -> Float
area (Circle r) = pi * r^2
area (Rect w h) = w * h
```

It is convenient and does no harm to add "`deriving (Eq,Show)`" to most algebraic data type definitions, and we'll do that for all of the remaining examples. But there is a problem in examples like
```
data T = C (Int -> Int)
```
since functions can't be tested for equality or printed.

The same goes for equality of shapes, which refers to equality on `Float`, and for conversion from `Shape` to `String`, which refers to show for `Float`:

```
eqShape :: Shape -> Shape -> Bool
eqShape (Circle r) (Circle r')  = (r == r')
eqShape (Rect w h) (Rect w' h') = (w == w') && (h == h')
eqShape _          _            = False

showShape :: Shape -> String
showShape (Circle r) = "Circle " ++ showF r
showShape (Rect w h) = "Rect " ++ showF w ++ " " ++ showF h

showF :: Float -> String
showF x | x >= 0    = show x
        | otherwise = "(" ++ show x ++ ")"
```

showF puts parentheses around negative numbers in order to make showShape produce Circle (-1.7) rather than Circle -1.7. Typing the latter into Haskell gives a type error.

Because the type definition includes "`deriving (Eq,Show)`", both of these function definitions are generated automatically and incorporated into the built-in `==` and `show` functions.

As we've already seen for functions on lists, patterns with variables make it possible to write simple and concise function definitions. The alternative is to define the following functions—using pattern matching—for discriminating between cases and extracting values:

```
isCircle :: Shape -> Bool
isCircle (Circle r) = True
isCircle (Rect w h) = False

isRect :: Shape -> Bool
isRect (Circle r) = False
isRect (Rect w h) = True

radius :: Shape -> Float
radius (Circle r) = r

width :: Shape -> Float
width (Rect w h) = w
```

The elegant combination of algebraic data types with pattern-matching function definitions was first introduced by Rod Burstall (1934–), a British computer scientist and Professor Emeritus at the University of Edinburgh, see ▶ https://en. wikipedia.org/wiki/Rod_Burstall.

```
height :: Shape -> Float
height (Rect w h) = h
```

and then using them to write the function area. Yuck!

```
area :: Shape -> Float
area s =
  if isCircle s then
    let
      r = radius s
    in
      pi * r^2
  else if isRect s then
    let
      w = width s
      h = height s
    in
      w * h
  else error "impossible"
```

This is the way that the computer interprets the two-line definition of area given above, where
let *var* = *exp* in *exp′* is another way of writing *exp′* where *var* = *exp*.

Tuples

Our first example of a polymorphic data type definition defines the type of pairs:

```
data Pair a b = Pair a b deriving (Eq,Show)
```

The type `Pair Int Bool` is the one that we write (`Int,Bool`) using Haskell's built-in type of pairs, with the value `Pair 3 True` of type `Pair Int Bool` being written (`3,True`).

The type variables a and b are used to indicate the polymorphism, meaning that the same definition also gives types/values like

```
Pair [1,2] 'b' :: Pair [Int] Char
```

Of course, the types used in place of a and b may be the same, as in

```
Pair 3.1 2.45 :: Pair Float Float
```

The variables a and/or b are only used once on the right-hand side of this type definition, but this is not a requirement: they may be used multiple times, or not at all. We'll see an example of multiple use when we define lists in the next section.

This type definition may look confusing because it includes an equation with the same thing on both sides! It's important to understand that these have different purposes: on the left-hand side, `Pair` a b defines the name of the *type*; on the right-hand side, `Pair` a b defines the name of the *constructor* used to produce *values* of that type. Another way of writing the definition would be to use different names for the type and the constructor:

```
data Pair a b = MkPair a b deriving (Eq,Show)
```

with `MkPair 3 True :: Pair Int Bool`, but the first definition is more in tune with Haskell's use of the same notation for types and values in examples like (`3,True`) :: (`Int,Bool`) and [1,2] :: [Int].

The following definitions of equality and the conversion from pairs to strings are generated automatically:

```
eqPair :: (Eq a, Eq b) => Pair a b -> Pair a b -> Bool
eqPair (Pair x y) (Pair x' y') = x == x' && y == y'

showPair :: (Show a, Show b) => Pair a b -> String
showPair (Pair x y) = "Pair " ++ show x ++ " " ++ show y
```

Since equality of pairs relies on equality of the pair's components, we can only check equality of a pair of type **Pair** *t s* when == works on both of the types *t* and *s*. That's what "(**Eq** a, **Eq** b) =>" in the type of eqPair means. A similar comment applies to the type of showPair, where what is required of the component types is that show can be used to convert their values to strings. This use of **Eq**, which is the same as we've seen earlier, will be explained in more detail in Chap. 24. Ditto for **Show**.

We need a separate definition for the type of triples:

```
data Triple a b c = Triple a b c deriving (Eq,Show)
```

and the same for *n*-tuples for any other *n*.

Lists

Recall the definition of Haskell's built-in type of lists from Chap. 10:

Definition. A **list** of type [*t*] is either

1. *empty*, written [], (empty list) or
2. *constructed*, written *x* : *xs*, with *head x* (an element of type *t*) and *tail xs* (a list of type [*t*]).

A definition of lists as a polymorphic algebraic data type says precisely the same thing in symbols, with the name **Nil** instead of [] and the name **Cons** instead of infix : to avoid clashes with built-in lists:

```
data List a = Nil
            | Cons a (List a)
  deriving (Eq,Show)
```

This example demonstrates the use of recursion in algebraic data type definitions, by using the name of the type being defined—in this case, **List**—on the right-hand side of the definition. How this works for building values of type **List** is exactly as explained earlier for built-in lists:

- **Nil** is a value of type **List Int**, since **Nil** :: **List** a;
- and so **Cons 8 Nil** is a value of type **List Int**, since **Cons** :: a -> **List** a -> **List** a;
- and so **Cons 6 (Cons 8 Nil)** is a value of type **List Int**;
- and so **Cons 4 (Cons 6 (Cons 8 Nil))** is a value of type **List Int**.

Compare this explanation of **Cons 4 (Cons 6 (Cons 8 Nil))** :: **List Int** with the explanation of why 4:(6:(8:[])) is a list of type [**Int**] on page 82.

As in our previous examples, **Nil** and **Cons** are constructors and so can be used in pattern-matching function definitions. For example, the following definition of append for **List** a is the same as the definition of ++ for [a] on page 96:

```
append :: List a -> List a -> List a
append Nil ys          = ys
append (Cons x xs) ys = Cons x (append xs ys)
```

Note once more how the pattern of recursion in the function definition mirrors the recursion in the type definition. We will see the same thing with definitions of other algebraic data types below that involve recursion.

What we don't get with the above data type definition is Haskell's special notations for lists, such as [`True`,`False`,`False`], [1..10], and list comprehension notation. We also don't get Haskell's identification of `String` with `List Char`, or string notation "Haskell!".

Optional Values

There are situations in which a function is required to return a value but there is no sensible value to return. An example is the division function, where division by zero is undefined.

One way of dealing with such a situation is to generate an error, which causes computation to halt:

```
> 3 `div` 0
*** Exception: divide by zero
```

The error in this example was produced by applying the Prelude function `error` to the string "divide by zero".

Another involves use of the built-in type `Maybe` to indicate that the function will return an **optional value** as its result, for which one of the possible values represents *absence* of a value. That probably sounds confusing, so let's look at how `Maybe` is defined and how it's used.

The definition is simple:

```
data Maybe a = Nothing | Just a
  deriving (Eq,Show)
```

`Maybe` is a polymorphic type, with `Maybe` *t* being for optional values of type *t*. There are two constructors: one (`Nothing`) indicating absence of a value; and one (`Just :: a -> Maybe a`) for when a value is present. To show how that works, here's a version of integer division that returns an optional value of type `Int`:

```
myDiv :: Int -> Int -> Maybe Int
myDiv n 0 = Nothing
myDiv n m = Just (n `div` m)
```

Then we get:

```
> 3 `myDiv` 0
Nothing
> 6 `myDiv` 2
Just 3
```

A list of pairs used in this way is called an **association list**, see ▶ https://en.wikipedia.org/wiki/Association_list.

Another example is the Prelude function `lookup` which searches a list of pairs for a pair whose first component matches a given key and returns its second component:

```
lookup :: Eq a => a -> [(a,b)] -> Maybe b
lookup key []                    = Nothing
lookup key ((x,y):xys) | key == x = Just y
                       | otherwise = lookup key xys
```

The result type is `Maybe b` rather than b because there may be no such pair in the list. In that case, `lookup` returns `Nothing`. If it does find a matching pair (*x*,*y*), it returns `Just` *y*.

`Maybe` can also be used in the type of an **optional function parameter**. This can be useful when there is a default value for the parameter which can be overridden by supplying a different value.

The following function raises the first parameter to the power given by the second parameter. When the first parameter is absent (`Nothing`), the default value of 2 is used:

15

```
power :: Maybe Int -> Int -> Int
power Nothing n  = 2 ^ n
power (Just m) n = m ^ n
```

Then:

```
> power Nothing 3
8
> power (Just 3) 3
27
```

The fact that Haskell's type system keeps careful track of the distinction between the type t and the type `Maybe` t means that some extra work is required when supplying and using optional values. Obviously, power won't work if you simply omit the first parameter, relying on Haskell to figure out what you mean:

```
> power 3
<interactive>:2:7: error:
    • No instance for (Num (Maybe Int)) arising from the literal '3'
    • In the first argument of 'power', namely '3'
      In the expression: power 3
      In an equation for 'it': it = power 3
```

You need to signal that there is no value for the first parameter by supplying the parameter `Nothing`, as above. Similarly, if you do want to supply a value for the first parameter then you need to apply the constructor `Just` to it, yielding a value of type `Maybe Int`, before passing it to power.

For the same reason, the result of a function that returns an optional result typically needs to be unpacked before it can be used. Forgetting to do so produces a type error. For instance, the following attempt to use the result of `myDiv` in an expression that requires an `Int`:

```
wrong :: Int -> Int -> Int
wrong n m = (n `myDiv` m) + 3
```

yields the error:

```
    • Couldn't match expected type 'Int' with actual type 'Maybe Int'
    • In the expression: (n `myDiv` m) + 3
      In an equation for 'wrong': wrong n m = (n `myDiv` m) + 3
Compilation failed.
```

because myDiv produces a value of `Maybe Int`, not `Int`.

Here is a corrected version of this example, in which a **case** expression is used to deal with the different possible results of myDiv:

```
right :: Int -> Int -> Int
right n m = case n `myDiv` m of
              Nothing -> 3
              Just r  -> r + 3
```

A **case** expression allows case analysis via pattern matching on values that are not function parameters. The syntax should be self-explanatory.

When the result is `Just` r, the value r `:: Int` is available for use in the rest of the computation.

Disjoint Union of Two Types

The type `Maybe` t combines values of t with a value that indicates the absence of a value. The built-in type `Either`, which has a similar definition, can be used to combine the values of two different types into a single type:

```
data Either a b = Left a | Right b
  deriving (Eq,Show)
```

The polymorphic type `Either` is used to combine two types, and so it has two constructors: one (`Left :: a -> Either a b`) for values of the first type; and one (`Right :: b -> Either a b`) for values of the second type. So values of the type `Either Int String` are either `Left` *n* for some *n* `:: Int` or `Right` *s* for some *s* `:: String`.

Here's an example of the use of `Either` to produce a list containing both integers and strings:

```
mylist :: [Either Int String]
mylist = [Left 4, Left 1, Right "hello", Left 2,
          Right " ", Right "world", Left 17]
```

It's important to understand that `mylist` is not a counterexample to the principle that all of the values in a Haskell list have the same type! All of the values of `mylist` do have the same type, namely `Either Int String`. The constructors `Left` and `Right`, which "inject" values of `Int` and `String` into that type, allow both types of values to belong to the same list.

To show how to write code that uses values of such a type, here's a function that adds together all of the integers in a list like `mylist`:

```
addints :: [Either Int String] -> Int
addints xs = sum [ n | Left n <- xs ]
```

and another function that concatenates all of the strings in such a list:

```
addstrs :: [Either Int String] -> String
addstrs xs = concat [ s | Right s <- xs ]
```

Then we get

```
> addints mylist
24
> addstrs mylist
"hello world"
```

The type `Either` *s t* can be thought of as the **union** of the types *s* and *t*, but note that it is actually the **disjoint union**. The constructors `Left` and `Right` distinguish whether a value of `Either` *s t* comes from *s* or *t*, even if *s* and *t* are the same type. Thus we can use it to produce a type containing two "copies" of the values of a single type, as in `Either Int Int` which contains values of the form `Left` *n* and `Right` *n* for every *n* `:: Int`.

It's tempting to omit the constructors `Left`/`Right`—or the constructor `Just` in the case of `Maybe`—and to hope that Haskell will somehow figure out what you mean. That is unfortunately not possible: the constructors are the key to making polymorphic typechecking work for algebraic data types. See ▶ https://en.wikipedia.org/wiki/Hindley-Milner_type_system for an entry to the relevant literature.

The definitions of `addints` and `addstrs` reveal a subtle difference between patterns in function definitions—where a missing case can lead to an error when the function is applied—and patterns in comprehensions, which can be used as here to select values from a generator that match the pattern, disregarding the rest.

15

Exercises

1. Consider the following declaration:

```
data Fruit = Apple String Bool
           | Orange String Int
    deriving (Eq,Show)
```

An expression of type `Fruit` is either an `Apple` or an `Orange`. We use a `String` to indicate the variety of the apple or orange, a `Bool` to say whether an apple has a worm, and an `Int` to count the number of segments in an orange. For example:

```
Apple "Bramley" False   -- a Bramley apple with no worm
Apple "Braeburn" True   -- a Braeburn apple with a worm
Orange "Moro" 10        -- a Moro orange with 10 segments
```

(a) Write a function `isBloodOrange :: Fruit -> Bool` which returns `True` for blood oranges and `False` for apples and other oranges. Blood orange varieties are: Tarocco, Moro and Sanguinello. For example:

```
isBloodOrange (Orange "Moro" 12) == True
isBloodOrange (Apple "Granny Smith" True) == False
```

(b) Write a function `bloodOrangeSegments :: [Fruit] -> Int` which returns the total number of blood orange segments in a list of fruit.

(c) Write a function `worms :: [Fruit] -> Int` which returns the number of apples that contain worms.

2. Extend `Shape` and `area :: Shape -> Float` to include further kinds of geometric shapes: triangles (defined by the lengths of two sides and the angle between them); rhombuses (defined by the lengths of the diagonals); and regular pentagons (defined by the length of a side).

3. "Wide lists" are like ordinary lists but with an additional constructor, `Append`, which creates a wide list by appending two wide lists.

(a) Define an algebraic data type `Widelist a`.

(b) Define the following functions on wide lists:

```
lengthWide :: Widelist a -> Int
nullWide :: Widelist a -> Bool
reverseWide :: Widelist a -> Widelist a
```

(c) Define a function

```
fixWide : Widelist a -> [a]
```

which converts a wide list to an ordinary Haskell list of its elements in the same order. Use `fixWide` to test that `lengthWide`, `nullWide` and `reverseWide` produce the same results as the corresponding functions on ordinary lists. To do the tests, you'll need to include the following code to generate random wide lists:

```
import Control.Monad
instance Arbitrary a => Arbitrary (Widelist a) where
  arbitrary = sized list
    where
      list n | n<=0 = return Nil
             | otherwise
                     = oneof [liftM2 Cons arbitrary sublist,
                              liftM2 Append sublist sublist]
          where sublist = list (n `div` 2)
```

(d) Define functions for mapping and folding over wide lists:

```
mapWide :: (a -> b) -> Widelist a -> Widelist b
foldrWide :: (a -> b -> b) -> b -> Widelist a -> b
```

4. Define a function `myInsertionSort :: Ord a => List a -> List a` corresponding to `insertionSort` on page 85.
Define a function `toList :: [a] -> List a` and use it to test that `myInsertionSort` produces the same results as `insertionSort`.

5. Write a data type definition for natural numbers (`Nat`) that reflects the definition of natural numbers on page 95. Then define functions

```
myPlus  :: Nat -> Nat -> Nat
myTimes :: Nat -> Nat -> Nat
myPower :: Nat -> Nat -> Nat
```

Define a function `toNat :: Int -> Nat` and use it to test that `myPlus` and `myTimes` produce the same results as `plus` and `times` on page 95 for parameters between 0 and 100. (You can check `myPower` against `power` too, but you will need to limit the range of parameters much more severely.)

6. Define a function

```
createLock :: Password -> a -> Locked a
```

where

```
type Password = String
type Locked a = Password -> Maybe a
```

The function `createLock` should take a password and some data and create a locked version of the data, such that the data in the locked version can only be accessed by supplying the correct password. For example:

```
> locked = createLock "secret" 12345
> locked "wrong"
Nothing
> locked "secret"
Just 12345
```

7. Define the function

```
mapMaybe :: (a -> Maybe b) -> [a] -> [b]
```

(in the **Data.Maybe** library module) which is analogous to map except that it throws away values of the list for which the function yields **Nothing** and keeps *v* when the function's result is of the form **Just** *v*.
Define composition for functions delivering a result of **Maybe** type:

```
(...) :: (b -> Maybe c) -> (a -> Maybe b) -> (a -> Maybe c)
```

8. Define the Prelude function

```
either :: (a -> c) -> (b -> c) -> Either a b -> c
```

that combines functions of types a -> c and b -> c to give a function of type **Either** a b -> c. Use either to define a function

```
join :: (a -> c) -> (b -> d) -> Either a b -> Either c d
```

Expression Trees

Contents

© The Author(s), under exclusive license to Springer Nature Switzerland AG 2021
D. Sannella et al., *Introduction to Computation*, Undergraduate Topics
in Computer Science, https://doi.org/10.1007/978-3-030-76908-6_16

Trees

Tree-like structures are ubiquitous in Informatics. They are used to provide conceptual models of situations and processes involving hierarchies, for representing the syntax of languages involving nesting, and for representing data in a way that is amenable to processing by recursive algorithms. We have already seen some pictures of trees in Chap. 12, in the explanation of `foldr` and `foldl`, and the sequent calculus proofs in Chap. 14 have a tree-like structure.

Algebraic data types are ideal for representing tree-like structures. The combination of algebraic data types for representing language syntax with pattern-matching function definitions for writing functions that operate on syntax can be regarded as a "killer app" of Haskell and similar functional programming languages.

We will now look at some examples of the use of algebraic data types for representing the syntax of simple languages. Our second example provides a link to the chapters on logic by showing how Haskell can be used to represent and manipulate logical expressions and to solve problems in logic.

Early uses of functional programming languages were in connection with computer-assisted theorem proving systems, where trees are used for representing the syntax of the logical language and for representing proofs, among other things. See for instance ▶ https://en. wikipedia.org/wiki/ Logic_for_Computable_Functions.

Arithmetic Expressions

We will begin by looking at simple arithmetic expressions involving integers, addition and multiplication. The extension with other arithmetic operations is left as an exercise.

```
data Exp = Lit Int
         | Add Exp Exp
         | Mul Exp Exp
   deriving Eq
```

This data type definition says that there are three kinds of values of type `Exp`:

- the constructor `Lit` (short for "literal") applied to an `Int`;
- the constructor `Add` (short for "addition") applied to two values of type `Exp`; and
- the constructor `Mul` (short for "multiplication") applied to two values of type `Exp`.

The last two cases involve recursion. Just as with lists, complicated values of type `Exp` are built from simpler values, starting with the simplest values of all, which involve the constructor `Lit` and no recursion. A difference with respect to lists is that there are two recursive cases, rather than just one (`Cons`). Another is that each of the recursive cases require two values of the type being defined, rather than one.

For this type, we'll use "`deriving Eq`" in place of "`deriving (Eq,Show)`" because we will provide a definition of `show` for `Exp` that is different from the one that Haskell generates automatically.

Here are some example values of type `Exp`:

```
e0 = Add (Lit 1) (Mul (Lit 2) (Lit 3))
e1 = Mul (Add (Lit 1) (Lit 2)) (Lit 3)
e2 = Add e0 (Mul (Lit 4) e1)
```

Unfortunately, it's a little hard to see what expressions we have just defined. The following function for converting from `Exp` to `String` produces a more familiar notation:

```
showExp :: Exp -> String
showExp (Lit n)   = show n
showExp (Add e f) = par (showExp e ++ " + " ++ showExp f)
showExp (Mul e f) = par (showExp e ++ " * " ++ showExp f)
```

```
par :: String -> String
par s = "(" ++ s ++ ")"
```

and then the following (to be explained in Chap. 24) incorporates this into the built-in show function:

```
instance Show Exp where
  show e = showExp e
```

which gives:

```
> e0
(1 + (2 * 3))
> e1
((1 + 2) * 3)
> e2
((1 + (2 * 3)) + (4 * ((1 + 2) * 3)))
```

Alternatively, combine the function definition with the **instance** declaration:

```
instance Show Exp where
  show (Lit n)
    = show n
  show (Add e f)
    = par (show e ++ " + " ++ show f)
  show (Mul e f)
    = par (show e ++ " * " ++ show f)
```

The definition of showExp uses full parenthesisation to show the structure of expressions.

The expressions e0 and e1 involve the same literals and the same operations in the same order, but it's obvious from a comparison of their parenthesised outputs that they have different structures. Drawing them as trees makes it clear how their nesting structures differ:

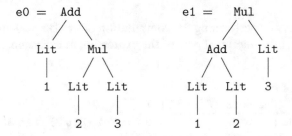

And drawing e2 as a tree shows how its sub-expressions e0 and e1 (drawn in boxes) contribute to the overall structure of the expression:

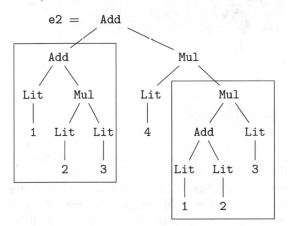

Evaluating Arithmetic Expressions

You learned when you were a child how to compute the value of expressions like e2. What you learned is expressed in Haskell by the following recursive function definition:

```
evalExp :: Exp -> Int
evalExp (Lit n)   = n
evalExp (Add e f) = evalExp e + evalExp f
evalExp (Mul e f) = evalExp e * evalExp f
```

Taking the three cases of evalExp in turn:

- the value of Lit n is just n :: Int;
- to get the value of Add $e\,f$, add the value of e and the value of f; and
- to get the value of Mul $e\,f$, multiply the value of e and the value of f.

This gives the following results:

```
> evalExp e0
7
> evalExp e1
9
> evalExp e2
43
```

Let's see how this works by performing the computation of evalExp e0 one step at a time, expanding the underlined part of the expression at each step:

evalExp e0
      Retrieving value of e0
= evalExp (Add (Lit 1) (Mul (Lit 2) (Lit 3)))
      Applying 2nd equation, with e = Lit 1 and f = Mul (Lit 2) (Lit 3)
= evalExp (Lit 1) + evalExp (Mul (Lit 2) (Lit 3))
      Applying 1st equation, with n = 1
= 1 + evalExp (Mul (Lit 2) (Lit 3))
      Applying 3rd equation, with e = Lit 2 and f = Lit 3
= 1 + (evalExp (Lit 2) * evalExp (Lit 3))
      Applying 1st equation, with n = 2
= 1 + (2 * evalExp (Lit 3))
      Applying 1st equation, with n = 3
= 1 + (2 * 3)
      Doing the addition and multiplication
= 7

Note the type: evalExp :: Exp -> Int. That is, evalExp takes an *expression*, and produces its *value*, which is an integer.

Values of type Exp represent the **syntax** of arithmetic expressions, where constructors like Add :: Exp -> Exp -> Exp build complicated expressions from simpler expressions. Values of type Int represent their **semantics**, with operations like (+) :: Int -> Int -> Int computing the values of complicated expressions from the values of their constituent sub-expressions.

This helps to understand the meaning of an equation like

```
evalExp (Add e f) = evalExp e + evalExp f
```

in the definition of evalExp. On the left-hand side, we have syntax: Add e f represents an arithmetic expression, with sub-expressions e and f. On the right-hand side, we have semantics: evalExp e + evalExp f uses addition to compute the value of the expression from the values evalExp e and evalExp f of its sub-expressions.

```
> e0
(1 + (2 * 3))
> e1
((1 + 2) * 3)
> e2
((1 + (2 * 3)) + (4 * ((1 + 2) * 3)))
```

16

evalExp can be viewed as a *homomorphism* between the algebraic structures (Exp, Add, Mul) and (Int, +, *), see ▶ https://en.wikipedia.org/wiki/Homomorphism. This is typical for functions that map syntax to semantics.

Arithmetic Expressions with Infix Constructors

Using infix notation for the constructors `Add` and `Mul`, by surrounding them with backticks, makes the same example a little easier to read:

```
data Exp = Lit Int
         | Exp `Add` Exp
         | Exp `Mul` Exp
  deriving Eq

e0 = Lit 1 `Add` (Lit 2 `Mul` Lit 3)
e1 = (Lit 1 `Add` Lit 2) `Mul` Lit 3
e2 = e0 `Add` ((Lit 4) `Mul` e1)

instance Show Exp where
  show (Lit n)     = show n
  show (e `Add` f) = par (show e ++ " + " ++ show f)
  show (e `Mul` f) = par (show e ++ " * " ++ show f)

evalExp :: Exp -> Int
evalExp (Lit n)     = n
evalExp (e `Add` f) = evalExp e + evalExp f
evalExp (e `Mul` f) = evalExp e * evalExp f
```

```
> e0
(1 + (2 * 3))
> e1
((1 + 2) * 3)
> e2
((1 + (2 * 3)) + (4 * ((1 + 2) * 3)))
```

We can go further, and use symbolic constructors:

```
data Exp = Lit Int
         | Exp :+: Exp
         | Exp :*: Exp
  deriving Eq
```

As we have seen, the names of constructors are required to begin with an upper case letter. But if their names are composed of symbols rather than letters, they are instead required to begin with a colon and are infix, as with `:+:` and `:*:` in this example, and `:` for Haskell's lists. (We could instead use `:+` and `:*`, but we add a closing colon for the sake of symmetry.)

Finishing the example:

```
e0 = Lit 1 :+: (Lit 2 :*: Lit 3)
e1 = (Lit 1 :+: Lit 2) :*: Lit 3
e2 = e0 :+: ((Lit 4) :*: e1)

instance Show Exp where
  show (Lit n)   = show n
  show (e :+: f) = par (show e ++ " + " ++ show f)
  show (e :*: f) = par (show e ++ " * " ++ show f)

evalExp :: Exp -> Int
evalExp (Lit n)   = n
evalExp (e :+: f) = evalExp e + evalExp f
evalExp (e :*: f) = evalExp e * evalExp f
```

Abstract syntax refers to the *structure* of a syntactic expression, taking into account the kind of expression (e.g. literal, addition or multiplication) and its sub-expressions. See ▶ https://en.wikipedia.org/wiki/Abstract_syntax. **Concrete syntax** refers to the representation of a syntactic expression in terms of *text*, including things like the names of the operations (e.g. + and ∗), whether or not they are infix, and the use of parentheses for grouping. Algebraic data types capture abstract syntax, while strings (as in the output of `show`) are used for concrete syntax. Using notation like `:+:` for constructors makes the code easier to read, but at the same time it blurs this important distinction while also risking confusion between syntax (`:+:`,`:*:`) and semantics (+, ∗).

Propositions

Our second example, which is a little more complicated, is for logical expressions (**propositions**). These expressions are built from variables and truth values using negation, conjunction, and disjunction.

The name "well-formed formula" (WFF) is often used instead of "proposition".

```
type Name = String
data Prop = Var Name
          | F
          | T
          | Not Prop
          | Prop :||: Prop
          | Prop :&&: Prop
   deriving Eq
```

This algebraic data type definition says that we have the following kinds of values of type `Prop`:

- the constructor `Var` (variable) applied to a `Name`, which is a `String` according to the type definition;
- the constructors `F` (false) and `T` (true);
- the constructor `Not` (negation) applied to a value of type `Prop`;
- the infix constructor `:||:` (disjunction) applied to two values of type `Prop`; and
- the infix constructor `:&&:` (conjunction) applied to two values of type `Prop`.

We use F and T to avoid conflict with `False` and `True`.

Here are some examples of propositions:

```
p0 = Var "a" :&&: Not (Var "a")
p1 = (Var "a" :&&: Var "b")
     :||: (Not (Var "a") :&&: Not (Var "b"))
p2 = (Var "a" :&&: Not (Var "b")
             :&&: (Var "c" :||: (Var "d" :&&: Var "b"))
       :||: (Not (Var "b") :&&: Not (Var "a")))
     :&&: Var "c"
```

p2 was used as an example for building truth tables on page 29.

As with `Exp`, the definition of show that Haskell generates automatically for `Prop` makes propositions hard to read, so we provide a different definition:

```
instance Show Prop where
  show (Var x)     = x
  show F           = "F"
  show T           = "T"
  show (Not p)     = par ("not " ++ show p)
  show (p :||: q) = par (show p ++ " || " ++ show q)
  show (p :&&: q) = par (show p ++ " && " ++ show q)
```

which gives

```
> p0
(a && (not a))
> p1
((a && b) || ((not a) && (not b)))
> p2
((((a && (not b)) && (c || (d && b))) || ((not b) && (not a))) && c)
```

As with arithmetic expressions, the structure of propositions is best understood when they are drawn as trees:

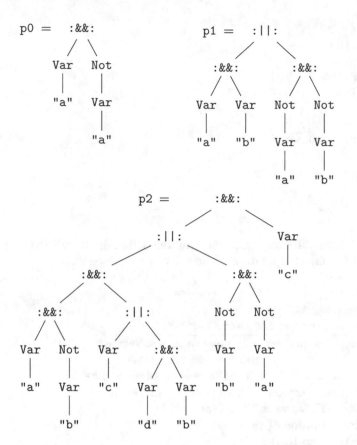

```
> p0
(a && (not a))
> p1
((a && b) || ((not a) && (not b)))
> p2
(((((a && (not b)) && (c || (d && b)))
  || ((not b) && (not a)))
&& c)
```

Evaluating Propositions

Evaluating a proposition should produce a Boolean value, either **True** or **False**. But the value of a proposition that contains variables will depend on the values of those variables. It makes no sense to ask for the value of a proposition without providing this information.

What we need is a **valuation** that associates the variables in a proposition with their values. We'll use a function from names to **Bool**:

```
type Valn = Name -> Bool
```

Now, if vn :: **Valn** such that vn "a" = True+, the value of the proposition Var "a" in vn will be **True**.

The definition of evaluation of propositions follows similar lines to evaluation of arithmetic expressions, with more cases (because there are more constructors for the type **Prop**) and with a valuation as an additional parameter to handle variables:

```
evalProp :: Valn -> Prop -> Bool
evalProp vn (Var x)   = vn x
evalProp vn F         = False
evalProp vn T         = True
evalProp vn (Not p)   = not (evalProp vn p)
evalProp vn (p :||: q) = evalProp vn p || evalProp vn q
evalProp vn (p :&&: q) = evalProp vn p && evalProp vn q
```

To check that this works, let's define a valuation that gives values to the variables named "a", "b", "c", and "d", which are all the variables that appear in our example propositions p0, p1, and p2:

```
> p0
(a && (not a))
> p1
((a && b) || ((not a) && (not b)))
> p2
((((a && (not b)) && (c || (d && b)))
   || ((not b) && (not a)))
   && c)
```

```
valn :: Valn
valn "a" = True
valn "b" = True
valn "c" = False
valn "d" = True
```

This gives

```
> evalProp valn p0
False
> evalProp valn p1
True
> evalProp valn p2
False
```

If you are having difficulty understanding the definition of `evalProp`, try studying the following computation of the value of `evalProp valn p0`:

```
evalProp valn p0
        Retrieving value of p0
= evalProp valn (Var "a" :&&: Not (Var "a"))
        Applying 6th equation, with vn = valn, p = Var "a"
                           and q = Not (Var "a")
= evalProp valn (Var "a") && evalProp valn (Not (Var "a"))
        Applying 1st equation, with vn = valn and x = "a"
= valn "a" && evalProp valn (Not (Var "a"))
        Applying definition of valn
= True && evalProp valn (Not (Var "a"))
        Applying 4th equation, with vn = valn and p = Var "a"
= True && not (evalProp valn (Var "a"))
        Applying 1st equation, with vn = valn and x = "a"
= True && not (valn "a")
        Applying definition of valn
= True && not True
        Doing the negation and conjunction
= False
```

Once again, notice how the structure of the definitions of `evalProp` and `show` for `Prop` follow the structure of the definition of `Prop`:

- there is one equation for each constructor;
- the equations for `Var`, `F` and `T` don't involve recursion;
- the equations for `Not`, `:||:` and `:&&:` are recursive, in just the same way that the definition of `Prop` is recursive in these cases.

The correspondence is so close that you can read off the "shape" of the definitions of `evalProp` and `show` directly from the form of the definition of `Prop`. You've seen this before for recursive definitions over lists.

Satisfiability of Propositions

Recall that a proposition is **satisfiable** if there is at least one combination of values of variables—a valuation—for which the proposition is true. Now that we have propositions available as values of type `Prop`, and a function `evalProp` for evaluating them, we are almost in a position to write a function to check satisfiability. The main thing missing is a list of all of the possible valuations

Satisfiability of Propositions

for the variables in the proposition that we want to test for satisfiability, for use as parameters of evalProp.

The first step is to compute the names of all of the variables in a proposition:

```
type Names = [Name]

names :: Prop -> Names
names (Var x)    = [x]
names F          = []
names T          = []
names (Not p)    = names p
names (p :||: q) = nub (names p ++ names q)
names (p :&&: q) = nub (names p ++ names q)
```

This uses the function nub from the **Data.List** library module which removes duplicates from a list. This is necessary because p and q may have variables in common in the last two cases of the definition. We get the expected results:

```
> names p0
["a"]
> names p1
["a","b"]
> names p2
["a","b","c","d"]
```

Now we need to compute the list of all possible valuations that give values to the variables in a given list.

```
empty :: Valn
empty y = error "undefined"

extend :: Valn -> Name -> Bool -> Valn
extend vn x b y | x == y    = b
                | otherwise = vn y

valns :: Names -> [Valn]
valns [] = [ empty ]
valns (x:xs)
        = [ extend vn x b | vn <- valns xs, b <- [True, False] ]
```

Here, empty is the empty valuation, which produces an error for every variable name; extend *vn x b* yields the valuation which gives the value *b* to the variable *x* without changing the values of other variables in *vn*.

This definition of valns is a little tricky. Let's start by looking at the recursive case. It combines all of the possible choices for the value of x—**False** and **True**—with all of the choices in valns xs for the values of the other variables.

In the base case, valns [] is defined to deliver a list containing just the empty valuation. You may think that including the empty valuation is pointless and so be tempted to replace this line of the definition with

```
valns [] = []
```

but the effect of that would be to make valns *xs* produce the empty list of valuations for any *xs* :: **Names**.

Make sure that you understand why!

Let's look at what valns produces, to check that it works. Haskell is unable to display functions, so we'll work it out by hand. Using the informal notation

$$\{x_0 \mapsto v_0, \ldots, x_n \mapsto v_n\}$$

for the function that maps x_j to v_j for $0 \le j \le n$, we have:

```
> p0
(a && (not a))
> p1
((a && b) || ((not a) && (not b)))
> p2
((((a && (not b)) && (c || (d && b)))
  || ((not b) && (not a)))
 && c)
```

```
> p0
(a && (not a))
> p1
((a && b) || ((not a) && (not b)))
> p2
((((a && (not b)) && (c || (d && b)))
  || ((not b) && (not a)))
 && c)
```

$$\begin{aligned}
\texttt{valns []} \quad &= [\{\textit{anything} \mapsto \textit{error}\}] \\
\texttt{valns ["b"]} \quad &= [\{\texttt{"b"} \mapsto \textbf{False}, \textit{anything else} \mapsto \textit{error}\}, \\
& \quad \{\texttt{"b"} \mapsto \textbf{True}, \textit{anything else} \mapsto \textit{error}\}] \\
\texttt{valns ["a","b"]} &= [\{\texttt{"a"} \mapsto \textbf{False}, \texttt{"b"} \mapsto \textbf{False}, \textit{anything else} \mapsto \textit{error}\}, \\
& \quad \{\texttt{"a"} \mapsto \textbf{False}, \texttt{"b"} \mapsto \textbf{True}, \textit{anything else} \mapsto \textit{error}\}, \\
& \quad \{\texttt{"a"} \mapsto \textbf{True}, \texttt{"b"} \mapsto \textbf{False}, \textit{anything else} \mapsto \textit{error}\}, \\
& \quad \{\texttt{"a"} \mapsto \textbf{True}, \texttt{"b"} \mapsto \textbf{True}, \textit{anything else} \mapsto \textit{error}\}]
\end{aligned}$$

As expected, there are 2^n possible valuations over n variables.

Now a function to test satisfiability is easy to write, using list comprehension and the Prelude function or :: [Bool] -> Bool:

```
satisfiable :: Prop -> Bool
satisfiable p = or [ evalProp vn p | vn <- valns (names p) ]
```

Let's see what happens for our examples:

```
> [ evalProp vn p0 | vn <- valns (names p0) ]
[False,False]
> satisfiable p0
False
> [ evalProp vn p1 | vn <- valns (names p1) ]
[True,False,False,True]
> satisfiable p1
True
> [ evalProp vn p2 | vn <- valns (names p2) ]
[False,False,True,True,False,False,False,False, ... etc.]
> satisfiable p2
True
```

Structural Induction

Recall structural induction on lists from Chap. 11:

Proof Method (**Structural Induction on lists**). To prove that a property P holds for all finite lists of type $[t]$:

Base case: Show that P holds for []; and

Induction step: Show that if P holds for a given list xs of type $[t]$ (the induction hypothesis), then it also holds for $x : xs$ for any value x of type t.

This was justified by the definition of lists:

Definition. A **list** of type $[t]$ is either

1. *empty*, written [], or
2. *constructed*, written $x : xs$, with *head* x (an element of type t) and *tail* xs (a list of type $[t]$).

which exactly captures the algebraic data type definition of List, but using Haskell notation in place of the constructors Nil and Cons:

deriving (Eq,Show)deriving (Eq,Show) is omitted because it isn't relevant to the present topic.

```
data List a = Nil
            | Cons a (List a)
```

The same justification yields a structural induction principal for other algebraic data types. For example, for the type Exp of arithmetic expressions:

```
data Exp = Lit Int
         | Exp `Add` Exp
         | Exp `Mul` Exp
```

we get:

Proof Method (Structural Induction on Exp). To prove that a property P holds for all finite values of type Exp:

Base case: Show that P holds for Lit n, for any n :: Int;

Induction step for Add: Show that if P holds for given e :: Exp and f :: Exp (the induction hypotheses), then it also holds for e `Add` f; and

Induction step for Mul: Show that if P holds for given e :: Exp and f :: Exp (the induction hypotheses), then it also holds for e `Mul` f.

Let's use this to prove that replacing all sub-expressions of the form e `Add` Lit 0 by e and Lit 0 `Add` f by f doesn't affect the values of expressions. We'll prove that

$$\text{evalExp}\,(\text{simplify}\,e) = \text{evalExp}\,e$$

by structural induction on e, where simplify is defined as follows:

```
simplify :: Exp -> Exp
simplify (Lit n)       = Lit n
simplify (e `Add` f)
  | e' == Lit 0      = f'
  | f' == Lit 0      = e'
  | otherwise        = e' `Add` f'
  where e' = simplify e
        f' = simplify f
simplify (e `Mul` f) = (simplify e) `Mul` (simplify f)
```

Notice that simplify will simplify Lit n `Add` (Lit 0 `Add` Lit 0) to Lit n, because Lit 0 `Add` Lit 0 simplifies to Lit 0 and then Lit n `Add` Lit 0 simplifies to Lit n.

Base case: simplify (Lit n) = Lit n so evalExp (simplify (Lit n)) = evalExp (Lit n)

Induction step for Add: Suppose that

$$\text{evalExp}\,(\text{simplify}\,e) = \text{evalExp}\,e$$
$$\text{evalExp}\,(\text{simplify}\,f) = \text{evalExp}\,f$$

for given e, f :: Exp. Then we need to show that

$$\text{evalExp}\,(\text{simplify}\,(e\,\text{`Add`}\,f)) = \text{evalExp}\,(e\,\text{`Add`}\,f)$$

Case 1, simplify e == Lit 0:
evalExp (simplify (e `Add` f))
  = evalExp (simplify f)
  = evalExp f (applying the second IH)
evalExp (e `Add` f)
  = evalExp e + evalExp f
  = evalExp (simplify e) + evalExp f (applying the first IH)
  = evalExp (Lit 0) + evalExp f
  = 0 + evalExp f
  = evalExp f

Case 2, simplify f == Lit 0: Similarly.

Case 3, the `otherwise` **case in** `simplify`:
evalExp (simplify (e `` `Add` `` f))
    = evalExp ((simplify e) `` `Add` `` (simplify f))
    = evalExp (simplify e) + evalExp (simplify f)
    = evalExp e + evalExp f (applying the IHs)
    = evalExp (e `` `Add` `` f)

Induction step for `Mul`: Suppose that

> evalExp (simplify e) = evalExp e
> evalExp (simplify f) = evalExp f

for given e, f `:: Exp`. Then we need to show that

> evalExp (simplify (e `` `Mul` `` f)) = evalExp (e `` `Mul` `` f)

evalExp (simplify (e `` `Mul` `` f))
    = evalExp ((simplify e) `` `Mul` `` (simplify f))
    = evalExp (simplify e) * evalExp (simplify f)
    = evalExp e * evalExp f (applying the IHs)
    = evalExp (e `` `Mul` `` f)

There is a direct correspondence between the definition of an algebraic data type and its structural induction principal:

1. there is a base case corresponding to each non-recursive case of the data type definition;
2. there is a separate induction step corresponding to each of the recursive cases of the definition; and
3. in each of those induction steps, there is an induction hypothesis corresponding to each of the recursive occurrences of the type being defined.

Well, almost every, there's a problem with any algebraic data type T having a constructor that requires a value of type $T \rightarrow t$, for some t. But such types are rare.

Every algebraic data type definition gives rise to a structural induction principal that can be read off from the definition by applying (1)–(3). For `Exp`, there is one base case (for `Lit`) by (1); there are two induction steps (for `Add` and `Mul`) by (2); and each induction step has two induction hypotheses by (3). For `List`, there is one base case (for `[]`, aka `Nil`); there is one induction step (for `:`, aka `Cons`); and that induction step has one induction hypothesis.

Mutual Recursion

Mutual recursion arises naturally in human language. For example, "the hair on her head" is a noun phrase, which includes the prepositional phrase "on her head", which in turn includes the noun phrase "her head".

You have seen how algebraic data types are useful for representing the syntax of simple languages of arithmetic expressions and propositions. Slightly more complicated examples of languages often involve **mutual recursion**, where the definition of one data type A refers to another data type B, whose definition refers back to A. The cycle of references may involve more than two types. To cope with this situation, Haskell allows mutually recursive algebraic data type definitions, as well as the mutually recursive function definitions that naturally arise in functions over such types.

Remember that the order of definitions doesn't matter in Haskell, so the forward reference to `Cond` in the definition of `Exp` isn't a problem.

Let's extend our language of arithmetic expressions by adding conditional expressions, where conditions compare the values of expressions:

```haskell
data Exp = Lit Int
         | Add Exp Exp
         | Mul Exp Exp
         | If Cond Exp Exp
    deriving Eq
```

```
data Cond = Eq Exp Exp
          | Lt Exp Exp
          | Gt Exp Exp
  deriving Eq
```

The types **Exp** and **Cond** are mutually recursive: the last case (**If**) of **Exp** refers to **Cond**, and all three cases of **Cond** refer to **Exp**.

Here are some example values of the types **Exp** and **Cond**:

```
e0 = Add (Lit 1) (Mul (Lit 2) (Lit 3))
e1 = Mul (Add (Lit 1) (Lit 2)) (Lit 3)
c0 = Lt e0 e1
e2 = If c0 e0 e1
```

Tree diagrams for e0 and e1 are on page 145. Here is e2 (which includes c0) as a tree:

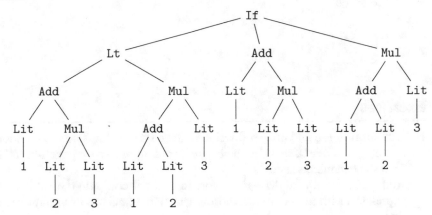

The definitions of the function show for **Exp** and show for **Cond** are mutually recursive: the last case of show for **Exp** calls show for **Cond**, and all of the cases of show for **Cond** call show for **Exp**. That's exactly the same pattern of recursion as in the definitions of the types **Exp** and **Cond**.

```
instance Show Exp where
  show (Lit n)    = show n
  show (Add e f)  = par (show e ++ " + " ++ show f)
  show (Mul e f)  = par (show e ++ " * " ++ show f)
  show (If c e f) = "if " ++ show c
                          ++ " then " ++ show e
                          ++ " else " ++ show f

instance Show Cond where
  show (Eq e f) = show e ++ " == " ++ show f
  show (Lt e f) = show e ++ " < " ++ show f
  show (Gt e f) = show e ++ " > " ++ show f
```

This gives:

```
> c0
(1 + (2 * 3)) < ((1 + 2) * 3)
> e2
if (1 + (2 * 3)) < ((1 + 2) * 3) then (1 + (2 * 3)) else ((1 + 2) * 3)
```

The definition of evaluation is also mutually recursive:

```
evalExp :: Exp -> Int
evalExp (Lit n)   = n
evalExp (Add e f) = evalExp e + evalExp f
```

```
evalExp (Mul e f)  = evalExp e * evalExp f
evalExp (If c e f) = if evalCond c then evalExp e
                                   else evalExp f

evalCond:: Cond -> Bool
evalCond (Eq e f) = evalExp e == evalExp f
evalCond (Lt e f) = evalExp e < evalExp f
evalCond (Gt e f) = evalExp e > evalExp f
```

This gives:

```
> evalExp e0
7
> evalExp e1
9
> evalCond c0
True
> evalExp e2
7
```

The grammars of languages used in Informatics are usually expressed in **Backus-Naur Form** (BNF), see ► https://en.wikipedia.org/wiki/Backus-Naur_form. Here is a BNF definition for our example of arithmetic expressions extended with conditional expressions:

⟨exp⟩ ::= ⟨int⟩
      | ⟨exp⟩ + ⟨exp⟩
      | ⟨exp⟩ * ⟨exp⟩
      | if ⟨cond⟩ then ⟨exp⟩ else ⟨exp⟩
⟨cond⟩ ::= ⟨exp⟩ == ⟨exp⟩
      | ⟨exp⟩ < ⟨exp⟩
      | ⟨exp⟩ > ⟨exp⟩

The notation for defining algebraic data types is deliberately similar to BNF.

Exercises

1. Add subtraction and integer division to the data type **Exp** and the functions showExp and evalExp. (Use the original version of **Exp** on page 144 rather than the extended version.)

    Add variables and re-define evalExp to make it compute the value of an expression with respect to an environment that associates its variables with their values.

2. Define a function

    ```
    ttProp :: Valn -> Prop -> [(Prop,Bool)]
    ```

    that computes the entries in one row of the truth table for a proposition. For example, ttProp valn p0 should produce

    ```
    [(a,True), ((not a),False), ((a && (not a)),False)]
    ```

    representing

    $$\begin{array}{c|c|c} a & \neg a & a \wedge \neg a \\ \hline 1 & 0 & 0 \end{array}$$

    and ttProp valn p1 should produce

    ```
    [(a,True), (b,True), ((not a),False), ((not b),False),
     ((a && b),True), (((not a) && (not b)),False),
     (((a && b) || ((not a) && (not b))),True)]
    ```

    representing

    $$\begin{array}{c|c|c|c|c|c|c} a & b & \neg a & \neg b & a \wedge b & \neg a \wedge \neg b & (a \wedge b) \vee (\neg a \wedge \neg b) \\ \hline 1 & 1 & 0 & 0 & 1 & 0 & 1 \end{array}$$

```
> p0
(a && (not a))
> p1
((a && b) || ((not a) && (not b)))
> p2
((((a && (not b)) && (c || (d && b)))
  || ((not b) && (not a)))
 && c)

valn "a" = True
valn "b" = True
valn "c" = False
valn "d" = True
```

16

Hint: Start with a function that produces a list of all the sub-expressions of a proposition. Then use the function sortOn from the **Data.List** library module together with a function for computing the size of a proposition to arrange the entries in order, starting with the simplest and ending with the entry for the proposition itself. Use the function nub from **Data.List** to remove duplicate entries. Then use evalProp to compute the values of all of the entries in the list.

3. Define a function to check whether a proposition is a tautology. Check that Not p0 is a tautology and that p1 is not a tautology.

   Define another function to check whether two propositions are equivalent. Check that p2 is equivalent to `Not (Var "b") :&&: Var "c"`, as our conversion of p2 to CNF on page 167 says it should be, and that p2 is not equivalent to p1.

4. A sequent is satisfiable if there is at least one valuation for which all of its antecedents are true and at least one of its succedents is true. Define a type `Sequent` for representing sequents $\Gamma \vDash \Delta$ where $\Gamma, \Delta :: $ `[Prop]`. Write a function `satSequent :: Sequent -> Bool` that checks if a sequent is satisfiable or not.

   (This is related to but different from Exercise 14.1, where the required function checks *validity* of sequents formed from predicates over the simple universe of shapes in Chap. 6, rather than *satisfiability* of sequents formed from propositions in `Prop`.)

5. A proposition is in **negation normal form** if the only use of negation is in applications directly to variables.

   Write a function `isNNF :: Prop -> Bool` to test whether a proposition is in negation normal form.

   Write a function `toNNF :: Prop -> Prop` that converts a proposition to negation normal form by applying the following equivalences:

   $$\neg(p \wedge q) = \neg p \vee \neg q$$
   $$\neg(p \vee q) = \neg p \wedge \neg q$$
   $$\neg\neg p = p$$

   Test that `toNNF` p produces a result that is in negation normal form and that is equivalent to p. To do the tests, you'll need to include the following code to generate random propositions:

   ```
   import Control.Monad
   instance Arbitrary Prop where
     arbitrary = sized prop
       where
         prop n | n<=0 = oneof [liftM Var arbitrary,
                                 return T,
                                 return F]
                | otherwise
                     = oneof [liftM Not subprop,
                              liftM2 (:||:) subprop subprop,
                              liftM2 (:&&:) subprop subprop]
           where subprop = prop (n `div` 2)
   ```

6. Mobiles are hanging sculptures made of rods, wires and pendants. At each end of every rod in a mobile is a wire from which is hung a pendant or another rod. Rods and wires are weightless; pendants have weight. Here are some examples of mobiles:

```
> p0
(a && (not a))
> p1
((a && b) || ((not a) && (not b)))
> p2
((((a && (not b)) && (c || (d && b)))
  || ((not b) && (not a)))
 && c)
```

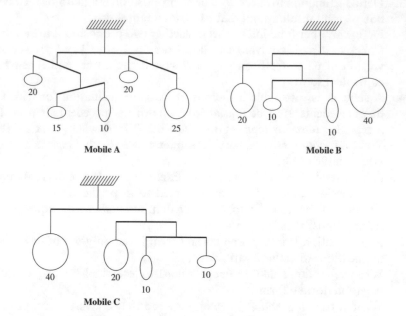

Mobile A

Mobile B

Mobile C

The *weight* of a mobile is the sum of the weights of the pendants attached to it. A mobile is *balanced* if the mobiles attached at the ends of each rod are of the same weight, and these mobiles are themselves balanced. In the pictures above, mobile A is not balanced but mobiles B and C are balanced.

(a) Define an algebraic data type for representing mobiles, and functions for computing the weight of a mobile and testing whether a mobile is balanced.

(b) When the wind blows, the rods and pendants of a mobile move in space. We can *reflect* a mobile, or any sub-mobile of a larger mobile, about its vertical axis: for a pendant, the reflection is just itself; for a rod, we swap the positions of the two mobiles hanging off its ends. Mobile m is *equal* to mobile m' if m can be transformed to yield m' by applying reflections to some or all of the rods in m. In the pictures above, mobiles B and C are equal. Define a function for testing equality of mobiles.

(c) Define a function `bmobile :: [Int] -> Maybe Mobile` that produces a balanced mobile, if there is one, from a list of pendant weights.
Hint: Start by writing a function

```
eqsplits :: [Int] -> [([Int],[Int])]
```

such that `eqsplits` *ns* produces a list of all partitions of *ns* into two lists of equal weight. So

```
eqsplits [1,2,3] = [([3], [1, 2]), ([1, 2], [3])]
eqsplits [1,2] = []
```

Then write `bmobile`, using `eqsplits` to produce candidate partitions for argument lists of length > 1.

7. Give structural induction principals for `Prop`, for `Bool`, and for `Mobile` from Exercise 6.

16

8. Define mutually recursive algebraic data types for noun phrases, prepositional phrases, and any other syntactic categories required to capture the following examples and others like them:

> the big dog
> the silly man with the tiny dog
> a dog with the silly man on the hill
> the cat in the hat
> the cat with the man in the hat on the hill

Comment on any ambiguities that arise, using tree diagrams to illustrate.

Karnaugh Maps

Contents

© The Author(s), under exclusive license to Springer Nature Switzerland AG 2021
D. Sannella et al., *Introduction to Computation*, Undergraduate Topics
in Computer Science, https://doi.org/10.1007/978-3-030-76908-6_17

Simplifying Logical Expressions

The symbols used in these circuit diagrams are explained on page 223.

Complex logical expressions like $(a \wedge \neg b \wedge (c \vee (d \wedge b)) \vee (\neg b \wedge \neg a)) \wedge c$ are hard to understand and hard to work with. The much simpler expression $\neg b \wedge c$, to which it is equivalent, is obviously an improvement. When logical expressions are used to design hardware circuits, simpler expressions produce circuits that are cheaper because they have fewer components. The left-hand diagram is a circuit for the complex expression, and the right-hand diagram is a circuit for the simpler equivalent expression.

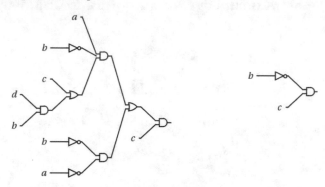

One way of simplifying a logical expression, that we will look at later, is to apply equivalences like the double negation law and De Morgan's laws:

$$\neg\neg a = a \qquad \neg(a \vee b) = \neg a \wedge \neg b \qquad \neg(a \wedge b) = \neg a \vee \neg b$$

Karnaugh is pronounced "karnaw". Karnaugh maps are due to the American physicist, mathematician and inventor Maurice Karnaugh (1924–), see ▶ https://en.wikipedia. org/wiki/Maurice_Karnaugh. Karnaugh maps can be used for expressions with more than four predicates but they are harder to draw and to understand.

in an attempt to reduce the depth of nesting, eliminate redundant terms, etc.

In this chapter we will look at a different method, called **Karnaugh maps**, that is useful for simplifying logical expressions that involve no more than four predicates. A Karnaugh map is a particular representation of the truth table for an expression from which a simplified form of the expression can be directly read off.

Conjunctive Normal form and Disjunctive Normal form

To explain conjunctive normal form, we first need to introduce some terminology.

- A **literal** is a predicate like p or a negated predicate like $\neg p$.
- A **clause** is a disjunction of literals, for instance $p \vee \neg q \vee r$

In Mathematics, a **normal form** is a standard way of presenting an expression. Usually, there is an algorithm for converting any expression into normal form. If the normal form of an expression is unique, meaning that equivalent expressions have the same normal form, then it is called a **canonical form**. Conversion of two expressions into canonical form is a way of checking whether or not they are equivalent.

A logical expressions is in **conjunctive normal form** (CNF) if it consists of a conjunction of clauses. Some examples of expressions in conjunctive normal form are

- $(a \vee \neg b) \wedge c \wedge (\neg a \vee d) \wedge (a \vee b \vee \neg c)$ (four clauses)
- $p \vee \neg q \vee r$ (one clause)

Every logical expression can be converted into an equivalent expression in conjunctive normal form. The fact that negation only appears at the level of literals and that nesting is strictly limited makes expressions in conjunctive normal form easy to understand.

17

Disjunctive normal form is similar, but with the roles of conjunction and disjunction reversed. While an expression in conjunctive normal form is a *conjunction of disjunctions* of literals, an expression in **disjunctive normal form** (DNF) is a *disjunction of conjunctions* of literals. An example of an expression in disjunctive normal form is $(a \wedge \neg b \wedge c) \vee (\neg a \wedge d) \vee (a \wedge \neg c)$. Again, every logical expression can be converted into an equivalent expression in disjunctive normal form.

Expressions in CNF and DNF are easy to understand and expressions in one of these forms are required for some purposes, as we will see later. However, converting an expression to CNF or DNF will sometimes produce an expression that is exponentially larger than the original expression.

In DNF, the word "clause" is sometimes used for a *conjunction* of literals, with a DNF expression being a disjunction of clauses. But we will reserve that word for a disjunction of literals as in CNF.

Karnaugh Maps

Consider the expression

$$(a \wedge \neg b \wedge (c \vee (d \wedge b))) \vee (\neg b \wedge \neg a)) \wedge c$$

Here is its truth table:

a	b	c	d	$c \vee (d \wedge b)$	$a \wedge \neg b \wedge (c \vee (d \wedge b))$	$\neg b \wedge \neg a$	$(a \wedge \neg b \wedge (c \vee (d \wedge b))) \vee (\neg b \wedge \neg a)) \wedge c$
0	0	0	0	0	0	1	0
0	0	0	1	0	0	1	0
0	0	1	0	1	0	1	1
0	0	1	1	1	0	1	1
0	1	0	0	0	0	0	0
0	1	0	1	1	0	0	0
0	1	1	0	1	0	0	0
0	1	1	1	1	0	0	0
1	0	0	0	0	0	0	0
1	0	0	1	0	0	0	0
1	0	1	0	1	1	0	1
1	0	1	1	1	1	0	1
1	1	0	0	0	0	0	0
1	1	0	1	1	0	0	0
1	1	1	0	1	0	0	0
1	1	1	1	1	0	0	0

A Karnaugh map is simply a 4×4 representation of the 16 entries in the last column of the truth table:

	cd 00	01	11	10
ab 00	0	0	1	1
01	0	0	0	0
11	0	0	0	0
10	0	0	1	1

An ordering of the binary numbers with this property is called a **Gray code**. They are used for error correction in digital communication, see ▶ https://en.wikipedia.org/wiki/Gray_code.

The values of a and b are given along the left edge of the table, and the values of c and d are given along the top edge. The order of those values are in the sequence 00, 01, 11, 10, which is different from the order in the truth table. The special feature of the Karnaugh map order is that just one digit changes between consecutive items. With truth table order, the sequence would be 00, 01, 10, 11, where both digits change when moving from 01 to 10.

Now, let's consider how to characterise the values of a, b, c, and d for which the entry in the Karnaugh map is 1. We start by collecting adjacent 1s into rectangular groups, in such a way that all of the 1s are included in some group. There are two groups:

$$cd$$

		00	01	11	10
	00	0	0	1	1
	01	0	0	0	0
ab	11	0	0	0	0
	10	0	0	1	1

Looking at the labels for a and b along the left edge of the table and the labels for c and d along the top edge, the first of these groups corresponds to a value of 0 for a and b and a value of 1 for c, while the value of d doesn't matter because there is a 1 when d is 0 as well as when d is 1. This combination of values is described by the expression $\neg a \wedge \neg b \wedge c$. The second group corresponds to $a = 1$, $b = 0$, and $c = 1$, which is described by $a \wedge \neg b \wedge c$. All of the other entries are 0, so the expression $(\neg a \wedge \neg b \wedge c) \vee (a \wedge \neg b \wedge c)$ describes all of the entries that are 1. It follows that

You may notice that $(\neg a \wedge \neg b \wedge c) \vee (a \wedge \neg b \wedge c)$ can be further simplified to give $\neg b \wedge c$. This example will be revisited later in order to show how that expression can be obtained directly from the Karnaugh map.

$$(\neg a \wedge \neg b \wedge c) \vee (a \wedge \neg b \wedge c)$$

is equivalent to $(a \wedge \neg b \wedge (c \vee (d \wedge b)) \vee (\neg b \wedge \neg a)) \wedge c$, the expression that we started with.

Let's look at a different example. Here's the Karnaugh map for the expression $((a \wedge \neg b) \vee c \vee (\neg d \wedge b) \vee a) \wedge \neg c$.

$$cd$$

		00	01	11	10
	00	0	0	0	0
	01	1	0	0	0
ab	11	1	1	0	0
	10	1	1	0	0

There are different ways of making groups of 1s, including the following:

17

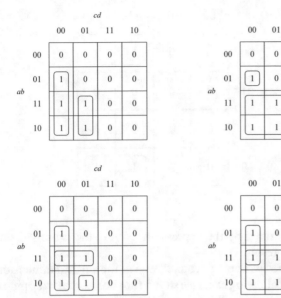

Any grouping is valid, but some groupings yield results that are more compact. The best result is obtained by taking large blocks, since they can be described by smaller expressions covering a large number of 1s. Overlapping blocks are okay, and large blocks that overlap produce better results than smaller blocks that are disjoint. We will see below why we will want to require that the number of entries in each group should be a power of 2, i.e. 1, 2, 4, 8, or 16. That eliminates the first and third of these groupings, since they each contain a group of size 3.

A block of size 2^n can be described by a conjunction of $4 - n$ literals.

The second grouping yields the expressions $a \wedge \neg c$ (for the big group) and $\neg a \wedge b \wedge \neg c \wedge \neg d$ (for the small group), giving a result of $(a \wedge \neg c) \vee (\neg a \wedge b \wedge \neg c \wedge \neg d)$. The fourth grouping yields the expressions $a \wedge \neg c$ and $b \wedge \neg c \wedge \neg d$, giving a result of $(a \wedge \neg c) \vee (b \wedge \neg c \wedge \neg d)$.

Converting Logical Expressions to DNF

The procedure that we have been following in the last two examples will always produce an expression in DNF. Provided the number of entries in a group of cells is a power of 2, it can be described by a conjunction of literals. Combining these expressions using disjunction—a cell contains 1 if it is in the first group, **or** the second group, etc.—gives an expression in DNF.

Let's look at what happens when the size of a group is not a power of 2, by looking at the group of size 3 in the first and third groupings for the last example. That group is described by the expression $(a \vee b) \wedge \neg c \wedge \neg d$, and there is no conjunction of literals that describes it. (Try it!) The same happens for all groups having a size that isn't a power of 2.

Let's look again at the order of the values along the left edge and top edge of a Karnaugh map: 00, 01, 11, 10. The fact that just one digit changes between consecutive items is what makes it possible to describe a group of adjacent items with a conjunction of literals, provided—as we have just seen—that the size of the group is a power of 2. This relationship also holds between the first item in the sequence, 00, and the last item, 10. That allows groups to "wrap around" from one edge of the Karnaugh map to the other. Revisiting our first example, we can group the 1s into a single group as follows:

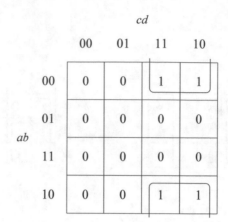

This group is described by the expression $\neg b \wedge c$, the promised simplified version of the earlier result.

In this example the "wrapping around" was in the vertical dimension of the Karnaugh map. The following example shows that horizontal wrapping around is also allowed, and even a combination of the two:

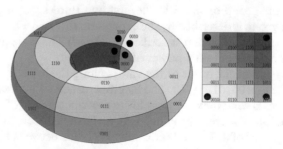

Other notations are sometimes used for these expressions, for example, when using logic for circuit design. There, $\neg a$ is written a' or $\bar{a}$, $a \vee b$ is written $a + b$, and $a \wedge b$ is written ab. So $(\neg b \wedge \neg d) \vee (\neg a \wedge b \wedge \neg d)$ becomes $b'd' + a'bd'$, with the terminology "sum of products" (SoP) for DNF and "product of sums" (PoS) for CNF.

This yields the DNF expression $(\neg b \wedge \neg d) \vee (\neg a \wedge b \wedge \neg d)$, where the group that includes all four corners corresponds to $(\neg b \wedge \neg d)$.

Vertical wrapping around corresponds geometrically to the Karnaugh map being on the surface of a cylinder, with the top edge joined to the bottom edge. Vertical *and* horizontal wrapping corresponds to it being on the surface of a torus (doughnut):

where the corner grouping is the four cells containing a blob.

Converting Logical Expressions to CNF

A similar procedure using Karnaugh maps will convert logical expressions to conjunctive normal form (CNF).

We start by writing expressions that describe *block of 0s* rather than blocks of 1s. Returning to our first example, for the expression $(a \wedge \neg b \wedge (c \vee (d \wedge b)) \vee (\neg b \wedge \neg a)) \wedge c$, we can group the 0s as follows:

	cd			
	00	01	11	10
ab 00	0	0	1	1
01	0	0	0	0
11	0	0	0	0
10	0	0	1	1

We then want to say that the 1s are in all of the places that are *not* in one of these blocks.

The horizontal block is described by the expression b, so the places *outside* that block are described by its negation, $\neg b$. The vertical block is described by $\neg c$ so the places outside it are described by c. The places that are outside *both* blocks are described by the *conjunction* of these two expressions, $\neg b \wedge c$.

Looking at our other example, we can group the 0s like this:

	cd			
	00	01	11	10
ab 00	0	0	0	0
01	1	0	0	0
11	1	1	0	0
10	1	1	0	0

The three groups of 0s are described by the expressions $\neg a \wedge \neg b$, $\neg a \wedge d$, and c. Negating these gives $\neg(\neg a \wedge \neg b)$, $\neg(\neg a \wedge d)$, and $\neg c$. So far so good, but the conjunction of these is $\neg(\neg a \wedge \neg b) \wedge \neg(\neg a \wedge d) \wedge \neg c$, which is not in CNF.

To give a result in CNF, we need to apply one of the De Morgan laws to the negated block descriptions to turn the negated conjunctions into disjunctions:

$$\neg(\neg a \wedge \neg b) = a \vee b$$
$$\neg(\neg a \wedge d) = a \vee \neg d$$

(The third block description can be used as it is.) The conjunction of the results is then $(a \vee b) \wedge (a \vee \neg d) \wedge \neg c$.

Exercises

1. Produce a truth table for the expression $((a \wedge \neg b) \vee c \vee (\neg d \wedge b) \vee a) \wedge \neg c$ (this part was Exercise 4.4(c)) and check that it yields the Karnaugh map given for this expression above.
2. Use a Karnaugh map to convert the expression $(c \vee \neg d) \wedge (b \vee \neg d) \wedge (\neg a \vee \neg b \vee c)$ into an equivalent expression in DNF.
3. Use a Karnaugh map to convert the expression $(\neg a \wedge \neg b) \vee (\neg a \wedge \neg c) \vee (\neg a \wedge \neg d) \vee (\neg b \wedge d) \vee (\neg b \wedge c)$ into an equivalent expression in CNF.
4. Recall the universe of discourse from Chap. 6:

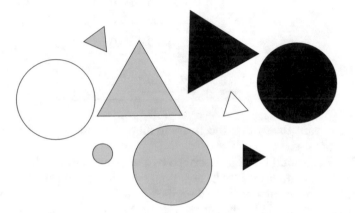

   Using the predicates $a = $ isWhite, $b = $ isBlack, $c = $ isSmall and $d = $ isDisc, produce a Karnaugh map that shows which combinations of these predicates and their negations are inhabited by something in the universe. Use it to produce DNF and CNF descriptions of this universe.
5. Using an example, explain why Karnaugh maps use the order 00, 01, 11, 10 for the values of a, b and c, d rather than truth table order.
6. A logical expression is in **full disjunctive normal form** if it is in disjunctive normal form and each clause contains all of the predicates in the expression. Let's require that no clause contains "complementary literals" (the same predicate both with and without negation); such a clause will always be false and so can be eliminated from the disjunction.

   Suppose that each clause of an expression in disjunctive normal form is represented as a set of literals, and then the expression itself is represented as a set of these clauses. Explain why that is a canonical form for logical expressions, in the sense that two logical expressions are equivalent whenever they have the same full disjunctive normal form.
7. Recall that a proof in sequent calculus reduces a sequent to an equivalent set of simple sequents. A simple sequent is equivalent to a disjunction of literals: for instance, $a \vDash b$ is equivalent to $\vDash \neg a, b$ by $\neg R$, which is equivalent to $\vDash \neg a \vee b$ by $\vee R$. It follows that a proof in sequent calculus of $\vDash exp$ yields a CNF expression—namely, the conjunction of the disjunctions corresponding to the simple sequents obtained from the proof—that is equivalent to exp.

   Use this procedure to convert $(a \wedge \neg b \wedge (c \vee (d \wedge b)) \vee (\neg b \wedge \neg a)) \wedge c$ to CNF, and compare the result with what was obtained on page 167 using Karnaugh maps.

Relations and Quantifiers

Contents

© The Author(s), under exclusive license to Springer Nature Switzerland AG 2021
D. Sannella et al., *Introduction to Computation*, Undergraduate Topics
in Computer Science, https://doi.org/10.1007/978-3-030-76908-6_18

Expressing Logical Statements

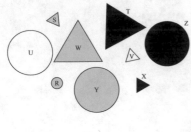

Our simple universe of discourse in Chap. 6 contained a number of things

```
data Thing = R | S | T | U | V | W | X | Y | Z deriving Show
things :: [Thing]
things = [R, S, T, U, V, W, X, Y, Z]
```

with predicates for describing features of things

```
type Predicate u = u -> Bool
```

including `isSmall :: Predicate Thing`, `isTriangle :: Predicate Thing`, etc.

We used these to express logical statements about the universe, for example "Every white triangle is small":

```
> and [ isSmall x | x <- things, isWhite x && isTriangle x ]
True
```

and "Some big triangle is grey":

```
> or [ isGrey x | x <- things, isBig x && isTriangle x ]
True
```

Later, we "lifted" operations on Booleans to operations on predicates. For instance, using `(&&) :: Bool -> Bool -> Bool`, we defined conjunction of predicates:

```
(&:&) :: Predicate u -> Predicate u -> Predicate u
(a &:& b) x = a x && b x
```

Don't confuse `&:&`, which is for combining predicates, with `:&&:`, which is a constructor for `Prop` (Chap. 16).

These operations on predicates make it more convenient to represent complex statements, for example "Every disc that isn't black is either big or not white":

```
> :{
| and [ (isBig |:| neg isWhite) x
|       | x <- things, (isDisc &:& neg isBlack) x ]
| :}
True
```

Normally, input in GHCi is restricted to a single line. To spread input over multiple lines, start with `:{` and finish with `:}`, as in this example. GHCi changes the prompt to a vertical bar when in multi-line mode.

Sequents are an alternative way of expressing the same statements. We can write "Every disc that isn't black is either big or not white" as

$$\text{isDisc}, \neg\text{isBlack} \vDash \text{isBig}, \neg\text{isWhite}$$

or equivalently

$$\text{isDisc \&:\& neg isBlack} \vDash \text{isBig |:| neg isWhite}$$

and "Some big triangle is grey" is

$$\text{isBig}, \text{isTriangle} \nvDash \neg\text{isGrey}$$

or equivalently

$$\text{isBig \&:\& isTriangle} \nvDash \text{neg isGrey}$$

Quantifiers

Yet another way to express statements involving "every" or "some" is to use **quantifiers**. You've probably already seen the **universal quantifier** ∀ and the **existential quantifier** ∃ in Mathematics. For example, the fact that for every natural number n there is a prime number p that is greater than n is written as

18

$$\forall n \in \mathbb{N}.\exists p \in \mathbb{N}.p \text{ is prime} \wedge p > n$$

Prime numbers are defined like this:

$$p \text{ is prime iff } \forall d \in \mathbb{N}.d \text{ divides } p \rightarrow d = 1 \vee d = p$$

where

$$d \text{ divides } p \text{ iff } \exists m \in \mathbb{N}.m \times d = p.$$

If you haven't seen the symbols $\forall$ and $\exists$ before then you have surely seen the same thing expressed in words.

We can code the universal and existential quantifiers in Haskell:

```
every :: [u] -> Predicate u -> Bool
every xs p = and [ p x | x <- xs ]


some :: [u] -> Predicate u -> Bool
some xs p = or [ p x | x <- xs ]
```

The first parameter of every/some is the domain of quantification: a list of things in the universe. The second parameter is a predicate, which is claimed to be satisfied for every/some item in the domain of quantification.

Then "Every white triangle is small" becomes

```
> every (filter (isWhite &:& isTriangle) things) isSmall
True
```

"Every disc that isn't black is either big or not white" becomes

```
> :{
| every (filter (isDisc &:& neg isBlack) things)
|       (isBig |:| neg isWhite)
| :}
True
```

and "Some big triangle is grey" becomes

```
> some (filter (isBig &:& isTriangle) things) isGrey
True
```

Notice how Haskell's filter function is used to express the list of things in the universe that have some feature or combination of features. This is similar to what we did in Chap. 6 to compute the list of big triangles

```
> [ x | x <- things, isBig x && isTriangle x ]
[T,W]
> filter (isBig &:& isTriangle) things
[T,W]
```

but using filter avoids writing out the list comprehension. That only saves a few characters, but it prevents us from making mistakes with the syntax of comprehension or getting one of the four uses of x wrong.

Sequents involve implicit quantification over the things in the universe of discourse: a sequent is valid if everything in the universe that satisfies all of the antecedent predicates satisfies at least one of the succedent predicates. An advantage of using explicit quantification, as in the last few examples, is that explicit quantifiers can be *nested*, which makes it possible to deal with more complicated examples.

$\forall n \in \mathbb{N}.\exists p \in \mathbb{N}.\ldots$ is pronounced "for all natural numbers n there exists a natural number p such that ...".

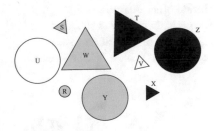

every and some are the same as the Prelude functions all and any, except for the order of parameters.

Remember: filter :: (a->Bool) -> [a] -> [a], where filter p xs yields the list of items in xs for which p :: a -> Bool produces True.

Our example with prime numbers ("for every natural number n there is a prime number p that is greater than n") demonstrates how complex nesting of quantification arises naturally in Mathematics: expanding the definition of a prime number and what it means for one number to divide another gives

This involves nesting of quantifiers inside other quantifiers ($\forall n \in \mathbb{N}.\exists p \in \mathbb{N}.\forall d \in \mathbb{N}. \ldots$), as well as nesting of quantifiers inside expressions (($\exists m \in \mathbb{N}. \ldots) \to \ldots$).

$$\forall n \in \mathbb{N}.\exists p \in \mathbb{N}.(\overbrace{(\forall d \in \mathbb{N}.(\underbrace{\exists m \in \mathbb{N}.m \times d = p}_{d \text{ divides } p}) \to d = 1 \vee d = p)}^{p \text{ is prime}} \wedge p > n$$

Relations

Predicates are limited to expressing features of individual things. Suppose that we want to express relationships *between* things?

One such relationship is the property of one thing being bigger than another thing. Our universe doesn't include a predicate that captures "bigger", but given the predicates `isBig` and `isSmall`, the only way that a can be bigger than b is if `isBig` a and `isSmall` b. We can use this fact to express the relationship of one thing being bigger than another as a Haskell function:

```
isBigger :: Thing -> Thing -> Bool
isBigger x y = isBig x && isSmall y
```

On the other hand, there are relationships that can't be obtained via predicates that we already have, for example one expressing the relative positions of things in the diagram. So this information would need to be added to our Haskell representation of the universe:

```
isAbove :: Thing -> Thing -> Bool
isAbove R _ = False
isAbove S x = x `elem` [R,U,V,X,Y]
...
```

`isBigger` and `isAbove` are **relations**: `Bool`-valued functions, like predicates, but taking two arguments of type `Thing` instead of just one, and delivering a result of `True` if the relationship between the first and second things holds.

We can define the type of relations as an abbreviation (see page 11):

```
type Relation u = u -> u -> Bool
```

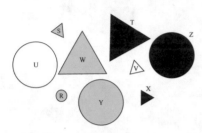

and then `isBigger :: Relation Thing` and `isAbove :: Relation Thing`. Haskell allows us to use these as infix functions, for instance, `U `isBigger` X` and `S `isAbove` R`, which avoids confusion about which thing is bigger than/above the other.

And now we can express "S is above every black triangle" like this:

```
> every (filter (isBlack &:& isTriangle) things) (S `isAbove`)
False
```

(This is false because **T** is a counterexample.)

This uses a **section** (`S `isAbove``) to partially apply `isAbove` to `S`. The section yields a predicate which can be applied to a `Thing` to give `True` (S is above that `Thing`) or `False` (it isn't).

The section (``isAbove` X`), which partially applies `isAbove` to its *second* argument, can be used to express "Every white disc is above **X**":

```
> every (filter (isWhite &:& isDisc) things) (`isAbove` X)
True
```

For a universe with more than one kind of thing—for example, people and dogs—we might instead define
```
type Relation u v =
  u -> v -> Bool
```
and then
```
bestFriend ::
  Relation Person Dog
```
with
```
Alice `bestFriend` Fido = True.
```

18

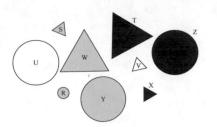

where (`` `isAbove` `` X) is the predicate that returns **True** when applied to a **Thing** that is above X.

Now consider the statement "Every black triangle is above every grey disc", which involves two quantifiers. To make the Haskell code for this more compact, let's first define

```
blackTriangles :: [Thing]
blackTriangles = filter (isBlack &:& isTriangle) things
greyDiscs :: [Thing]
greyDiscs = filter (isGrey &:& isDisc) things
```

and then we have

```
> every blackTriangles (\x -> every greyDiscs (x `isAbove`))
False
```

(This statement is false because X isn't above R or Y.)

This uses a **lambda expression**

```
\x -> every greyDiscs (x `isAbove`)
```

to express the predicate "is above every grey disc". This has type **Predicate Thing**. It's applied to items from blackTriangles, so the variable x :: **Thing** is drawn from that list. The result of the lambda expression is then **True** if x is above every item in the list greyDiscs.

You might need to read this twice to understand why this gives "is above every grey disc" rather than "every grey disc is above".

We could instead define a named function

```
isAboveEveryGreyDisc :: Predicate Thing
isAboveEveryGreyDisc x = every greyDiscs (x `isAbove`)
```

and then

```
every blackTriangles (\x -> every greyDiscs (x `isAbove`))
```

becomes

```
> every blackTriangles isAboveEveryGreyDisc
False
```

which is pretty close to the English formulation of the statement.

A similar statement is "Every big black triangle is above every grey disc". After defining

```
bigBlackTriangles :: [Thing]
bigBlackTriangles =
  filter (isBig &:& isBlack &:& isTriangle) things
```

we have

```
> every bigBlackTriangles isAboveEveryGreyDisc
True
```

Another Universe

Let's look at another universe of discourse, containing people

```
data Person = Angela | Ben | Claudia | Diana | Emilia
                    | Fangkai | Gavin | Hao | Iain
people :: [Person]
people = [Angela, Ben, Claudia, Diana, Emilia,
               Fangkai, Gavin, Hao, Iain]
```

and one relation on `Person`

```
loves :: Relation Person
Angela `loves` Ben = True
Angela `loves` _   = False
...
```

The `loves` relation between the people in the universe is given by the following diagram, where an arrow pointing from *a* to *b* means that *a* `loves` *b*:

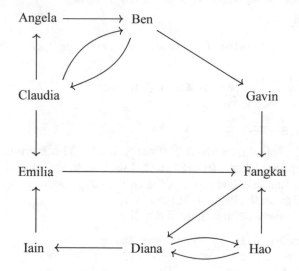

We can express statements like "Angela loves somebody":

```
> some people (Angela `loves`)
True
```

"Everybody loves Ben":

```
> every people (`loves` Ben)
False
```

"Somebody loves themself":

```
> some people (\x -> x `loves` x)
False
```

and "Somebody loves somebody who loves them":

```
> some people (\x -> some people (\y -> x `loves` y && y `loves` x))
True
```

Dependencies

Some subtleties arise in statements involving nested quantifiers, relating to dependencies between "inner" and "outer" quantified variables.

Consider the expanded version of our example with prime numbers:

$$\forall n \in \mathbb{N}.\exists p \in \mathbb{N}.(\overbrace{\forall d \in \mathbb{N}.(\underbrace{\exists m \in \mathbb{N}.m \times d = p}_{d \text{ divides } p}) \to d = 1 \vee d = p}^{p \text{ is prime}}) \wedge p > n$$

This begins with $\forall n \in \mathbb{N}$, which means that the statement is required to hold for every n; let's pick $n = 69$. Then we have $\exists p \in \mathbb{N}$. In order for the statement to hold, p must be chosen to be a prime number that is greater than n, for example 73. That is, the choice of p **depends on** n. Then, d must be chosen to be a divisor of p, so the choice of d (73 or 1) depends on p. Finally, the choice of m (1 or 73) depends on p and d.

Now let's look at an example in English: "Everybody loves somebody". This can be interpreted in two ways, according to whether or not the choice of the loved person depends on the choice of the loving person, or not. The distinction between these is not clearly reflected in informal English.

The first interpretation, "For every person, there is some person who they love", allows the choice of the person who is loved to depend on the choice of the person who is doing the loving. It can be expressed in Haskell as:

```
> every people (\y -> some people (y `loves`))
True
```

In this interpretation of the statement, the person loved by Angela might be different from the person loved by Ben. (This is indeed the case: Angela loves Ben and Ben loves Claudia and Gavin.) The dependency is conveyed in the Haskell code by the fact that the existential quantifier is inside the predicate

```
\y -> some people (y `loves`)
```

of the universal quantifier.

The second interpretation is "There is some person who everybody loves", which requires a fixed choice of the person who is loved. In Haskell, this is

```
> some people (\x -> every people (`loves` x))
False
```

Now, the universal quantifier is inside the predicate

```
\x -> every people (`loves` x)
```

of the existential quantifier. The fact that the same person needs to be loved by everybody means that, in our universe, the statement is false.

These dependencies come about because the inner quantifiers are within the **scope** of the outer quantified variables. This means that, in fact, even more dependencies are possible: d is in the scope of n as well as p, so d can depend on both n and p. Likewise, m can depend on n, p, and d.

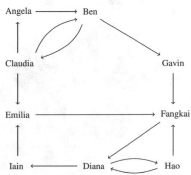

Consider the related function

```
\y -> head (filter (y `loves`)
people) of type Person -> Person.
```

Provided the existentially quantified statement is true, this computes a so-called "witness": a person that satisfies the predicate. The dependency means that it needs to be a *function*, with the output depending on the value taken by the universally quantified variable, rather than a constant. This is known as a **Skolem function**, after the Norwegian mathematician Thoralf Skolem (1887–1963), see ▶ https://en. wikipedia.org/wiki/Thoralf_Skolem.

Exercises

1. Define a Haskell version of the $\exists!$ quantifier ("there exists a unique") and give an example of its use.
   Hint: Use the fact that $\exists!x.P(x)$ is equivalent to

   $$\exists x.P(x) \land (\forall x, y.P(x) \land P(y) \rightarrow x = y).$$

   In order to use equality in the definition, you will need to use the type

   ```
   existsUnique :: Eq u => [u] -> Predicate u -> Bool
   ```

   And then, to give an example of its use involving the type **Thing**, you will need to change its definition to allow use of equality:

   ```
   data Thing = R | S | T | U | V | W | X | Y | Z
     deriving (Eq,Show)
   ```

2. Express the following Haskell codings of English statements from above as sequents:

   (a) "S is above every black triangle"
   ```
   every (filter (isBlack &:& isTriangle) things) (S `isAbove`)
   ```
   (b) "Every white disc is above X"
   ```
   every (filter (isWhite &:& isDisc) things) (`isAbove` X)
   ```

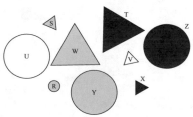

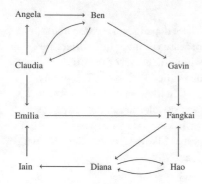

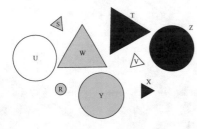

See ▶ https://www.youtube.com/ watch?v=z-2_OstpR5c and ▶ https://www.seeing-stars.com/ ImagePages/ DeanMartinGravePhoto.shtml.

(c) "Angela loves somebody"

```
some people (Angela `loves`)
```

(d) "Everybody loves Ben"

```
every people (`loves` Ben)
```

(e) "Somebody loves themself"

```
some people (\x -> x `loves` x)
```

3. Translate the following Haskell code into English:

    (a) `some people (\x -> every people (x `loves`))`
    (b) `every people (\x -> some people (`loves` x))`

4. One way of coding the statement "Every white disc is bigger than some triangle" in Haskell is as

```
every (filter (isWhite &:& isDisc) things) isBig
  && some (filter isTriangle things) isSmall
```

which says that every white disc is big and some triangle is small.
Can you express this as a sequent, without use of explicit quantification?
What about "There is some person who everybody loves", coded as

```
some people (\x -> every people (`loves` x))
```

5. The first line of the song "Everybody Loves Somebody" is "Everybody loves somebody sometime".

    (a) This statement is ambiguous. List the different possible readings.
    (b) Pick a reading and express it in Haskell, after explaining how the universe of people would need to be enriched with additional information.

6. In the Haskell coding of "Every white disc is bigger than some triangle" in Exercise 4, the choice of the small triangle doesn't depend on the choice of the big white disc.
In "Every white thing is bigger than something having the same shape", there is a clear dependency. Express this statement in Haskell: once using just predicates, and once using the relation `isBigger`.

7. Express the statements

    (a) For everything white, there is some black triangle that it is above.
    (b) There is some black triangle that everything white is above.

    in Haskell.

18

Checking Satisfiability

Contents

Satisfiability

You've seen how to use sequent calculus to check whether a sequent is universally true or has a counterexample. We're now going to look into the problem of checking **satisfiability**: whether a logical expression is true for at least one combination of values for the variables, or predicates, that the expression contains.

In Chap. 4, you saw how to check satisfiability using truth tables. In Chap. 16, we did it in Haskell using the same method. There's a way of checking satisfiability using sequent calculus, see Exercise 1 below. In this chapter, you're going to learn about an algorithm called DPLL that checks satisfiability efficiently, but only for expressions that are in conjunctive normal form (CNF).

Recall that a logical expression is in CNF if it consists of a conjunction of *clauses*, where each clause is a disjunction of *literals*. A literal is a predicate or variable, or the negation of a predicate or variable. Any logical expression can be converted into CNF. In Chap. 17, you learned how to convert expressions to CNF using Karnaugh maps, for propositions with no more than four variables. For expressions with more variables, there's a conversion method that uses sequent calculus, see Exercise 17.7. There's also a conversion algorithm that works by iteratively applying the laws of Boolean algebra, see Exercise 2.

Many hard combinatorial problems can be represented using logic, and then satisfiability checking can be used to find solutions. Examples of practical applications of satisfiability checking include: verification of hardware, software, and communication protocols; AI planning and scheduling; AI diagnosis; circuit design; and genomics.

Representing CNF

We can represent propositions that are in CNF using the algebraic data type `Prop` from Chap. 16. For example, the CNF expression

$$(\neg A \vee \neg B \vee C) \wedge (\neg A \vee D \vee F) \wedge (A \vee B \vee E) \wedge (A \vee B \vee \neg C)$$

corresponds to the following expression of type `Prop`:

```
cnf =
  (Not (Var "A") :||: Not (Var "B") :||: (Var "C"))
  :&&: (Not (Var "A") :||: (Var "D") :||: (Var "F"))
  :&&: ((Var "A") :||: (Var "B") :||: (Var "E"))
  :&&: ((Var "A") :||: (Var "B") :||: Not (Var "C"))
```

Drawing this as a tree gives:

19

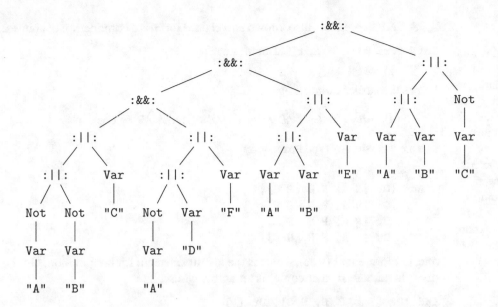

The type **Prop** can express all propositions, including propositions that are not in CNF. For an algorithm that works only on CNF expressions—like the DPLL algorithm that we're going to focus on in the rest of this chapter—it's convenient to use a simpler representation that can handle CNF expressions and nothing else.

We start with **atoms**, from which we will build literals. Atoms may be variables taken from a fixed set, for instance:

```
data Var = A | B | C | D | E | F | G | H
  deriving (Eq,Show)
```

or they may be applications of the predicates that were introduced in Chap. 6:

```
data PredApp = IsSmall Thing
             | IsWhite Thing
             | ...
  deriving (Eq,Show)
```

or something else. We'll regard them as arbitrary in our representation of CNF expressions. This will allow a free choice of atoms, including **Var** and **PredApp**.

A **literal** is an atom (which we'll call a *positive* literal) or a negated atom (a *negative* literal):

```
data Literal atom = P atom | N atom
  deriving Eq
```

The type **Literal** is polymorphic in the type of atoms. Thus **Literal Var** would be the type of literals over atoms of type **Var**, with values like **P A** (representing A) and **N B** (representing ¬B).

Then, a **clause** is a disjunction of literals:

```
data Clause atom = Or [Literal atom]
  deriving Eq
```

Examples of expressions of type **Clause Var** are **Or [P A, N B, N C]**, representing A ∨ ¬B ∨ ¬C, and **Or [N A]** (the single literal ¬A). The empty clause **Or []** represents 0 (false), which is the identity element for disjunction.

A CNF expression, also known as a **clausal form**, is a conjunction of clauses:

The algebraic data types `Clause` and `Form` share the special property of having just one constructor that takes just one parameter. Such types are essentially new names for the type that is the parameter of the constructor, with the constructor (for example `Or :: [Literal atom] -> Clause atom`) providing the translation in one direction, and pattern matching (`\(Or ls) -> ls`) providing the translation in the other direction. Replacing "`data`" by "`newtype`" in their definitions improves the performance of programs involving these types.

```
data Form atom = And [Clause atom]
  deriving Eq
```

Here's the earlier example again:

$$(\neg A \vee \neg B \vee C) \wedge (\neg A \vee D \vee F) \wedge (A \vee B \vee E) \wedge (A \vee B \vee \neg C)$$

as an expression of type `Form Var`:

```
cnf' =
  And [Or [N A, N B, P C],
       Or [N A, P D, P F],
       Or [P A, P B, P E],
       Or [P A, P B, N C]]
```

`And []` represents 1 (true), which is the identity element for conjunction. On the other hand, a `Form` that contains an empty clause, like

```
And [Or [P A, N B, N C], Or []]
```

In the scientific literature, a logical expression in CNF is often written as simply a set (the CNF) of sets (the clauses) of literals. We use `And` and `Or` to remind ourselves of which is which, and so that Haskell's typechecker can help catch our mistakes.

is equivalent to 0 because an empty clause is 0 and $a \wedge 0 = 0$.

Here are functions for converting values of these types to `String`, to make them easier to read. The definition of `show` for `Clause` uses the function `intercalate` from the `Data.List` library module to intersperse disjunction symbols between the disjuncts in the clause, and similarly for the definition of `show` for `Form`.

```
instance Show a => Show (Literal a) where
  show (P x) = show x
  show (N x) = "not " ++ show x

instance Show a => Show (Clause a) where
  show (Or ls) = "(" ++ intercalate " || " (map show ls) ++ ")"

instance Show a => Show (Form a) where
  show (And ls) = intercalate " && " (map show ls)
```

This gives:

```
> cnf'
(not A || not B || C) && (not A || D || F) && (A || B || E) && (A || B || not C)
```

The DPLL Algorithm: Idea

The Davis–Putnam–Logemann–Loveland (DPLL) algorithm takes a CNF expression e as input and searches for sets of literals Γ that, if true, imply that e is true, i.e. $\Gamma \vDash e$. It produces a list of all such sets of literals. If the list is non-empty then e is satisfiable; otherwise it is not satisfiable.

It is important that Γ is a **consistent** set of literals, meaning that it doesn't contain both a and $\neg a$ for any atom a. (Why?) That property is ensured by DPLL.

We'll develop the idea behind the algorithm using simple CNF expressions of type `Form Var`, and then use that understanding to develop the full algorithm in Haskell, which works for any type of atoms.

The heart of the algorithm is a function that takes a CNF expression e and an atom a and produces a simpler CNF expression that is equivalent to e if a is true. Let's look at our example

$$(\neg A \vee \neg B \vee C) \wedge (\neg A \vee D \vee F) \wedge (A \vee B \vee E) \wedge (A \vee B \vee \neg C)$$

19

to see how that works. We start by regarding it as a list of four clauses:

$$\neg A \vee \neg B \vee C$$
$$\neg A \vee D \vee F$$
$$A \vee B \vee E$$
$$A \vee B \vee \neg C$$

that the expression requires to be true.

Now, let's assume that A is true. How does that affect our list of clauses?

$\neg A \vee \neg B \vee C$: If A is true then $\neg A$ can't be true. So, under the assumption that A is true, $\neg A \vee \neg B \vee C$ reduces to $\neg B \vee C$.

$\neg A \vee D \vee F$: Similarly, under the assumption that A is true, $\neg A \vee D \vee F$ reduces to $D \vee F$.

$A \vee B \vee E$: On the other hand, if A is true then $A \vee B \vee E$ is true; we don't care whether B and E are true or not.

$A \vee B \vee \neg C$: Similarly, if A is true then $A \vee B \vee \neg C$ is true.

The following diagram summarises this:

$$
\begin{array}{ccc}
\neg A \vee \neg B \vee C & & \neg B \vee C \\
\neg A \vee D \vee F & \xrightarrow{\ A\ } & D \vee F \\
A \vee B \vee E & & \\
A \vee B \vee \neg C & &
\end{array}
$$

The clauses that are true drop out: if A is true then

$$(\neg A \vee \neg B \vee C) \wedge (\neg A \vee D \vee F) \wedge (A \vee B \vee E) \wedge (A \vee B \vee \neg C)$$

is equivalent to

$$(\neg B \vee C) \wedge (D \vee F)$$

Now, let's suppose that we were wrong about A and it's actually false, that is, $\neg A$ is true. What do we get, starting from our original list of four clauses?

$\neg A \vee \neg B \vee C$: If $\neg A$ is true then $\neg A \vee \neg B \vee C$ is true.

$\neg A \vee D \vee F$: Similarly, if $\neg A$ is true then $\neg A \vee D \vee F$ is true.

$A \vee B \vee E$: If $\neg A$ is true (i.e. A is false) then $A \vee B \vee E$ reduces to $B \vee E$.

$A \vee B \vee \neg C$: Similarly, if $\neg A$ is true then $A \vee B \vee \neg C$ reduces to $B \vee \neg C$.

Putting together both cases gives the following diagram, with the original list of clauses in the middle and the two alternatives—A is true and A is false—on the right and left, respectively:

$$
\begin{array}{ccccc}
& & \neg A \vee \neg B \vee C & & \neg B \vee C \\
& \xleftarrow{\ \neg A\ } & \neg A \vee D \vee F & \xrightarrow{\ A\ } & D \vee F \\
B \vee E & & A \vee B \vee E & & \\
B \vee \neg C & & A \vee B \vee \neg C & &
\end{array}
$$

We can now consider which atoms other than A might be true or false. But looking into the clauses on the left of the diagram, we can see that they'll both drop out if we assume that B is true:

$$
\begin{array}{cccccc}
& & & \neg A \vee \neg B \vee C & & \neg B \vee C \\
\xleftarrow{\ B\ } & & \xleftarrow{\ \neg A\ } & \neg A \vee D \vee F & \xrightarrow{\ A\ } & D \vee F \\
& B \vee E & & A \vee B \vee E & & \\
& B \vee \neg C & & A \vee B \vee \neg C & &
\end{array}
$$

What this sequence of steps shows is that the original CNF expression is true if both $\neg A$ and B are true. That is, $\Gamma = \neg A, B$ is a solution to the satisfiability problem for $(\neg A \vee \neg B \vee C) \wedge (\neg A \vee D \vee F) \wedge (A \vee B \vee E) \wedge (A \vee B \vee \neg C)$. It's easy to see that another solution is $\Gamma = \neg A, E, \neg C$:

$$\xleftarrow{\neg C} \begin{matrix} \\ B \vee \neg C \end{matrix} \xleftarrow{E} \begin{matrix} B \vee E \\ B \vee \neg C \end{matrix} \xleftarrow{\neg A} \begin{matrix} \neg A \vee \neg B \vee C \\ \neg A \vee D \vee F \\ A \vee B \vee E \\ A \vee B \vee \neg C \end{matrix} \xrightarrow{A} \begin{matrix} \neg B \vee C \\ D \vee F \end{matrix}$$

There are four more solutions involving A being true instead of false: $A, \neg B, D$; $A, \neg B, F$; A, C, D; and A, C, F. These are found by reducing the list of clauses on the right-hand side of the diagram.

The DPLL Algorithm: Implementation

cs << l is pronounced "cs given l".

The operation that we've identified as the heart of the DPLL algorithm is a function `cs << l` that takes a list `cs` of clauses and a literal `l` and returns a simplified list of clauses that is equivalent to `cs` if `l` is true. It's expressed in Haskell like this:

```
(<<) :: Eq atom => [Clause atom] -> Literal atom -> [Clause atom]
cs << l = [ Or (delete (neg l) ls)
          | Or ls <- cs, not (l `elem` ls) ]

neg :: Literal atom -> Literal atom
neg (P a) = N a
neg (N a) = P a
```

This definition works for any type of atom that can be tested for equality. It uses the function `delete` from the **Data.List** library module, which deletes an item from a list.

Actually, delete only deletes the first occurrence of the item. Using this means that our clauses must not contain repeated literals.

To show how the << function works, here's another example that's a little more complicated than the one above. The clauses in `cs` are shown in the leftmost column, and the other two columns show the effect of `cs << l` in two steps, for $l = $ `P A` `::` `Literal Var`, which represents the positive literal A:

cs	[Or ls | Or ls <- cs, not (A \`elem\` ls)]	[Or (delete ¬A ls) | Or ls <- cs, not (A \`elem\` ls)]
$\neg A \vee C \vee D$	$\neg A \vee C \vee D$	$C \vee D$
$\neg B \vee F \vee D$	$\neg B \vee F \vee D$	$\neg B \vee F \vee D$
$\neg B \vee \neg F \vee \neg C$	$\neg B \vee \neg F \vee \neg C$	$\neg B \vee \neg F \vee \neg C$
$\neg D \vee \neg B$	$\neg D \vee \neg B$	$\neg D \vee \neg B$
$B \vee \neg C \vee \neg A$	$B \vee \neg C \vee \neg A$	$B \vee \neg C$
$B \vee F \vee C$	$B \vee F \vee C$	$B \vee F \vee C$
$B \vee \neg F \vee \neg D$	$B \vee \neg F \vee \neg D$	$B \vee \neg F \vee \neg D$
$A \vee E$		
$A \vee F$		
$\neg F \vee C \vee \neg E$	$\neg F \vee C \vee \neg E$	$\neg F \vee C \vee \neg E$
$A \vee \neg C \vee \neg E$		

In the first step, we select the clauses from `cs` that don't contain `l` (that is, A). This is the effect of the part of the comprehension after the vertical bar, where the generator `Or ls <- cs` extracts a list `ls` of literals from each of the clauses in `cs`. In the example, this removes three clauses.

In the second step, we remove the negation of 1 (that is, $\neg A$) from the remaining clauses, if it appears. That's the effect of the part of the comprehension before the vertical bar. In the example, it simplifies two clauses and leaves the others unchanged.

Continuing the example, here's the result of cs $<<$ A from the rightmost column above together with the result of cs $<<$ $\neg A$, where N A, the negation of 1, is assumed to be truc:

cs	cs $<<$ A	cs $<<$ $\neg A$
$\neg A \vee C \vee D$	$C \vee D$	
$\neg B \vee F \vee D$	$\neg B \vee F \vee D$	$\neg B \vee F \vee D$
$\neg B \vee \neg F \vee \neg C$	$\neg B \vee \neg F \vee \neg C$	$\neg B \vee \neg F \vee \neg C$
$\neg D \vee \neg B$	$\neg D \vee \neg B$	$\neg D \vee \neg B$
$B \vee \neg C \vee \neg A$	$B \vee \neg C$	
$B \vee F \vee C$	$B \vee F \vee C$	$B \vee F \vee C$
$B \vee \neg F \vee \neg D$	$B \vee \neg F \vee \neg D$	$B \vee \neg F \vee \neg D$
$A \vee E$		E
$A \vee F$		F
$\neg F \vee C \vee \neg E$	$\neg F \vee C \vee \neg E$	$\neg F \vee C \vee \neg E$
$A \vee \neg C \vee \neg E$		$\neg C \vee \neg E$

The main function of the DPLL algorithm takes a CNF expression e and produces all sets (represented as lists) of literals that, if true, imply that e is true. Here's our first version of its definition, which omits some details that will be supplied soon:

```
dpll :: Eq atom => Form atom -> [[Literal atom]]
dpll (And [])              = ...
dpll (And (Or [] : cs))    = ...
dpll (And (Or (l:ls) : cs)) =
  [ l : ls | ls <- dpll (And (cs << l)) ]
  ++
  [ neg l : ls | ls <- dpll (And (Or ls : cs << neg l)) ]
```

The definition of dpll is recursive. It computes all solutions that are obtained by assuming that the literal 1 is true, and (separately) by assuming that its negation neg 1 is true, and then appends them. The recursive calls will do the same thing for other literals, continuing the process until one of the base cases is reached. This is another example of a *divide and conquer* algorithm, like quicksort on page 86: we split a problem into two simpler problems and then combine the solutions to those simpler problems to give a solution to the original problem.

In this version of the definition of dpll, the literal 1 is chosen to be the first literal in the first clause, Or (l:ls). When 1 is assumed to be true, that clause doesn't contribute any further to the result (that is, the recursive call is dpll (And (cs $<<$ 1))) because it contains 1 and is therefore true. On the other hand, when its negation neg 1 is assumed to be true, the rest of that clause is taken into account when computing the result (that is, the recursive call is dpll (And (Or ls : cs $<<$ neg 1))) since in that case, Or (l:ls) reduces to Or ls.

The definition of dpll is missing the right-hand side of its two base cases. The first is for the case of a CNF expression containing no clauses. As explained earlier, such an expression is true: the conjunction of the empty set of conjuncts is 1. The second is for the case where the first clause is empty. In that case, the expression is false: the disjunction of the empty set of disjuncts is 0, and conjoining 0 with anything gives 0. We can, therefore, complete the above definition as follows:

These reductions come directly from the rules of the sequent calculus. For example,

$$\frac{A \vDash C \vee D}{A \vDash \neg A \vee C \vee D} \neg R$$

and

$$\frac{}{A \vDash A \vee E} I$$

The DPLL algorithm applied to a CNF expression e can, therefore, be regarded as searching for both a consistent set of literals Γ and a proof in sequent calculus that $\Gamma \vDash e$. The fact that e is in CNF makes the proof search easy.

The DPLL algorithm was introduced by Martin Davis, George Logemann and Donald Loveland in 1962 as a refinement of an earlier algorithm due to Davis and Hilary Putnam, see ▶ https://en.wikipedia.org/wiki/DPLL_algorithm. The Rolling Stones's first international number one hit "(I Can't Get No) Satisfaction" is a homage to the DPLL algorithm. Just kidding!

```
dpll :: Eq atom => Form atom -> [[Literal atom]]
dpll (And [])            = [[]]   -- one trivial solution
dpll (And (Or [] : cs))  = []     -- no solution
dpll (And (Or (l:ls) : cs)) =
  [ l : ls | ls <- dpll (And (cs << l)) ]
  ++
  [ neg l : ls | ls <- dpll (And (Or ls : cs << neg l)) ]
```

Note the difference between the results for the two base cases! In the first base case, the expression is true so the empty set of literals is a solution. In the second base case, the expression is false so there are no solutions. The fact that there are no solutions to this sub-problem doesn't necessarily mean that there are no solutions to the original problem. It just means that this particular path in the search, under which a particular set of assumptions have been made about the truth of literals, doesn't lead to a solution.

Let's continue the example, looking into the branch where $\neg A$ has been assumed to be true. It isn't in the code for dpll, but we can take a little shortcut by observing that there are two so-called *unit clauses*, each containing a single literal: E and F. A solution always has to require that such literals are true; assuming that one is false would lead immediately to an empty clause, with no solution. The subsequent choice of assuming that C is true is forced for the same reason:

cs $\ll \neg A$	cs $\ll \neg A$ $\ll E \ll F$	cs $\ll \neg A$ $\ll E \ll F$ $\ll C$
$\neg B \vee F \vee D$		
$\neg B \vee \neg F \vee \neg C$	$\neg B \vee \neg C$	$\neg B$
$\neg D \vee \neg B$	$\neg D \vee \neg B$	$\neg D \vee \neg B$
$B \vee F \vee C$		
$B \vee \neg F \vee \neg D$	$B \vee \neg D$	$B \vee \neg D$
E		
F		
$\neg F \vee C \vee \neg E$	C	
$\neg C \vee \neg E$	$\neg C$	[]

The empty clause in the last step means that there's no solution if $\neg A$ is assumed to be true. According to Exercise 3 below, there's no solution if A is assumed to be true. So there's no solution to the satisfiability problem: the original CNF expression is not satisfiable. This is confirmed by running the dpll function:

```
> dpll example2
[]
```

As already mentioned, we didn't completely follow the implementation of dpll in this example. Instead of always using the first literal in the first clause for the case split, which is what the function definition says, we used the literal in a unit clause if one was present. This refinement is incorporated in the following (final) version of dpll, which reorders the clauses before making the choice of literal:

```
dpll :: Eq atom => Form atom -> [[Literal atom]]
dpll f =
  case prioritise f of
    []             -> [[]]  -- the trivial solution
    Or [] : cs     -> []    -- no solution
    Or (l:ls) : cs ->
      [ l : ls | ls <- dpll (And (cs << l)) ]
```

19

This strategy is called *unit propagation*, see ▶ https://en.wikipedia.org/wiki/Unit_propagation.

```
   ++
   [ neg l : ls | ls <- dpll (And (Or ls : cs << neg l)) ]
```

```
prioritise :: Form atom -> [Clause atom]
prioritise (And cs) = sortOn (\(Or ls) -> length ls) cs
```

The function `prioritise` uses the `sortOn` function from the `Data.List` library module to sort the clauses in order of increasing length so that the ones with the fewest literals, including unit clauses if any, are first. The improvement is dramatic: for the Sudoku example below, it makes the difference between producing a result in a few seconds and running overnight with no result. Other choices for `prioritise` are possible.

In practice, DPLL is often very efficient. But it's very slow in some cases, taking time that is exponential in the size of its input, which is the same as checking satisfiability using truth tables. No algorithm for satisfiability is known that doesn't share this property. The efficiency of DPLL and other satisfiability algorithms boils down to how their heuristics—in the case of DPLL, the choice of the function `prioritise`—perform in problems that arise in practice.

In fact, satisfiability is a so-called *NP-complete* problem, meaning that if it can be solved efficiently in every case then a very large class of difficult algorithmic problems also have efficient solutions, and vice versa, see ▶ https://en.wikipedia.org/wiki/NP-completeness. The question of whether there is an efficient solution to any NP-complete problem is the so-called "P versus NP" problem. There is a $1 million prize for the first person to prove either that P=NP—an algorithm for satisfiability that is *always* efficient would suffice—or P≠NP.

Application: Sudoku

As an interesting and fun application of DPLL, we're going to use it to solve sudoku puzzles. A sudoku puzzle is a 9×9 grid with some cells filled with digits from 1 to 9. For example:

9								2
			9		4	6	3	
3		6			8	1		
6			9			3		
9			8		2			1
		2			7			5
		3	5			7		4
5	1	7		8				
4						1		

The aim is to fill in the rest of the cells with digits so that each column, each row, and each of the outlined 3×3 regions contain all of the digits from 1 to 9. Here is a solution to the puzzle above, which is a very difficult one to solve manually:

See ▶ https://sudokuguy.com/ for lots of sudoku tutorials.

1	9	4	3	7	6	8	5	2
7	5	8	2	9	1	4	6	3
3	2	6	4	5	8	1	9	7
6	7	1	9	4	5	3	2	8
9	4	5	8	3	2	6	7	1
8	3	2	1	6	7	9	4	5
2	6	3	5	1	9	7	8	4
5	1	7	6	8	4	2	3	9
4	8	9	7	2	3	5	1	6

Solving a sudoku puzzle can be viewed as a satisfiability problem, that we can solve using DPLL. The starting configuration of the puzzle together with all of the rules about the allowed placement of digits can be formulated as a very large CNF expression. Then a solution to the puzzle is a placement of the missing digits which satisfies that expression.

We'll use triples of integers as the atoms in our representation. The triple (i, j, n) represents the cell in the ith row and jth column being occupied by the digit n. So the following code represents the requirement that every cell in the puzzle is filled:

```
allFilled :: Form (Int,Int,Int)
allFilled = And [ Or [ P (i,j,n) | n <- [1..9] ]
                | i <- [1..9], j <- [1..9] ]
```

This produces a CNF expression with 81 clauses, one for each cell. The first clause is

```
Or [P (1,1,1), P (1,1,2), P (1,1,3), P (1,1,4), P (1,1,5),
    P (1,1,6), P (1,1,7), P (1,1,8), P (1,1,9)]
```

which says that the cell in row 1 and column 1 must contain one of the digits 1–9.

Another requirement is that no cell is filled twice:

```
noneFilledTwice :: Form (Int,Int,Int)
noneFilledTwice = And [ Or [ N (i, j, n), N (i, j, n') ]
                      | i <- [1..9], j <- [1..9],
                        n <- [1..9], n' <- [1..(n-1)]]
```

This produces a CNF expression with 2916 clauses, 36 for each cell. Each of the clauses for the cell in row 1 and column 1 is of the form `Or [N (1,1,`n`)`, `N (1,1,`n'`)]` for $1 \leq n' < n \leq 9$, saying that n and n' can't both be in that cell.

The requirement that each row contains all of the digits 1–9 is:

```
rowsComplete :: Form (Int,Int,Int)
rowsComplete = And [ Or [ P (i, j, n) | j <- [1..9] ]
                   | i <- [1..9], n <- [1..9] ]
```

which produces 81 clauses, 9 for each row. The first clause for row 1 is

```
Or [P (1,1,1), P (1,2,1), P (1,3,1), P (1,4,1), P (1,5,1),
    P (1,6,1), P (1,7,1), P (1,8,1), P (1,9,1)]
```

which says that the digit 1 much be in row 1 and column j for some $1 \leq j \leq 9$. Similar code—see Exercise 4 below—expresses the requirements that each column and each of the outlined 3×3 squares contains all of the digits. Each of those requirements corresponds to 81 clauses with 9 literals each.

Finally, we need to express the starting configuration of the puzzle. The one above is given by a CNF expression composed of 30 unit clauses, one for each entry:

```
sudokuProblem =
  And [Or [P (1,2,9)], Or [P (1,9,2)], Or [P (2,5,9)],
       Or [P (2,7,4)], Or [P (2,8,6)], ... etc ...]
```

The following function is useful for forming the conjunction of CNF expressions:

```
(<&&>) :: Form a -> Form a  -> Form a
And xs <&&> And ys = And ( xs ++ ys )
```

and then the entire specification is

```
sudoku =
  allFilled <&&> noneFilledTwice <&&> rowsComplete
  <&&> columnsComplete <&&> squaresComplete
  <&&> sudokuProblem
```

for a total of 3270 clauses containing 8778 literals.

Unfortunately, although this is a complete specification of the problem, to make it tractable for our implementation of `dpll` we need to add three more constraints, each containing a large number of clauses. The additional constraints are actually consequences of the ones above, and adding them dramatically increases the size of the specification. But the effect of adding them is to restrict the search space explored by `dpll` by reducing the number of false avenues.

The first additional constraint says that each row contains no repeated digits:

```
rowsNoRepetition :: Form (Int,Int,Int)
rowsNoRepetition = And [ Or [ N (i, j, n), N (i, j', n) ]
                         | i <- [1..9], n <- [1..9],
                           j <- [1..9], j' <- [1..(j-1)] ]
```

This adds 2916 clauses, 324 for each row, each containing two literals. Each of the clauses for the first row is of the form `Or [N (1,`j`,`n`), N (1,`j'`,`n`)]` for $1 \le j' < j \le 9$, saying that n can't be in both cell $(1,j)$ and cell $(1,j')$. See Exercise 4 below for the other two additional constraints, which express the analogous non-repetition conditions for the columns and outlined 3×3 squares.

The revised specification

```
sudoku =
  allFilled <&&> noneFilledTwice <&&> rowsComplete
  <&&> columnsComplete <&&> squaresComplete
  <&&> sudokuProblem <&&> rowsNoRepetition
  <&&> columnsNoRepetition <&&> squaresNoRepetition
```

is composed of 12018 clauses containing 26274 literals, and can be solved by `dpll` in a few seconds:

```
> :set +s
> dpll sudoku
[[P (1,2,9), P (1,9,2), P (2,5,9), ... etc ...]]
(21.94 secs, 7,178,847,560 bytes)
```

The command `:set +s` tells Haskell to display elapsed time and space usage after evaluation.

There's only one solution—as usual for sudoku problems—and it's the one given above.

Exercises

1. Use sequent calculus to check whether or not the expressions $(a \vee b) \wedge (\neg a \wedge \neg b)$ and $((a \wedge \neg b) \vee c \vee (\neg d \wedge b) \vee a) \wedge \neg c$ are satisfiable. Use the following observations:

   - A proof which shows that $\vDash \neg e$ is universally valid amounts to a proof that e is not satisfiable.
   - A proof which shows that $\vDash \neg e$ is not universally valid amounts to a proof that e is satisfiable, and the simple sequents that arise from the proof can be used to give values for the atoms in e that makes it true.

2. Use the following laws of Boolean algebra to convert the expression $(a \wedge \neg b \wedge (c \vee (d \wedge b)) \vee (\neg b \wedge \neg a)) \wedge c$ to CNF:

$$\neg(a \vee b) = \neg a \wedge \neg b \qquad \neg 0 = 1 \qquad \neg\neg a = a \qquad \neg 1 = 0 \qquad \neg(a \wedge b) = \neg a \vee \neg b$$
$$a \vee (b \wedge c) = (a \vee b) \wedge (a \vee c) \qquad a \vee 1 = 1 = \neg a \vee a \qquad (a \wedge b) \vee c = (a \vee c) \wedge (b \vee c)$$
$$a \wedge (b \vee c) = (a \wedge b) \vee (a \wedge c) \qquad a \wedge 0 = 0 = \neg a \wedge a \qquad (a \vee b) \wedge c = (a \wedge c) \vee (b \wedge c)$$
$$a \vee a = a = 0 \vee a \qquad a \vee (a \wedge b) = a \qquad a \vee b = b \vee a \qquad a \vee (b \vee c) = (a \vee b) \vee c$$
$$a \wedge a = a = 1 \wedge a \qquad a \wedge (a \vee b) = a \qquad a \wedge b = b \wedge a \qquad a \wedge (b \wedge c) = (a \wedge b) \wedge c$$

3. Complete the example that began on page 182 by completing the branch where A has been assumed to be true, to conclude that the problem has no solution in that case. Unlike the branch on page 184 where $\neg A$ has been assumed to be true, there's more than one sub-case.

4. Write Haskell code to express the requirements on sudoku puzzles that

   • Each column contains all of the digits 1–9
   • Each of the outlined 3 × 3 squares contains all of the digits 1–9
   • Each column contains no repeated digits:
   • Each of the outlined 3 × 3 squares contains no repeated digit

   as CNF expressions.

5. Write a function (`<||>`) `:: Form a -> Form a -> Form a` that produces the disjunction of two CNF expressions.

6. Write a function `toProp :: Show a => Form a -> Prop` that converts an expression in clausal form to a proposition of type `Prop`, using `show` to convert the atoms to variable names. Use `toProp` to compare the performance of `not . null . dpll` with the function `satisfiable` on page 151.

7. Improve the final version of `dpll` on page 184 by adding special treatment of unit clauses and clauses containing two literals.

8. Use DPLL to solve some sudoku puzzles from newspapers and online sources. Try to determine whether the difficulty of the puzzle affects the time to find a solution.

To do a proper comparison, you should compile your Haskell code, see ► https://downloads.haskell.org/ ~ghc/8.2.1/docs/html/users_guide/ usage.html, and use Haskell's profiler to measure performance, see ► https://downloads.haskell.org/ ~ghc/8.2.1/docs/html/users_guide/ profiling.html.

19

Data Representation

Contents

D. Sannella et al., *Introduction to Computation*, Undergraduate Topics
in Computer Science, https://doi.org/10.1007/978-3-030-76908-6_20

Four Different Representations of Sets

Whenever you write a program that requires a kind of data that isn't already built into Haskell, you need to decide how to represent it in terms of the existing types. Sometimes you'll be able to find something appropriate in one of Haskell's library modules. Other times, you will decide to use some combination of existing types, or you'll define a new algebraic data type.

The way that your data is represented has a major impact on your code. For example, if you define a new algebraic data type, then some or all of your functions will be defined by cases using the new type's constructors. And you might need to define helper functions on your new types that would already be provided if you had decided to use built-in types instead.

In this chapter, we're going to look at four different ways of representing sets of integers, to show how the choice of representation of data affects your code. To make it easy to compare the different choices, we'll use the same type name and implement the same functions for each choice of representation:

A list of types and function names with their types, intended as a declaration of what is provided by a software component, is called an **API**, which stands for "application program interface", or **interface** for short. See ▶ https://en.wikipedia.org/wiki/API.

```
type Set
empty :: Set
singleton :: Int -> Set
set :: [Int] -> Set
union :: Set -> Set -> Set
element :: Int -> Set -> Bool
equal :: Set -> Set -> Bool
```

These should be self-explanatory, except perhaps for `set`, which takes a list of elements and produces a set containing those elements.

One of the things that your choice of data representation affects is the run time and space consumption of your code. The four examples below provide a good opportunity to study how the run time of programs can be characterised, and to see how the choice of data representation affects it.

Before proceeding, it is very important for you to understand that—at least at this point in your study of programming, and most of the time later on—you should strive to make your code simple, clear and above all **correct**, and forget about efficiency!

❭❭ Premature optimization is the root of all evil.
    Donald Knuth, 1974 Turing Award winner

For example, efficiency is obviously important in Google's indexing systems, which index hundreds of billions of web pages, see ▶ https://www.google.com/search/howsearchworks/crawling-indexing/.

That said, efficiency is sometimes important in programs that manipulate large amounts of data. In such cases, the best way to achieve the required efficiency is via timing and/or space usage measurements on the running simple/clear/correct version of the program, in order to identify where most of the time or space is being used. Usually, it will turn out—often to the complete surprise of the code's author!—that the problem is in a few small critical parts of the program, and then effort can be focused on improving just those parts.

Rates of Growth: Big-O Notation

Consider the Prelude function `elem`:

```
elem :: Eq a => a -> [a] -> Bool
m `elem` []     = False
m `elem` (n:ns) = m==n || m `elem` ns
```

What is a sensible answer to the question "What's the run time of `elem`?" An answer like "13.6 milliseconds" might be accurate but it isn't very helpful because the exact run time will depend on the hardware used. More important, it will depend on the length of the second parameter. The time for a list with 10 elements will be very different from the time for a list with 10 billion elements. It will also depend on where—and whether—the item in question appears in the list.

We need a way to characterise the run time of functions that is dependent on the size of the input. The best way of dealing with issues like where the item appears in the list, in the case of `elem`, is to focus on **worst-case** run time, which is when the item is not in the list at all. Best-case run time (where the item appears at the beginning of the list) is uninteresting. Average-case run time is interesting but turns out to be much harder to determine.

The run time of a function for small inputs is usually unimportant. Much more important is the run time for large and very large inputs, where the difference between an efficient and inefficient function might be huge. In the end, what matters is this: *How does the worst-case run time of a function increase for increasing sizes of input?* From the answer to this question, we can work out if it's feasible to use the function on large inputs.

Big-O notation is used to describe rates of growth. For a list of length n, the worst-case run time for `elem` will be $an + b$, where a is the time taken to check one item in the list, including the time to compute the disjunction and the overhead of the recursive call, and b is the overhead involved in starting the computation. In big-O notation, the worst-case run time of `elem` is $O(n)$. This deliberately glosses over the "constant factors" a and b, since they will vary according to the hardware used and the version of Haskell, and focuses on the fact that the rate of growth of the run time with increasing size of input is **linear**.

The idea of big-O notation is that a function f is $O(g)$ if g is an upper bound for f, for big enough inputs, ignoring constant factors. Formally:

Definition. f is $O(g)$ if there are constants c and m such that $|f(n)| \leq cg(n)$ for all $n \geq m$.

For example, $2n$ is $O(\frac{1}{2}n^2)$ because $2n \leq \frac{1}{2}n^2$ for all $n \geq 4$. And $3n + 17$ is $O(n)$ because $3n + 17 \leq 4n$ for all $n \geq 17$.

Because constant factors don't matter, and neither does the base of logarithms or lower order terms in polynomials (see Exercise 1), big-O notation normally uses simple functions: $O(1)$, $O(\log n)$, $O(n)$, $O(n^2)$, $O(n^3)$, $O(2^n)$, etc. And the function used is the slowest-growing one that provides an upper bound: $2n$ is $O(n^2)$ but it is also $O(n)$, so we use the latter. We use the terminology: **constant time**, **logarithmic time**, **linear time**, **quadratic time**, **cubic time**, **exponential time**, etc. In big-O notation, n always refers to the size of the parameter, whatever "size" means for the kind of data in question. In the case of lists, n refers to the length of the list. For the trees that will come up later in this chapter, it refers to the number of nodes in the tree.

The reason for distinguishing between logarithmic time, linear time, quadratic time, etc. but not (for example) between n and $n/2$ is shown by looking at the graphs of these functions:

In the case of `elem`, is the average case when the item is halfway down the list? If items have type `Int`, perhaps it should be when the item is not in the list, since no matter how long the list is, almost all values of type `Int` will be absent?

$O(n)$ is pronounced "big-O of n". The name "big-O" is used to distinguish it from "little-o notation", see ▶ https://en.wikipedia.org/wiki/Big_O_notation.

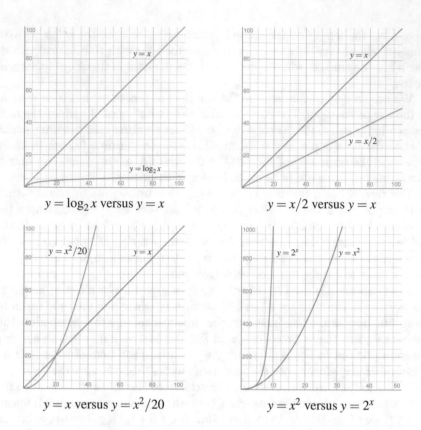

$y = \log_2 x$ versus $y = x$ $y = x/2$ versus $y = x$

$y = x$ versus $y = x^2/20$ $y = x^2$ versus $y = 2^x$

The first graph compares logarithmic time with linear time. The logarithm function grows *very* slowly. A function that takes logarithmic time can process twice as much data in only one additional unit of time. In comparison, a function that takes linear time will require twice as much time to process twice as much data. For a million items of data, the difference is between 20 units of time and a million units of time.

The second graph compares two versions of linear time: $n/2$ versus n. Here the difference is a constant factor. The second one will always take twice as long as the first one. That's a difference, but not a really substantial one in relation to the other graphs: for a million items of data, the difference is between half a million and a million units of time.

The third graph compares linear time with fast quadratic time, $n^2/20$. Up to 20 items, quadratic time is faster, but after that linear time is faster. For a million items of data, the difference is between a million and 50 billion units of time.

The final graph compares quadratic time with exponential time, and the scales on the axes have been adjusted to show that the exponential is not just a straight line. The exponential grows *very* quickly, taking twice as much time to process each additional data item. For 100 items, the difference is between 10,000 and 1,267,650,600,228,229,401,496,703,205,376 units of time.

With `elem`, we've seen an example of a linear time function. We can determine that it is $O(n)$ by looking at the structure of the code:

```
elem :: Eq a => a -> [a] -> Bool
m `elem` []     = False
m `elem` (n:ns) = m==n || m `elem` ns
```

which includes a single recursive call on a list that is one item shorter. All further function calls are to constant-time functions. So the time required is a constant for each recursive call, with one recursive call for each item in the list.

20

Here's an example of a quadratic time function:

```
subset :: Eq a => [a] -> [a] -> Bool
ms `subset` ns = and [ m `elem` ns | m <- ms ]
```

This function calls `elem` for each element of `ms`. We already know that `elem` is linear, so the time to do that will be $O(n^2)$. (Actually $O(mn)$, if `ms` has length m and `ns` has length n, but in doing such calculations one often assumes that all parameters have the same size.) There is a call to the linear-time function `and` on a list of size n that will take time $O(n)$, but that's less than $O(n^2)$ so the result is $O(n^2)$.

Finding all of the divisors of an n-digit binary number by exhaustive search takes exponential time. Given an n-digit binary number m, testing all of the binary numbers that are less than m to see if they are divisors of m will take time $O(2^n)$, since there are 2^n n-digit binary numbers. Each additional digit doubles the number of potential divisors to check.

Representing Sets as Lists

We'll start with the simplest representation of all, and define a set of integers to be a list containing the elements in the set:

```
type Set = [Int]
```

So the set $\{1, 2, 3\}$ can be represented by the list `[1,2,3]` or `[3,2,1]` or `[1,1,3,2,3]`, among others. The empty set is just the empty list, and `singleton` n is just `[n]`. The `set` function doesn't need to do anything to convert a list to a set, because it already *is* a set. We can use `++` for the `union` function. And the `element` function is just `elem` from the Prelude:

```
empty :: Set
empty = []

singleton :: Int -> Set
singleton n = [n]

set :: [Int] -> Set
set ns = ns

union :: Set -> Set -> Set
ms `union` ns = ms ++ ns

element :: Int -> Set -> Bool
n `element` ns = n `elem` ns
```

Testing equality of sets using equality on `[Int]` would give the wrong answer, because the same elements might be listed in a different order or some elements might appear more than once. For example:

```
> [1,2,1,1] == [2,1]
False
```

even though `[1,2,1,1]` and `[2,1]` contain the same elements, and are therefore equal as sets. So we define `subset` as a helper function, and define `equal` using `subset`:

```
equal :: Set -> Set -> Bool
ms `equal` ns = ms `subset` ns && ns `subset` ms
  where
  ms `subset` ns = and [ m `elem` ns | m <- ms ]
```

Since `singleton` amounts to a single application of : and `set` is just the identity function, they run in constant time, $O(1)$. The function `element` is just `elem`, which we have already seen is linear time, $O(n)$, and the function `union` is just `++`:

```
(++) :: [a] -> [a] -> [a]
[] ++ ys     = ys
(x:xs) ++ ys = x : (xs ++ ys)
```

which is linear time by the same reasoning as for `elem`. On the other hand, `equal` is quadratic, since it calls `subset` twice, which we have already seen is $O(n^2)$.

Representing Sets as Ordered Lists Without Duplicates

An easy improvement to our representation of sets using lists is to insist that the lists are in ascending (or descending) order, and that there are no duplicated elements. Putting the list in order makes it easier to search for elements, and when duplicates are eliminated as well then testing set equality is easy.

We'll need the `nub` and `sort` functions from Haskell's **Data.List** library module, so we start by importing them:

```
import Data.List(nub,sort)
```

Sets are still represented as lists of integers, but we're only interested in lists that satisfy an **invariant**: they need to be in ascending order and contain no duplicates. We'll spell out the invariant as a Haskell function, but it won't be used in the code so any other notation would be just as good:

An advantage of defining the invariant in Haskell is that it can then be used in testing.

```
type Set = [Int]

invariant :: Set -> Bool
invariant ns = and [ m < n | (m,n) <- zip ns (tail ns) ]
```

Requiring that each element in the list is *strictly* less than the next one excludes duplicates.

The empty set is still just the empty list and `singleton` n is still $[n]$. But `set` needs to sort the list of elements into ascending order and remove any duplicates to convert a list of elements to a set, while `union` needs to merge the two sets, eliminating any duplicates and retaining the order to ensure that the result satisfies the invariant:

```
empty :: Set
empty = []

singleton :: Int -> Set
singleton n = [n]

set :: [Int] -> Set
set ns = nub (sort ns)

union :: Set -> Set -> Set
ms `union` []                     = ms
[] `union` ns                     = ns
(m:ms) `union` (n:ns) | m == n    = m : (ms `union` ns)
                      | m < n     = m : (ms `union` (n:ns))
                      | otherwise = n : ((m:ms) `union` ns)
```

Let's check that these work:

```
> set [1,42,2,7,1,3,2]
[1,2,3,7,42]
> set [3,1,2] `union` set [1,7,4,1]
[1,2,3,4,7]
```

Testing whether or not an integer is in a set can now take advantage of the fact that the list representation of the set is in ascending order. The definition of `element` starts at the beginning of the list and checks consecutive elements until it either finds the integer it's looking for or else reaches an element that is larger, meaning that the required element is absent:

```
element :: Int -> Set -> Bool
m `element` []                     = False
m `element` (n:ns) | m < n     = False
                   | m == n    = True
                   | m > n     = m `element` ns
```

Finally, testing equality of sets represented as ordered lists without duplicates is the same as testing ordinary list equality:

```
equal :: Set -> Set -> Bool
ms `equal` ns = ms == ns
```

The change of data representation changes the run time of some functions in the API. All good sorting functions require time $O(n \log n)$, but `nub` requires quadratic time, so `set` requires quadratic time $O(n^2)$, while `union` still requires linear time $O(n)$ and `singleton` is still $O(1)$.

Although taking advantage of the order of the list means that `element` is able to return a result of **False** when it reaches an value that is larger than the one it is looking for, rather than continuing all the way to the end of the list, its worst-case run time is still linear, $O(n)$. On the other hand, the run time of `equal` is linear, which is much faster than before (to be precise, it's linear in the length of the shorter of the two parameter lists). In terms of worst-case run time, that's the only improvement with respect to the unordered list representation. The run time of `invariant` doesn't matter because it's just documentation.

Sorting a list so that duplicates are eliminated can be done in $O(n \log n)$ time, see Exercise 2. And, if *ms* has length *m* and *ns* has length *n* then *ms* `union` *ns* is $O(m + n)$. But if we assume—as is often done in these calculations—that $m = n$, then $O(m + n) = O(2n)$ which is $O(n)$.

Representing Sets as Ordered Trees

We can do better than ordered lists by using a binary tree representation. If done properly, checking if an integer is in a set will then take logarithmic time. We will be able to check just one path in the tree from the root to the location of the integer that we're looking for. The length of such a path is at most the *depth* of the tree, which is $\log_2 n$ for a tree containing *n* values.

The first step is to define an appropriate algebraic data type for sets:

```
data Set = Nil | Node Set Int Set
  deriving Show
```

This defines a binary tree with integer labels at each node. A set is either empty (`Nil`) or is a `Node` with a left sub-tree, an integer (its "label"), and a right sub-tree. Here's an example:

```
Node (Node (Node (Node Nil 1 Nil) 2 Nil)
           6
           (Node Nil 10 (Node Nil 12 Nil)))
     17
     (Node (Node (Node Nil 18 Nil)
                 20
                 (Node Nil 29 (Node Nil 34 Nil)))
```

The depth might be greater than $\log_2 n$ if the tree is unbalanced, see page 198.

In a tree, the points from which sub-trees hang—which in our trees contain integer **labels**—are called **nodes**. The node at the top of a tree is called its **root**. The sub-trees along the bottom—which in our trees are all `Nil`—are called **leaves**. The **depth** of a tree is the length of the longest path from the root to a leaf. A tree is **binary** if every node has at most two sub-trees. See ▶ https://en.wikipedia.org/wiki/Binary_tree.

```
     35
(Node (Node Nil 37 Nil)
      42
      (Node (Node Nil 48 Nil) 50 Nil)))
```

When drawn as a diagram, the terminology ("tree", "node", "label", "sub-tree", etc.) becomes clearer:

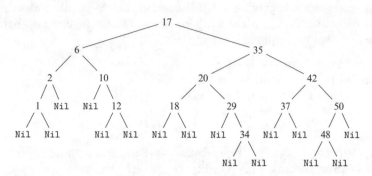

Ordered binary trees are also called **binary search trees**, see ▶ https://en.wikipedia.org/wiki/Binary_search_tree.

We're going to be interested in *ordered* binary trees in which the value at each node is greater than all of the values in its left sub-tree, and less than all of the values in its right sub-tree. This property—which the example above obeys—is expressed by the following invariant. It uses a helper function called `list` that produces a list containing all of the elements in a set by visiting all of its nodes and recording their labels:

```
invariant :: Set -> Bool
invariant Nil          = True
invariant (Node l n r) =
  and [ m < n | m <- list l ] &&
  and [ m > n | m <- list r ] &&
  invariant l && invariant r


list :: Set -> [Int]
list Nil        = []
list (Node l n r) = list l ++ [n] ++ list r
```

Maintaining the invariant is what makes this representation so efficient. For example, when searching for an integer in a set (`element`), each comparison of the integer in question with the integer at a node will allow us to ignore half of the remaining values in the tree, on average. Discarding half of the tree at each stage reduces a large tree very quickly—in $\log_2 n$ steps, where n is the number of nodes—to a single node.

The empty set is represented by `Nil` and `singleton` yields a tree containing a single node. The functions `set` and `union` require a helper function `insert`, which adds an integer to a tree. It uses the values in each node to decide whether the new element belongs to its left or right, and then recursively inserts the new element in that sub-tree, leaving the other sub-tree untouched. If it eventually finds that the element is already there then it does nothing; otherwise, it creates a new node:

```
empty :: Set
empty = Nil

singleton :: Int -> Set
singleton n = Node Nil n Nil

insert :: Int -> Set -> Set
insert m Nil = Node Nil m Nil
```

```
insert m (Node l n r)
  | m == n    = Node l n r
  | m < n     = Node (insert m l) n r
  | m > n     = Node l n (insert m r)
```

Here's a picture of what inserting 11 into the above set does:

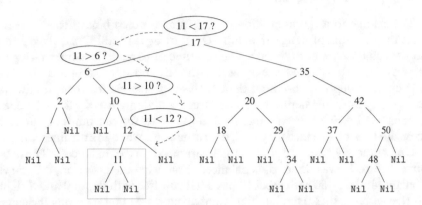

Note that the result of `insert` is the whole tree representing the set, including a new node containing the new element, if it wasn't already there. The result is *not* merely the new node.

To construct a set from a list, the items in the list are inserted one by one, starting with an empty set. We can express that using `foldr` and `insert`:

```
set :: [Int] -> Set
set = foldr insert empty
```

Similarly, the `union` function uses `insert` to add each of the elements in one set (computed using `list`) to the other set:

```
union :: Set -> Set -> Set
ms `union` ns = foldr insert ms (list ns)
```

The definition of `element` is similar to the definition of `insert` with respect to the way that it finds its way to the node that might contain the value being sought:

```
element :: Int -> Set -> Bool
m `element` Nil = False
m `element` (Node l n r)
  | m == n     = True
  | m < n      = m `element` l
  | m > n      = m `element` r
```

Here's a picture of what `element` does when checking whether or not 15 is in the above set:

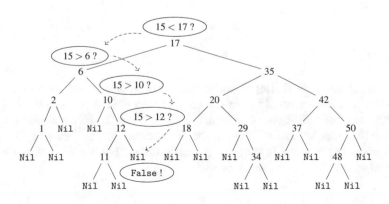

Make sure that you understand this point! If necessary, write out the computation step-by-step on a smaller example.

This is known as an **inorder traversal**, see ▶ https://en.wikipedia.org/wiki/Tree_traversal.

Since list traverses the tree from left to right, making a list of the values it encounters as it goes, and all trees representing sets respect the above invariant, the result will be in ascending order without duplicates. So checking equality of two sets can just use list equality on the resulting lists:

```
equal :: Set -> Set -> Bool
s `equal` t = list s == list t
```

The change to a representation of sets using ordered binary trees doesn't affect the run time of singleton (still $O(1)$) or equal (still $O(n)$, since list and list equality are both linear). The run time of element becomes $O(\log n)$ for trees that are **balanced**, meaning that the left and right sub-trees of every node differ in depth by no more than 1. (It's not possible for them to always have exactly the same depth, except for trees containing exactly $2^d - 1$ nodes for some d.) Likewise for insert. And in that case, the run times of set and union are $O(n \log n)$, since they both do n insertions, while list is linear, $O(n)$.

Unfortunately, our trees are not guaranteed to be balanced. The tree above, before 11 was inserted, is balanced. Following the insertion, it became unbalanced: the node with label 10 has a left sub-tree with depth 0 and a right sub-tree with depth 2. Here's a highly unbalanced tree that contains the same integers as the tree we started with above:

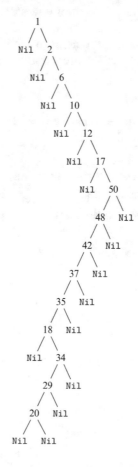

Inserting 11 involves visiting 6 nodes instead of 4, which isn't so bad. Nor is checking whether or not 15 is in the set, for which element needs to visit 7 nodes instead of 5. But checking whether or not 22 is in the set would require visiting 16 nodes instead of 5. So in the worst case, the run times of element and equal are the same as for the representation using ordered lists, while for set and union they're worse. Bummer!

Representing Sets as Balanced Trees

The good news is that it's possible to make sure that the ordered trees we use to represent sets are always balanced, using a clever data structure called **AVL trees** that is a variation on ordered binary trees. Balance is maintained by performing a re-balancing step after every insertion. The clever part is that it's possible to apply the re-balancing only to the nodes that are visited while doing the insertion, rather than having it affect the whole tree, and that each of the re-balancing steps requires only a "local" rearrangement that takes constant time.

AVL trees are named after the Russian mathematician and computer scientist Georgy Maximovich Adelson-Velsky, see ▶ https://en.wikipedia.org/wiki/Georgy_Adelson-Velsky, and the Russian mathematician Evgenii Mikhailovich Landis, see ▶ https://en.wikipedia.org/wiki/Evgenii_Landis.

We start by defining the representation of sets to be the same as for ordered binary trees, except that each node contains the depth of the tree with its root at that node as well as its left and right sub-trees and its label:

```
type Depth = Int
data Set = Nil | Node Set Int Set Depth
  deriving Show

depth :: Set -> Int
depth Nil = 0
depth (Node _ _ _ d) = d
```

Keeping track of the depth of each node is key to making re-balancing after insertion efficient. Computing the depth of a tree is linear in the size of the tree. If the depth of nodes involved in re-balancing needed to be recomputed, re-balancing would require linear time as well instead of—as we will see—logarithmic time.

Given an integer label and two sets (trees), the node function builds a node with that label and those two sub-trees, computing the depth for the node from the depths of the sub-trees:

```
node :: Set -> Int -> Set -> Set
node l n r = Node l n r (1 + (depth l `max` depth r))
```

As invariant, we will require that trees are ordered in the same sense as before. In addition, they are required to be balanced, and the depth information at each node is required to be accurate:

```
invariant :: Set -> Bool
invariant Nil            = True
invariant (Node l n r d) =
  and [ m < n | m <- list l ] &&
  and [ m > n | m <- list r ] &&
  abs (depth l - depth r) <= 1 &&
  d == 1 + (depth l `max` depth r) &&
  invariant l && invariant r
```

The definition of insert is the same as for ordered lists, except that the new function rebalance is applied to each node that is encountered while looking for the node at which to do the insertion, since those are the ones that might potentially become unbalanced:

```
insert :: Int -> Set -> Set
insert m Nil = node empty m empty
insert m (Node l n r _)
  | m == n  = node l n r
  | m < n   = rebalance (node (insert m l) n r)
  | m > n   = rebalance (node l n (insert m r))
```

See ► https://en.wikipedia.org/wiki/
AVL_tree for an animation showing
how re-balancing works.

Re-balancing is best understood using a few pictures which capture the ways that insertion of a single value below a node that obeys the balance property might cause it to become unbalanced. It turns out that there are only two ways, with their symmetric variants adding another two cases.

The first kind of unbalanced tree is shown by the tree on the left below. The picture is meant to indicate that trees B and C have the same depth, with tree A being 1 deeper, leading to a difference in depths of 2 at the node labelled y:

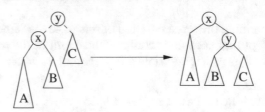

Balance is restored by replacing the tree on the left with the tree on the right. Crucially, the order of the tree is preserved. In the tree on the left, we know that all of the values in A are less than x, which is less than all of the values in B, that all of those values are less than y, and that y is less than all of the values in C. Those properties imply that the tree on the right is still ordered.

The second kind of unbalanced tree is shown on the left of the following diagram. Here, the tree at the node labelled x is 2 deeper than D:

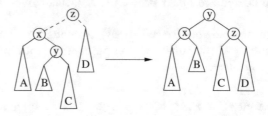

Again, balance is restored by adding the sub-trees A, B, C, and D under nodes with labels x, y, and z, in such a way that the order invariant is preserved.

The following definition of `rebalance` performs these re-balancing steps (cases 1 and 3) and their symmetric variants (cases 2 and 4). Case 5 is for the case where no re-balancing is required:

```
rebalance :: Set -> Set
rebalance (Node (Node a m b _) n c _)
  | depth a >= depth b && depth a > depth c
        = node a m (node b n c)
rebalance (Node a m (Node b n c _) _)
  | depth c >= depth b && depth c > depth a
        = node (node a m b) n c
rebalance (Node (Node a m (Node b n c _) _) p d _)
  | depth (node b n c) > depth d
        = node (node a m b) n (node c p d)
rebalance (Node a m (Node (Node b n c _) p d _) _)
  | depth (node b n c) > depth a
        = node (node a m b) n (node c p d)
rebalance a = a
```

Note that the definition of `rebalance` is not recursive: each call performs a single re-balancing step, which takes constant time. Re-balancing is done at each node that `insert` visits, but there are at most $\log n$ of those because the tree is balanced.

Representing Sets as Balanced Trees

Here's a picture of what inserting 11 into the balanced tree on page 196 does, leaving out re-balancing steps that make no changes:

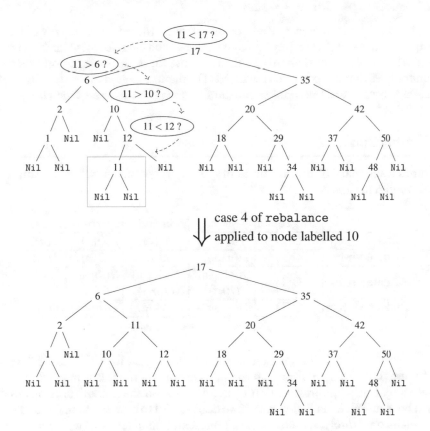

The definitions of empty, set, union and equal remain unchanged, while singleton needs to add depth information. The definitions of list and element are unchanged except that the pattern for **Node** needs to accommodate the presence of depth information:

```
empty :: Set
empty = Nil

singleton :: Int -> Set
singleton n = Node Nil n Nil 1

list :: Set -> [Int]
list Nil           = []
list (Node l n r _) = list l ++ [n] ++ list r

set :: [Int] -> Set
set = foldr insert empty

union :: Set -> Set -> Set
ms `union` ns = foldr insert ms (list ns)

element :: Int -> Set -> Bool
m `element` Nil = False
m `element` (Node l n r _)
  | m == n      = True
  | m < n       = m `element` l
  | m > n       = m `element` r
```

```
equal :: Set -> Set -> Bool
s `equal` t = list s == list t
```

Changing the representation of sets from ordered trees to AVL trees improves the worst-case run times of insert (and therefore set and union) and element to make them match the run times for those functions on ordered binary trees that happen to be balanced. The main change is to the definition of insert. Since rebalance takes constant time, insert has run time $O(\log n)$.

Comparison

Here's a summary of the run times of our five functions on sets in all four of our representations:

	singleton	set	union	element	equal
Lists	$O(1)$	$O(1)$	$O(n)$	$O(n)$	$O(n^2)$
Ordered lists	$O(1)$	$O(n \log n)$	$O(n)$	$O(n)$	$O(n)$
Ordered trees	$O(1)$	$O(n \log n)^*$ $O(n^2)^\dagger$	$O(n \log n)^*$ $O(n^2)^\dagger$	$O(\log n)^*$ $O(n)^\dagger$	$O(n)$
Balanced trees	$O(1)$	$O(n \log n)$	$O(n \log n)$	$O(\log n)$	$O(n)$

* average case   /   † worst case

According to this comparison, balanced trees appear to be the most efficient of these four representations of sets in terms of run time. But in general, the decision of which representation to use depends on the mix of functions that will be required. Even in this case, if we don't expect to ever use equal and to use element only rarely, then the simple representation in terms of unordered lists might be the best on grounds of the efficiency of set and union, and simplicity.

It's possible to implement union on balanced trees in linear time, see ▶ https://en.wikipedia.org/wiki/AVL_tree.

Polymorphic Sets

This chapter has been devoted to four different ways of representing sets of integers. But the choice of integers as elements was just to keep things simple. The same code, with minor adjustments to the types of functions, will work for sets of any kind of elements.

In the simplest representation of sets, as unordered lists, the functions element and equal can't be implemented without an equality test on set elements. So a requirement that the element type supports equality needs to be added to the polymorphic version of their types. This gives the following API:

```
type Set a
empty :: Set a
singleton :: a -> Set a
set :: [a] -> Set a
union :: Set a -> Set a -> Set a
element :: Eq a => a -> Set a -> Bool
equal :: Eq a => Set a -> Set a -> Bool
```

In each of the other three representations of sets—as ordered lists without duplicates, as ordered trees, and as balanced ordered trees—some functions additionally require an ordering on set elements:

```
type Set a
empty :: Set a
singleton :: a -> Set a
set :: Ord a => [a] -> Set a
union :: Ord a => Set a -> Set a -> Set a
element :: Ord a => a -> Set a -> Bool
equal :: Eq a => Set a -> Set a -> Bool
```

Despite the fact that the unordered list representation doesn't require an order relation on set elements, the types in this more restrictive version of the API will work for that representation as well. First, an instance of `Ord` is also an instance of `Eq`. Second, just as a non-polymorphic type signature can be provided for a polymorphic function definition, a type signature with a superfluous requirement—in this case, the `Ord` requirement for `set`, `union` and `element`, for unordered lists—will match a definition that doesn't make use of the required functions.

Exercises

1. Show that:

   (a) $an + b$ is $O(n)$
   (b) $n^d + n^{d-1} + \cdots n + 1$ is $O(n^d)$ for any integer $d \geq 0$
   (c) $\log_b n$ is $O(\log_d n)$ for any b, d
   (d) 3^n is not $O(2^n)$

2. In our representation of sets as ordered lists without duplicates, converting a list to a set required quadratic time:

   ```
   set :: [Int] -> Set
   set xs = nub (sort xs)
   ```

   This definition of `set` removes duplicates *after* sorting, and `nub` requires quadratic time. Instead, they can be removed *during* sorting.

   (a) Give a modified version of `quicksort` from page 86 that does this.
   (b) Unfortunately, although the average-case run time of `quicksort` is $O(n \log n)$, its worst-case run time is $O(n^2)$. What is the worst case?
   (c) Give a modified version of merge sort (Exercise 10.8) that removes duplicates, which will have worst-case run time of $O(n \log n)$. (**Hint:** The code for `union` on page 194 will come in handy.)
   (d) Define a version of `nub` that only removes duplicates when they are consecutive. Would that help?

3. Show that the two AVL tree re-balancing steps

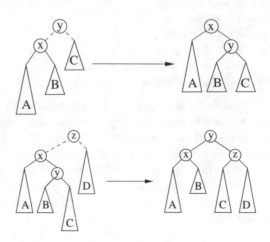

    preserve the order invariant.

4. Try representing sets as predicates:

    ```
    type Set = Int -> Bool
    ```

    What goes wrong?

5. Sets of integers may be represented using lists of *intervals*: the interval $[a, b]$ for $a \leq b$ represents the set of integers between a and b inclusive, where $[a, a]$ represents the set $\{a\}$. A list containing several intervals represents the union of the sets represented by the intervals. If the intervals don't overlap or "touch" then this representation is space-efficient; if they're kept in ascending order then manipulation of sets can be made time-efficient. Here's one example of a set represented this way, and four non-examples, where the interval $[a, b]$ is represented as the pair (a, b) `:: (Int,Int)`:

    ```
    [(1,3),(7,7),(10,11)]
    ```
    represents $\{1, 2, 3, 7, 10, 11\}$
    ```
    [(2,1),(5,6)]
    ```
    is invalid: $[2, 1]$ isn't a valid interval
    ```
    [(1,4),(3,6)]
    ```
    is invalid: intervals overlap
    ```
    [(1,4),(5,6)]
    ```
    is invalid: intervals "touch"
    ```
    [(3,4),(1,1)]
    ```
    is invalid: intervals aren't in ascending order

    Give a representation of sets as lists of intervals. What are the run times of the functions?

6. Add a function `delete :: Int -> Set -> Set` to the four representations of sets in this chapter, plus lists of intervals (Exercise 5). What is its run time?

20

Data Abstraction

Contents

Modular Design

A modern car contains about 100 million lines of code, including central functionality (brakes, engine), navigation, entertainment, etc. It is estimated that a fully autonomous car would contain about 1 billion lines of code. See ▶ https://spectrum.ieee.org/ transportation/self-driving/ accelerating-autonomous-vehicle- technology.

So far you've been writing small programs. Imagine for a moment that— perhaps 10 years from now—you're a member of a team building the software for an autonomous robotaxi. At that point, you'll be working on *much* larger programs in collaboration with other team members.

Any significant software system is built from components that are—or could have been—developed by different people. Each component is designed to work together with other components, with the connections between components mediated by well-defined **interfaces**. An interface is designed to enable simple access to the functionality that the component provides, in such a way that components can be combined without detailed knowledge of their internal workings. For example, the brakes of the robotaxi should work regardless of how the directional signals work. Some components provide functionality that is likely to be useful outside the system that they were designed for and may, therefore, be designed for future re-use.

See ▶ https://en.wikipedia.org/wiki/ Modular_programming.

How to make this way of building systems work well is the topic of **modular design**. As well as interface design and design for reuse, issues include means of combining components, decomposition of components into smaller pieces, component hierarchies, large-scale system architecture, specification versus implementation, abstraction and refinement, separation of concerns, and information hiding.

In Haskell, software components consisting of a number of types and functions are called **modules**, defined using the following syntax:

```
module MyModule where
  import YourModule

  type MyType = ...

  myFunction :: ...
  myFunction x y z | ... = ...
  etc.
```

It's standard practice to put the code for each module in a separate file, with the code for `MyModule` in a file called **MyModule.hs**.

All module names begin with an upper case letter. Every module should be relatively small and self-contained, so that it can be easily understood as a unit.

We've seen before how to **import** a module, using `import` as above. The Haskell Prelude is imported automatically into all modules. To attach an interface to a module, just give the names of the types and functions that it **exports** for use by other code, when it is imported:

```
module MyModule
  ( MyType, myFunction, ... ) where
  import YourModule
  ...
```

If no interface is given, then all of the types and functions are exported. An example of a function that should probably not be exported is a helper function whose only use is in the definition of another function.

Using `import` leads to a situation where one module builds on top of the functionality provided by one or more other modules. When a system is composed of a large collection of modules, any given module may depend on not just the modules that it directly imports, but also on the modules that *they* import, and so on. Thus a small change to one module may affect many other parts of the overall system, leading to a need for many further changes. The point of interfaces is partly to isolate modules from one another so that such knock-on effects of changes are minimised. Ideally, it should be possible to view modules as **black boxes** that provide a certain functionality, and can be

21

replaced by another black box providing the same functionality without *any* changes elsewhere in the system.

In this chapter, we're going to study a very important special case of this situation, in which the module being replaced provides the representation of a type, like sets of integers in Chap. 20. Such a change might be motivated by the need to remove a performance bottleneck. Using the same example, you'll see that defining such modules without appropriate interfaces doesn't isolate them adequately from modules that import them. The secret is to define them as **abstract data types**, making their interface just large enough but no larger, hiding internal details of the representation from external meddling, so that all interaction with the representation goes via well-controlled routes.

Sets as Unordered Lists

We'll start with our representation of sets of integers as unordered lists from page 193, packaged as a module:

```
module SetAsList
  ( Set, empty, singleton, set, union, element, equal ) where

  type Set = [Int]

  empty :: Set
  empty = []

  singleton :: Int -> Set
  singleton n = [n]
  ...
```

We'll use the type and functions in the API at the beginning of Chap. 20 as its interface.

Then we build another module on top of `SetAsList` by importing it, and including functions that refer to the functions that it exports:

```
module MyModule where
  import SetAsList

  intersect :: Set -> Set -> Set
  ms `intersect` ns = [ m | m <- ms, m `elem` ns ]

  intersect_prop :: Set -> Set -> Bool
  intersect_prop ns ms =
   (ns `intersect` ms) `equal` (ms `intersect` ns)
  ...
```

Suppose that `MyModule` includes the following function for selecting an integer from a non-empty set:

```
select :: Set -> Int
select ns = head ns
```

This definition is well typed: we have `ns :: Set` but `SetAsList` says

```
type Set = [Int]
```

so `ns :: [Int]`, and therefore `head ns :: Int`, as required by the type signature of `select`.

But now we have a problem: `select` is not a function, because it doesn't always take equal inputs to equal results! For example, `intersect` is commutative, as expressed by the QuickCheck property `intersect_prop`, so:

That is: `select` doesn't always take equal inputs to == results.

```
> ([1,2] `intersect` [3,2,1]) `equal` ([3,2,1] `intersect` [1,2])
True
```

but:

```
> select ([1,2] `intersect` [3,2,1])
1
> select ([3,2,1] `intersect` [1,2])
2
```

The problem doesn't just affect the intersect function. For instance:

```
> set [1,2] `equal` set [2,1]
True
```

but:

```
> select (set [1,2])
1
> select (set [2,1])
2
```

This is a problem because it shows how the results of computations involving functions from SetAsList can be affected by changes in code that should be innocuous. It is easy to imagine you, or a member of your team, replacing one expression producing a result of type Set with another expression that produces a value that is equal to it, for example in order to simplify the code. Suddenly, a function somewhere else produces a different result! And why has the left rear directional signal started blinking faster? Oops, better go back and spend a month re-testing all of the code that depends on that module!

As we'll see after looking at the other representations of sets, the cause of this problem is in the interface of SetAsList. The definition of select takes advantage of the fact that sets are implemented as lists. We need a way of keeping that information internal to the module definition, so that code elsewhere can't use it.

Sets as Ordered Lists Without Duplicates

Let's move on to our representation of sets of integers as ordered lists without duplicates from page 194. Again, we put the code in a module:

```
module SetAsOrderedList
  ( Set, empty, singleton, set, union, element, equal ) where

  import Data.List(nub,sort)

  type Set = [Int]

  invariant :: Set -> Bool
  invariant ns = and [ m < n | (m,n) <- zip ns (tail ns) ]
  ...

  union :: Set -> Set -> Set
  ms `union` []                     = ms
  [] `union` ns                     = ns
  (m:ms) `union` (n:ns) | m == n    = m : (ms `union` ns)
                        | m < n     = m : (ms `union` (n:ns))
                        | otherwise = n : ((m:ms) `union` ns)
  ...

  element :: Int -> Set -> Bool
  m `element` []                    = False
```

21

```
m `element` (n:ns) | m < n    = False
                   | m == n   = True
                   | m > n    = m `element` ns
...
```

Remember the invariant, expressed as a Haskell function: each element in a list used to represent a set will be strictly less than the next one. The invariant is very important. First, the definitions of union, element, and equal don't work for lists that don't satisfy it. Second, the functions **preserve** it: for example, if *ms* and *ns* satisfy the invariant then *ms* `union` *ns* will also satisfy the invariant.

Again, we build another module on top of `SetAsOrderedList`:

```
module MyModule where
  import SetAsOrderedList

  intersect :: Set -> Set -> Set
  ms `intersect` ns = [ m | m <- ms, m `elem` ns ]

  intersect_prop :: Set -> Set -> Bool
  intersect_prop ns ms =
   (ns `intersect` ms) `equal` (ms `intersect` ns)
  ...
```

Let's look at that select function again:

```
select :: Set -> Int
select ns = head ns
```

The definition is still well typed, for the same reason. And the good news is that it is no longer problematic: the head of any list that satisfies the invariant will be the smallest element of the list. Therefore,

ms `equal` *ns* yields True

implies that

select *ms* == select *ns* yields True

But that's the only good news.

The first problem is that there is nothing to prevent the functions in `SetAsOrderedList` from being applied to values that don't satisfy the invariant. Because the code for union, element, and equal depends on their inputs satisfying the invariant, the results will most likely be wrong. For example,

```
> [4,1,3] `union` [2,4,5,6]
[2,4,1,3,5,6]
> 1 `element` ([4,1,3] `union` [2,4,5,6])
False
```

despite the fact that 1 is in the list [2,4,1,3,5,6].

The second problem is that functions that don't preserve the invariant can be defined in modules that build on `SetAsOrderedList`. An example is the following alternative definition of union:

```
badUnion :: Set -> Set -> Set
ms `badUnion` ns = ms ++ ns
```

The elements in its result are the union of the elements in its inputs, but they need not be arranged in ascending order and duplicates are possible. For example:

```
> set [3,1]
[1,3]
```

```
> set [2,1,2]
[1,2]
> set [3,1] `badUnion` set [2,1,2]
[1,3,1,2]
```

Again, applying `element` and `equal` to such sets gives the wrong answers, for example:

```
> 2 `element` (set [3,1] `badUnion` set [2,1,2])
False
```

despite the fact that 2 is in the list `[1,3,1,2]`.

This shows that direct access to the representation of sets as lists causes problems, not just for the reason we saw in `SetAsList`, but also because there's nothing that requires functions built on top of `SetAsOrderedList` to respect its invariant. Later in this chapter, you'll see how to prevent such functions from being defined.

Sets as Ordered Trees

We'll now look into sets represented as ordered binary trees, as on page 195. Here's that code packaged as a module.

```
module SetAsOrderedTree
  ( Set (Nil, Node), empty, singleton, set, union,
    element, equal ) where

  data Set = Nil | Node Set Int Set
    deriving Show

  invariant :: Set -> Bool
  invariant Nil          = True
  invariant (Node l n r) =
    and [ m < n | m <- list l ] &&
    and [ m > n | m <- list r ] &&
    invariant l && invariant r
  ...

  insert :: Int -> Set -> Set
  insert m Nil = Node Nil m Nil
  insert m (Node l n r)
    | m == n   = Node l n r
    | m < n    = Node (insert m l) n r
    | m > n    = Node l n (insert m r)
  ...
```

To be able to manipulate trees in modules that import `SetAsOrderedTree`, we need to export the constructors of the algebraic data type `Set`, and not just the type name. Including `Set (Nil, Node)` in the export list, rather than just `Set`, has that effect.

Again, as with sets represented as ordered lists, there is an invariant. In this case, it says that at each node, the node label is greater than all of the values in its left sub-tree, and less than all of the values in its right sub-tree. Again, the invariant is preserved by the functions in `SetAsOrderedTree` and it needs to be obeyed by inputs to functions like `element` and `equal`.

Let's build another module on top of `SetAsOrderedTree`, and try a different way of defining a function for selecting an integer from a non-empty set:

```
module MyModule where
  import SetAsOrderedTree
```

```
select :: Set -> Int
select (Node _ n _) = n
...
```

This definition of `select` has the same problem as the `select` function defined in terms of sets represented as lists on page 207: it isn't a function, because it doesn't always take equal inputs to equal results:

```
> :{
| (set [0,3] `union` set [1,0,2])
|    `equal` (set [1,0,2] `union` set [0,3])
| :}
True
> select (set [0,3] `union` set [1,0,2])
3
> select (set [1,0,2] `union` set [0,3])
2
```

The module `SetAsOrderedTree` also has the same problems as `SetAsOrderedList` had earlier. First, there's nothing to prevent its functions from being applied to values that don't satisfy the invariant, giving wrong results:

```
> 1 `element` (Node Nil 2 (Node Nil 1 Nil))
False
```

despite the fact that 1 appears in the right sub-tree. Second, functions that don't preserve the invariant can be defined in modules that build on `SetAsOrderedTree`.

Sets as AVL Trees

Following the same procedure for our final representation of sets, we can package the code for AVL trees on page 199 as a module too:

```
module SetAsAVLTree
  ( Set (Nil, Node), empty, singleton, set, union,
          element, equal ) where

  type Depth = Int
  data Set = Nil | Node Set Int Set Depth
    deriving Show
  ...

  invariant :: Set -> Bool
  invariant Nil              = True
  invariant (Node l n r d) =
    and [ m < n | m <- list l ] &&
    and [ m > n | m <- list r ] &&
    abs (depth l - depth r) <= 1 &&
    d == 1 + (depth l `max` depth r) &&
    invariant l && invariant r
  ...

  insert :: Int -> Set -> Set
  insert m Nil = node empty m empty
  insert m (Node l n r _)
    | m == n    = node l n r
```

```
   | m < n    = rebalance (node (insert m l) n r)
   | m > n    = rebalance (node l n (insert m r))
...
```

Because an AVL tree is a kind of ordered tree, the same problems arise with this module as with `SetAsOrderedTree` above, and there are no new problems. In fact, the part of the invariant related to balance is only important to ensure that the worst-case run time of `insert` and `element` are $O(\log n)$. All of the functions in `SetAsAVLTree` will still produce correct results if given ordered but unbalanced trees, but they may fail to meet their specification for run time.

Abstraction Barriers

In the examples above, you've seen some of the problems that can arise when building systems using Haskell modules. Summarising, it's dangerous to allow unrestricted access to the representation of data in a module. This holds especially when the module defines an invariant on a data representation that its functions require inputs to satisfy and that functions are required to preserve.

We started with a vision in which a module defining one representation of a type could ideally be replaced in a large system by a module defining a different representation that provides the same functionality, without making any changes elsewhere in the system, and the system would continue to work as before. The problems that we've seen with using functions in an imported module, and with defining new functions in terms of them, show that achieving that vision isn't easy. All of those problems would prevent smooth replacement of one module by another. Moreover, consider the `intersect` function on page 207: this doesn't even *typecheck* if we replace `SetAsList` by `SetAsOrderedTree`, because it would attempt to use list comprehension on trees!

We need to erect an **abstraction barrier** which prevents the kinds of abuses that we have seen in the examples above. The barrier should make it impossible for code outside a module to make inappropriate use of the type that is used in the data representation or the way that data is represented using that type. It should be impossible to supply inappropriate inputs to functions in the module, and impossible to define inappropriate functions over the types defined in the module. But how?

The key is in the interface: it controls what the module exports for external use. And in the interfaces for `SetAsOrderedTree` and `SetAsAVLTree` there was a hint: we needed to export the constructors for `Set` in order to be able to manipulate trees outside the module. Was that actually a good idea?

Abstraction refers to removing details from something in order to concentrate attention on its essence. See ▶ https://en.wikipedia.org/wiki/Abstraction_(computer_science). An abstraction barrier should prevent access to those details.

Abstraction Barriers: `SetAsOrderedTree` and `SetAsAVLTree`

Let's see what happens when the constructors are removed from the interface of `SetAsOrderedTree`. We'll call the result `AbstractSetAsOrderedTree`:

```
module AbstractSetAsOrderedTree
  ( Set, empty, singleton, set, union, element, equal ) where

  data Set = Nil | Node Set Int Set
  ...
```

Then, when we try the same definition of `select`:

```
module AbstractMyModule where
  import AbstractSetAsOrderedTree
```

The definition of `Set` in `SetAsOrderedTree` included "`deriving Show`", which incorporates a definition of `show :: Set -> String` into the built-in `show` function. We've left it out here on the principle that internal details of the data representation should be kept hidden.

21

```
select :: Set -> Int
select (Node _ n _) = n
...
```

Haskell doesn't accept it, because the code makes reference to the constructor `Node`, which is no longer exported:

```
AbstractMyModule.hs:7:11-14: error:
    Not in scope: data constructor 'Node'
```

That's a good thing! The attempted definition of `select` isn't a function, so we *want* it to be rejected.

But suppose that we really really want a function that selects an integer from a set? There's no reasonable way to compute it using the functions that are supplied by the API. So we're forced to change the API to include it. Then we add a definition of `select` *inside* **AbstractSetAsOrderedTree**—and the modules packaging our other representations of sets—being careful to check that select s == select t whenever s `equal` t. The following definition, which delivers the smallest element in the set, satisfies that condition:

But there's an *unreasonable* way: given a set s, search through all values of type `Int`, returning the first n such that n `element` s.

```
select :: Set -> Int
select s = head (list s)
```

Once we've added `select` to the API, we could add functions `delete` and `isEmpty` as well, to give us a way to iterate over sets:

See Exercise 20.6 for definitions of `delete` for other set representations.

```
delete :: Int -> Set -> Set
delete m Nil          = Nil
delete m (Node l n r)
  | m < n                   = Node (delete m l) n r
  | m > n                   = Node l n (delete m r)
  | m == n && isEmpty l = r
  | m == n && isEmpty r = l
  | otherwise              = Node l min (delete min r)
                      where min = select r


isEmpty :: Set -> Bool
isEmpty Nil            = True
isEmpty (Node _ _ _) = False
```

Adding these three functions yields the following API:

```
type Set
empty :: Set
isEmpty :: Set -> Bool          -- new
singleton :: Int -> Set
select :: Set -> Int            -- new
set :: [Int] -> Set
union :: Set -> Set -> Set
delete :: Int -> Set -> Set     -- new
element :: Int -> Set -> Bool
equal :: Set -> Set -> Bool
```

A good API needs to balance the benefits of compactness with the requirements of modules that will use the functionality provided. For some purposes, our original design was too compact: there are enough functions for building sets, but only `element` and `equal` for accessing sets once built.

Hiding the constructors means that we can only build sets using the functions in the API. So applying functions to values built using the constructors, whether they satisfy the invariant or not, is impossible:

In a real-life software project, changes to APIs are rare because they typically require changes to modules that have already been finished and tested. At a minimum, such changes need to be negotiated with and approved by other team members.

```
> 1 `element` (Node Nil 2 (Node Nil 1 Nil))
<interactive>:1:14-17: error:
    Data constructor not in scope: Node :: t0 -> Integer -> t1 -> Set
<interactive>:1:19-21: error: Data constructor not in scope: Nil
etc.
```

Finally, if all of the functions in `AbstractSetAsOrderedTree` preserve the invariant, and since subsequent code can only produce sets by applying those functions, then all of the functions in modules that build on top of `AbstractSetAsOrderedTree` will preserve the invariant as well. The upshot is that there's no way to produce sets that don't satisfy the invariant.

The simple idea of omitting the constructors from the interface works well for both `SetAsOrderedTree` and `SetAsAVLTree`. The functions that are defined *inside* the module need to have full access to the details of the data representation. Those functions can be trusted to preserve the invariant. The functions that are defined *outside* the module should have no access to the data representation. Leaving the constructors out of the interface means that the only access for these external functions is via the functions defined inside the module. They are well behaved and preserve the invariant, and—provided the API provides adequate functionality—are expressive enough for use in defining other functions.

Abstraction Barriers: `SetAsList` and `SetAsOrderedList`

But what about `SetAsList` and `SetAsOrderedList`? They have no constructors, except for the built-in constructors on lists. What should we remove from their interfaces?

The answer is that we need to make their representation types into algebraic data types too, with constructors that aren't exported. For example:

We'll start with the original API, not including the functions `select`, `isEmpty` and `delete`.

```
module AbstractSetAsList
  ( Set, empty, singleton, set, union, element, equal ) where

  data Set = MkSet [Int]

  empty :: Set
  empty = MkSet []
  ...

  union :: Set -> Set -> Set
  MkSet ms `union` MkSet ns = MkSet (ms ++ ns)
  ...
```

Here we've changed the type definition

```
type Set = [Int]
```

to

```
data Set = MkSet [Int]
```

but we haven't exported the constructor `MkSet`, in order to prevent external access to the representation. This change means that functions inside `AbstractSetAsList` that take values of type `Set` as input need to use a pattern like `MkSet ns` to get access to the representation of sets in terms of lists. Functions that produce values of type `Set` need to apply `MkSet` to package a list as a `Set`. Both of these can be seen in the definition of union.

What happens when we try to define `select`?

```
module AbstractMyModule where
  import AbstractSetAsList
```

```
select :: Set -> Int
select ns = head ns
```

This produces a type error, because ns isn't a list:

```
AbstractMyModule.hs:5:20-21: error:
    • Couldn't match expected type '[Int]' with actual type 'Set'
    • In the first argument of 'head', namely 'ns'
      In the expression: head ns
      In an equation for 'select': select ns = head ns
```

Okay, so let's try using **MkSet** to get access to the list:

```
module AbstractMyModule where
  import AbstractSetAsList

  select :: Set -> Int
  select (MkSet ns) = head ns
```

Nope, **MkSet** isn't exported from **AbstractSetAsList** so we can't use it:

```
AbstractMyModule.hs:5:11-15: error:
    Not in scope: data constructor 'MkSet'
```

What happens with our earlier definition of the intersect function?

```
module AbstractMyModule where
  import AbstractSetAsList

  intersect :: Set -> Set -> Set
  ms `intersect` ns = [ m | m <- ms, m `elem` ns ]
```

Again, there are type errors because ms and ns aren't lists, and because the type of intersect requires the result to be a set and not a list:

```
AbstractMyModule.hs:6:23-49: error:
    • Couldn't match expected type 'Set' with actual type '[t1]'
    • In the expression: [m | m <- ms, m `elem` ns]
      In an equation for 'intersect':
          ms `intersect` ns = [m | m <- ms, m `elem` ns]
AbstractMyModule.hs:6:34-35: error:
    • Couldn't match expected type '[t1]' with actual type 'Set'
    • In the expression: ms
      In a stmt of a list comprehension: m <- ms
      In the expression: [m | m <- ms, m `elem` ns]
AbstractMyModule.hs:6:47-48: error:
    • Couldn't match expected type 't0 t1' with actual type 'Set'
    • In the second argument of 'elem', namely 'ns'
      In the expression: m `elem` ns
      In a stmt of a list comprehension: m `elem` ns
```

Using **MkSet** to get access to the underlying lists and to package the result as a list doesn't work for the same reason as for select above. Our earlier definition of badUnion, in terms of ++, will fail as well.

Remember that the definition of select needs to ensure that select s == select t whenever s `equal` t. The following definition does the trick:

```
select :: Set -> Int
select (MkSet ns) = minimum ns
```

(Using maximum would work just as well.)

Once the API is extended to include the functions select, isEmpty and delete, they can be used to define intersect:

```
intersect :: Set -> Set -> Set
s `intersect` t
  | isEmpty s = empty
  | choice `element` t
            = singleton choice `union` (rest `intersect`  t)
  | otherwise = rest `intersect`  t
  where choice = select s
        rest = delete choice s
```

This definition works for all representations of sets, not just lists. And since it's defined in terms of the functions in the API, it preserves the relevant invariant, if any.

Exactly the same idea works with **SetAsOrderedList**: we change the type definition from

```
type Set = [Int]
```

to

```
data Set = MkSet [Int]
```

change the function definitions inside **SetAsOrderedList** to use **MkSet** to get access to the representation of sets as ordered lists without duplicates, and omit **MkSet** from the interface.

These examples show how to define an **abstract data type**, in which the representation of data is kept private to the functions that need to access it while being protected from inappropriate external access. We always use an algebraic data type to define the representation, even if only one constructor is required, and then omit the constructor from the interface so that it is only accessible to those internal functions. All external access is forced to take place via the well-behaved functions that are provided for that purpose. As a result, another representation that provides the same functionality can be swapped in without changing anything else in the system!

Abstract data types were first proposed by Barbara Liskov (1939–), winner of the 2008 Turing Award (see ▶ https://en.wikipedia.org/wiki/Barbara_Liskov) and Stephen Zilles in connection with the design of the CLU programming language, see ▶ https://en.wikipedia.org/wiki/CLU_(programming_language).

Note that this method doesn't just make it *difficult* to discover information about the data representation. The representation is *completely hidden* by the type system. There's no way to penetrate the abstraction barrier: it's *impossible* to write code that gets access to the representation. Even if we manage to guess what it is, there's no way to take advantage of that information.

Testing

There's a tension between the need to test the functions of a system to make sure that they work, and the need to break a large system up into modules with interfaces: hiding information tends to make testing more difficult.

An example is with the invariants in the abstract data types above that represent sets in terms of ordered lists, ordered trees, and AVL trees. It's convenient to express these in Haskell for use by QuickCheck in testing that all of the functions that manipulate the representation preserve the invariant. But it's inappropriate to export the implementation details that are required to express the invariant for use outside the abstract data type, and that makes it awkward to do the testing from outside the module.

In testing, it's best to take a bottom-up approach, and attach an interface to a module that hides information only after it has been thoroughly tested and you're confident that the internal details are stable and correct. Such tests on individual functions in a module are known as **unit tests**. You can do that one module at a time, starting from the "bottom" of the import hierarchy. Once you reach the point of assembling modules into a system that provides functionality to an end-user, you do **system tests** to check that the overall behaviour of the system is correct.

It's important to keep the unit tests with the module that they relate to, so that they can be reused whenever that module is updated. Best is to keep them inside the body of the module, where they have access to the internal details that they require to run. Once you have them there, you can include a function in the interface that runs them and reports success or failure, for use in diagnosing faults if anything goes wrong later.

In modular design, working **top-down** means starting with the modules that provide user-level functionality and then proceeding to the modules that are needed in order to implement those, working "downwards" through the import hierarchy. Working **bottom-up** means starting with the modules that provide the basic data structures and algorithms that will be required and then building higher level functionality step by step on top of those. See ► https://en.wikipedia.org/wiki/Top-down_and_bottom-up_design.

Exercises

1. Give an example of a list that doesn't satisfy the invariant for ordered lists in `SetAsOrderedList`, for which `element` and `equal` produce the wrong result.

2. Use `QuickCheck` to check that `set` for ordered trees produces trees that satisfy the invariant, and that `insert` for ordered trees preserves the invariant. Use `QuickCheck` to check that the definition of `union` for ordered trees is commutative and that it preserves the invariant.
   Hint: Excluding inputs that violate the invariant can be done using a conditional test, but a randomly generated tree is very unlikely to satisfy the invariant. Instead, apply `set` to randomly generated lists to produce trees that satisfy the invariant, once you've done the first test to check that this is the case.

3. Check that the definition of `intersect` given on page 215 works in a module which imports any of `AbstractSetAsList`, `AbstractSetAsOrderedList`, `AbstractSetAsOrderedTree`, or `AbstractSetAsAVLTree`. It should be commutative, produce a set that contains the intersection of the values in its input sets, and preserve the invariant.

4. Define `AbstractSetAsOrderedList` as an abstract data type, with an API that includes `isEmpty`, `select`, and `delete`.

5. The invariants for `SetAsOrderedList`, `SetAsOrderedTree`, and `SetAsAVLTree` forbid duplicates. Investigate what happens when the functions in these modules are supplied with inputs of type `Set` which contain duplicates but otherwise satisfy the invariant. Does anything go wrong?

6. Define an abstract data type `AbstractNat` for natural numbers represented as non-negative integers with the following API:

```
type Nat
fromInt :: Int -> Nat
isZero :: Nat -> Bool
plus :: Nat -> Nat -> Nat
minus :: Nat -> Nat -> Nat
```

The functions `fromInt` and `minus` should produce an error when the result would otherwise be negative.
Define another module that imports `AbstractNat` and defines functions

```
times :: Nat -> Nat -> Nat
toInt :: Nat -> Int
```

It isn't possible to give efficient implementations of `times` and `toInt` in terms of the functions in the API given. In the real world, the module structure would be altered to include them in `AbstractNat`.

7. (a) Define an abstract data type `AbstractPolySetsAsList` for polymorphic sets represented as unordered lists, with an API that includes the functions

```
mapSet :: (a -> b) -> Set a -> Set b
foldSet :: (a -> a -> a) -> a -> Set a -> a
```

which should behave analogously to `map` and `foldr` for lists.

   (b) Define a module on top of `AbstractPolySetsAsList` that defines the function `intersect :: Set a -> Set a -> Set a` in terms of `mapSet` and `foldSet`, and check that it satisfies the properties in Exercise 3.

   (c) Since we can't depend on the order of elements in the list used to represent a set or the lack of duplicates, the function f passed to `foldSet` should be associative, commutative, and **idempotent** (x `f` x == x). Prove that if these properties are satisfied then `foldSet` takes equal arguments to equal results. Give counterexamples to show that this may not hold if one of these properties isn't satisfied.

The property that the starting value supplied to `foldSet` is an identity element for f is also desirable; it's required for `foldSet` in other representations of sets, but in none of the ones we have considered.

Efficient CNF Conversion

Contents

CNF Revisited

You've learned several methods for converting logical expressions to conjunctive normal form (CNF), starting with Karnaugh maps in Chap. 17. We'll look more closely at one of those methods, using the laws of Boolean algebra, later in this chapter. Conversion to CNF produces an expression that is in a simple regular form, and it's the starting point for using the DPLL algorithm to check satisfiability.

Unfortunately, converting an expression to CNF using any of these methods will sometimes produce an expression that is *much* larger than the expression you started with. In the worst case, converting an expression of size n to CNF will produce an expression of size 2^n. For example, if we start with the expression

See ▶ https://en.wikipedia.org/wiki/
Conjunctive_normal_form.

$$(X_1 \wedge Y_1) \vee (X_2 \wedge Y_2) \vee \cdots \vee (X_n \wedge Y_n)$$

then the equivalent CNF expression contains 2^n clauses, each having n literals. If $n = 20$, that's 1,048,576 clauses! You've seen that DPLL is pretty efficient, but it won't be able to cope with an input like that.

In this chapter, you're going to learn about the Tseytin transformation, a method for CNF conversion that produces much smaller expressions at the cost of introducing additional variables. The result is a CNF expression that isn't *equivalent* to the original expression, because of the extra variables, but is **equisatisfiable**, meaning that it's satisfiable if and only if the original expression was satisfiable. So it can be used as input to DPLL to get solutions to the original expression.

The Tseytin transformation is used in digital circuit design, so we will look at it first in that context and then at how to apply it to logical expressions.

Implication and Bi-implication

Back in Chap. 4, you learned about the implication ($\rightarrow$) connective, which captures the meaning of "if a then b", but it hasn't been used much since then. To refresh your memory, here is its truth table again:

a	b	$a \rightarrow b$
0	0	1
0	1	1
1	0	0
1	1	1

Implication can easily be expressed in terms of the other connectives: $a \rightarrow b = \neg a \vee b$. Here's a truth table that proves the equivalence:

a	b	$\neg a$	$\neg a \vee b$	$a \rightarrow b$
0	0	1	1	1
0	1	1	1	1
1	0	0	0	0
1	1	0	1	1

Translating $a \rightarrow b$ to $\neg a \vee b$ is sometimes useful to take advantage of everything that you have learned about negation and disjunction.

Treating 0 and 1 as numbers instead of as the mathematical symbols for false and true, it's easy to see that $a \rightarrow b$ is also equivalent to $a \leq b$. That's one reason for choosing those numbers to represent false and true. This equivalence will turn out to be crucial in Chap. 23.

22

Related to implication is **bi-implication** (↔) or **logical equivalence**:

a	b	a ↔ b
0	0	1
0	1	0
1	0	0
1	1	1

↔ is pronounced "if and only if". Some books use ⇔ instead of ↔.

Bi-implication is used to express "a if and only if b", which is often shortened to "a iff b". For example: x is even and prime iff x is 2. It's equivalent to two-way implication: $a \leftrightarrow b = (a \rightarrow b) \land (b \rightarrow a)$:

a	b	a → b	b → a	(a → b) ∧ (b → a)	a ↔ b
0	0	1	1	1	1
0	1	1	0	0	0
1	0	0	1	0	0
1	1	1	1	1	1

Another use for bi-implication $a \leftrightarrow b$ is to express the mathematical statement that "a is a necessary and sufficient condition for b". This combines two implications.

First, "a is a **necessary** condition for b" means $b \rightarrow a$. That is, if something is b then it must be a. For instance, being able to read is a necessary condition for being an accountant, but other conditions are also required.

Second, "a is a **sufficient** condition for b" means $a \rightarrow b$. That is, if something is a then it is guaranteed to be b. For instance, being outdoors in the open without an umbrella when it's raining hard is a sufficient condition for getting wet, but there are other ways to get wet.

It's straightforward to extend the type `Prop` of propositions from Chap. 16 to include implication and bi-implication:

Mathematical definitions are properly expressed using bi-implication, for example:

> An integer is even *iff* dividing it by 2 leaves no remainder.

But often the word "if" is used instead:

> An integer is even *if* dividing it by 2 leaves no remainder.

which, in the context of a definition, is understood to mean iff.

```haskell
type Name = String
data Prop = Var Name
          | F
          | T
          | Not Prop
          | Prop :||: Prop
          | Prop :&&: Prop
          | Prop :->: Prop
          | Prop :<->: Prop
  deriving Eq
```

All of the functions on `Prop` need to be extended to deal with the two new cases. Here is an extended version of `evalProp` from page 149. The first six cases are unchanged, and the cases for the new constructors `:->:` and `:<->:` follow the same pattern as the cases for the other constructors. Implication a `:->:` b corresponds to "`if` a `then` b `else True`" in Haskell, and bi-implication corresponds to equality on `Bool`:

The comparison functions `<`, `>`, `<=` and `>=` work on `Bool` as well as on `Int` and other numeric types, with `False <= True == True` and `True <= False == False`. Therefore, a `:->:` b also corresponds to a `<=` b.

```haskell
evalProp :: Valn -> Prop -> Bool
evalProp vn (Var x)     = vn x
evalProp vn F           = False
evalProp vn T           = True
evalProp vn (Not p)     = not (evalProp vn p)
evalProp vn (p :||: q)  = evalProp vn p || evalProp vn q
evalProp vn (p :&&: q)  = evalProp vn p && evalProp vn q
evalProp vn (p :->: q)  = not (evalProp vn p) || evalProp vn q
evalProp vn (p :<->: q) = evalProp vn p == evalProp vn q
```

$$\frac{}{\Gamma, a \vDash a, \Delta} \, I$$

$$\frac{\Gamma \vDash a, \Delta}{\Gamma, \neg a \vDash \Delta} \, \neg L \qquad \frac{\Gamma, a \vDash \Delta}{\Gamma \vDash \neg a, \Delta} \, \neg R$$

$$\frac{\Gamma, a, b \vDash \Delta}{\Gamma, a \wedge b \vDash \Delta} \, \wedge L \qquad \frac{\Gamma \vDash a, \Delta \quad \Gamma \vDash b, \Delta}{\Gamma \vDash a \wedge b, \Delta} \, \wedge R$$

$$\frac{\Gamma, a \vDash \Delta \quad \Gamma, b \vDash \Delta}{\Gamma, a \vee b \vDash \Delta} \, \vee L \qquad \frac{\Gamma \vDash a, b, \Delta}{\Gamma \vDash a \vee b, \Delta} \, \vee R$$

$$\frac{\Gamma \vDash a, \Delta \quad \Gamma, b \vDash \Delta}{\Gamma, a \to b \vDash \Delta} \, {\to} L \qquad \frac{\Gamma, a \vDash b, \Delta}{\Gamma \vDash a \to b, \Delta} \, {\to} R$$

$$\frac{\Gamma, a \to b, b \to a \vDash \Delta}{\Gamma, a \leftrightarrow b \vDash \Delta} \, {\leftrightarrow} L$$

$$\frac{\Gamma \vDash a \to b, \Delta \quad \Gamma \vDash b \to a, \Delta}{\Gamma \vDash a \leftrightarrow b, \Delta} \, {\leftrightarrow} R$$

The sequent calculus with $\to$ and $\leftrightarrow$

The **converse** of $a \to b$ (If I eat too much then I feel sick) is $b \to a$ (If I feel sick then I ate too much). The truth of the converse does *not* follow from the truth of the original statement (because there are other reasons why I might feel sick), see
▶ https://en.wikipedia.org/wiki/Affirming_the_consequent.

Extending `showProp :: Prop -> String` and `names :: Prop -> Names` is easy. (Do it!)

In order to do proofs with sequents that contain implication and bi-implication, we need to add the following rules to the sequent calculus:

$$\frac{\Gamma \vDash a, \Delta \quad \Gamma, b \vDash \Delta}{\Gamma, a \to b \vDash \Delta} \, {\to} L \qquad \frac{\Gamma, a \vDash b, \Delta}{\Gamma \vDash a \to b, \Delta} \, {\to} R$$

$$\frac{\Gamma, a \to b, b \to a \vDash \Delta}{\Gamma, a \leftrightarrow b \vDash \Delta} \, {\leftrightarrow} L \qquad \frac{\Gamma \vDash a \to b, \Delta \quad \Gamma \vDash b \to a, \Delta}{\Gamma \vDash a \leftrightarrow b, \Delta} \, {\leftrightarrow} R$$

The $\to L$ rule is a little difficult to understand on an intuitive level. It comes from the rules for negation and disjunction, using the equivalence $a \to b = \neg a \vee b$. You'll get used to it after doing a few proofs involving implication.

The $\to R$ rule expresses the proof strategy that to prove $a \to b$ you assume a and then try to prove b. Notice that if Γ and Δ are empty, the rule becomes just

$$\frac{a \vDash b}{\vDash a \to b}$$

which says that the meaning of $a \to b$, at the level of logical expressions, is the same as $a \vDash b$, at the level of sequents.

Finally, the $\leftrightarrow L$ and $\leftrightarrow R$ rules follow directly from the equivalence $a \leftrightarrow b = (a \to b) \wedge (b \to a)$.

Here's a proof that $a \to b \vDash \neg b \to \neg a$ is universally valid using the above rules, together with the rest of the rules of the sequent calculus:

$$\frac{\dfrac{\dfrac{}{\neg b, a \vDash a} \, I}{\neg b \vDash a, \neg a} \, \neg R \quad \dfrac{\dfrac{}{b \vDash b, \neg a} \, I}{b, \neg b \vDash \neg a} \, \neg L}{\dfrac{a \to b, \neg b \vDash \neg a}{a \to b \vDash \neg b \to \neg a} \, {\to} R} \, {\to} L$$

Substituting $\neg b$ for a and $\neg a$ for b gives a proof of $\neg b \to \neg a \vDash a \to b$, which shows that $a \to b$ and its **contrapositive**, $\neg b \to \neg a$, are equivalent.

Here is an example in English: the statement "If I eat too much then I feel sick" is true iff its contrapositive "If I don't feel sick then I didn't eat too much" is true.

Boolean Algebra

Exercise 19.2 asked you to use the laws of Boolean algebra to convert a logical expression to CNF. Let's look at that method again, adding laws to deal with implication and bi-implication:

$$
\begin{array}{lll}
a \leftrightarrow b = (a \to b) \wedge (b \to a) & & a \to b = \neg a \vee b \\
\neg(a \vee b) = \neg a \wedge \neg b & \neg 0 = 1 \quad \neg\neg a = a \quad \neg 1 = 0 & \neg(a \wedge b) = \neg a \vee \neg b \\
a \vee (b \wedge c) = (a \vee b) \wedge (a \vee c) & a \vee 1 = 1 = \neg a \vee a & (a \wedge b) \vee c = (a \vee c) \wedge (b \vee c) \\
a \wedge (b \vee c) = (a \wedge b) \vee (a \wedge c) & a \wedge 0 = 0 = \neg a \wedge a & (a \vee b) \wedge c = (a \wedge c) \vee (b \wedge c) \\
a \vee a = a = 0 \vee a & a \vee (a \wedge b) = a \quad a \vee b = b \vee a & a \vee (b \vee c) = (a \vee b) \vee c \\
a \wedge a = a = 1 \wedge a & a \wedge (a \vee b) = a \quad a \wedge b = b \wedge a & a \wedge (b \wedge c) = (a \wedge b) \wedge c
\end{array}
$$

There are a lot of equations here! The most difficult part of using them to convert an expression to CNF is knowing which equation to apply at each step. The following three-phase strategy works best:

1. First, use the laws in line 1 to eliminate all occurrences of $\leftrightarrow$ and $\rightarrow$.
2. Next, use the laws in line 2 to push $\neg$ inwards.
3. Finally, use the laws for distributivity of $\vee$ over $\wedge$ in line 3 to push $\vee$ inside $\wedge$.

The expression is now in CNF. Optionally, the laws $a \vee 1 = 1 = \neg a \vee a$, $a \wedge 0 = 0 = \neg a \wedge a$, $a \vee a = a = 0 \vee a$, $a \wedge a = a = 1 \wedge a$ and $a \wedge (a \vee b) = a$ can be used to simplify the result.

Let's now apply that strategy to convert the expression $r \leftrightarrow (s \wedge t)$ to CNF. As usual, in each step the part of the expression that changes is underlined.

Phase 1: eliminate $\leftrightarrow$ and $\rightarrow$

$\underline{r \leftrightarrow (s \wedge t)}$
$\qquad$ Applying $a \leftrightarrow b = (a \rightarrow b) \wedge (b \rightarrow a)$
$= \underline{(r \rightarrow s \wedge t)} \wedge (s \wedge t \rightarrow r)$
$\qquad$ Applying $a \rightarrow b = \neg a \vee b$
$= (\neg r \vee (s \wedge t)) \wedge \underline{(s \wedge t \rightarrow r)}$
$\qquad$ Applying $a \rightarrow b = \neg a \vee b$
$= (\neg r \vee (s \wedge t)) \wedge (\neg(s \wedge t) \vee r)$

Phase 2: push $\neg$ inwards

$(\neg r \vee (s \wedge t)) \wedge (\underline{\neg(s \wedge t)} \vee r)$
$\qquad$ Applying $\neg(a \wedge b) = \neg a \vee \neg b$
$(\neg r \vee (s \wedge t)) \wedge (\neg s \vee \neg t \vee r)$

Phase 3: push $\vee$ inside $\wedge$

$\underline{(\neg r \vee (s \wedge t))} \wedge (\neg s \vee \neg t \vee r))$
$\qquad$ Applying $a \vee (b \wedge c) = (a \vee b) \wedge (a \vee c)$
$(\neg r \vee s) \wedge (\neg r \vee t) \wedge (\neg s \vee \neg t \vee r)$

This example shows how more than one step is required in phase 1 to eliminate all occurrences of $\leftrightarrow$ and $\rightarrow$, even though the starting expression contained only one occurrence. In many examples, more than one step is also required in phase 2 to move all negations inwards as far as they will go, and in phase 3 to push all occurrences of $\vee$ inside $\wedge$ until an expression in CNF is obtained. But once a phase is finished, there is no need to revisit it.

In some of the previous chapters, starting with Chap. 4, you have been working with logical expressions built from *variables* which take the values 0 and 1. In other chapters, starting with Chap. 6, you have been working with logical expressions built from *predicates* which are `Bool`-valued functions:

```
type Predicate u = u -> Bool
```

The reason for not picking one of these domains of logical expressions and sticking to it consistently throughout is that *it doesn't matter*. Since both of them obey the laws of Boolean algebra, all of the ways of manipulating expressions that you have learned will work for both!

Logical Circuits

By now, you're used to writing logical expressions using names of variables or predicates and connectives like $\neg$, $\wedge$ and $\vee$. An alternative is to draw **logical circuit diagrams** using symbols like these

This is exactly the same as the way that the laws of ordinary algebra can be applied without worrying whether the expressions are interpreted using rational numbers, or real numbers, or complex numbers, or something else: everything works, provided the domain of interpretation is a **field**, which is a set together with two operations $+$ and $\times$ that satisfy certain laws, see ▶ https://en.wikipedia.org/wiki/Field_(mathematics).

Other logic gates are

NAND   NOR   XOR

where x NAND $y = \neg(x \wedge y)$,
x NOR $y = \neg(x \vee y)$ and XOR is
"exclusive or", x XOR $y = x \neq y$.
See ▶ https://en.wikipedia.org/wiki/
Logic_gate.

AND   OR   NOT

representing **logic gates**, and lines connecting them representing wires. For example, the following circuit

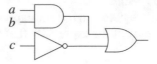

is another way of writing the expression $(a \wedge b) \vee \neg c$.

Expressions written using symbols always take the form of a tree, as you saw when we used algebraic data types to represent arithmetic expressions and propositions. You can see the circuit diagram above as a tree if you rotate it by 90°, with the sub-trees corresponding to $(a \wedge b)$ and $\neg c$ joined by an OR gate.

There is more freedom in circuit diagrams than in logical expressions, because wires can route signals to multiple gates. For example, the circuit

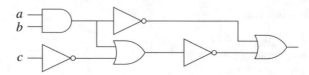

is equivalent to $\neg(a \wedge b) \vee \neg((a \wedge b) \vee \neg c)$. Writing that expression requires repetition of the sub-expression $a \wedge b$ because the wire from the output of the AND gate, which computes $a \wedge b$, goes to both a NOT gate and the first OR gate.

Here's where that expression comes from. If we label the "internal" wires in the circuit diagram

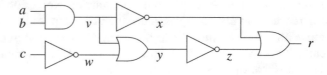

then it can be represented by the following set of logical equivalences, one for each gate:

$$v \leftrightarrow a \wedge b$$
$$w \leftrightarrow \neg c$$
$$x \leftrightarrow \neg v$$
$$y \leftrightarrow v \vee w$$
$$z \leftrightarrow \neg y$$
$$r \leftrightarrow x \vee z$$

Solving for the output r by successively substituting for variables gives an expression that is equivalent to the circuit:

$$
\begin{aligned}
r &\leftrightarrow x \vee z \\
&\leftrightarrow \neg v \vee \neg y \\
&\leftrightarrow \neg(a \wedge b) \vee \neg(v \vee w) \\
&\leftrightarrow \neg(a \wedge b) \vee \neg((a \wedge b) \vee \neg c)
\end{aligned}
$$

22

The Tseytin Transformation

You have just seen how a circuit diagram can be represented by a set of logical equivalences. Each of the equivalences corresponds to one of the gates in the circuit: $r \leftrightarrow (s \wedge t)$ for AND, $r \leftrightarrow (s \vee t)$ for OR, and $r \leftrightarrow \neg s$ for NOT.

Earlier you saw how to convert the expression for an AND gate, $r \leftrightarrow (s \wedge t)$, into CNF, yielding the result:

$$r \leftrightarrow (s \wedge t) : \quad (\neg r \vee s) \wedge (\neg r \vee t) \wedge (\neg s \vee \neg t \vee r)$$

You can do the same thing for the expressions for OR and NOT gates, obtaining:

$$r \leftrightarrow (s \vee t) : \quad (\neg s \vee r) \wedge (\neg t \vee r) \wedge (\neg r \vee s \vee t)$$
$$r \leftrightarrow \neg s : \quad (s \vee r) \wedge (\neg r \vee \neg s)$$

It follows that we can replace the list of logical equivalences that represent the above circuit with the following list of CNF expressions:

$$
\begin{aligned}
v \leftrightarrow a \wedge b : & \quad (\neg v \vee a) \wedge (\neg v \vee b) \wedge (\neg a \vee \neg b \vee v) \\
w \leftrightarrow \neg c : & \quad (c \vee w) \wedge (\neg w \vee \neg c) \\
x \leftrightarrow \neg v : & \quad (v \vee x) \wedge (\neg x \vee \neg v) \\
y \leftrightarrow v \vee w : & \quad (\neg v \vee y) \wedge (\neg w \vee y) \wedge (\neg y \vee v \vee w) \\
z \leftrightarrow \neg y : & \quad (y \vee z) \wedge (\neg z \vee \neg y) \\
r \leftrightarrow x \vee z : & \quad (\neg x \vee r) \wedge (\neg z \vee r) \wedge (\neg r \vee x \vee z)
\end{aligned}
$$

If we now take the conjunction of these:

$$
\begin{aligned}
& (\neg v \vee a) \wedge (\neg v \vee b) \wedge (\neg a \vee \neg b \vee v) \\
\wedge \; & (c \vee w) \wedge (\neg w \vee \neg c) \\
\wedge \; & (v \vee x) \wedge (\neg x \vee \neg v) \\
\wedge \; & (\neg v \vee y) \wedge (\neg w \vee y) \wedge (\neg y \vee v \vee w) \\
\wedge \; & (y \vee z) \wedge (\neg z \vee \neg y) \\
\wedge \; & (\neg x \vee r) \wedge (\neg z \vee r) \wedge (\neg r \vee x \vee z)
\end{aligned}
$$

require the output r to be 1, and simplify:

$$
\begin{aligned}
& (\neg v \vee a) \wedge (\neg v \vee b) \wedge (\neg a \vee \neg b \vee v) \\
\wedge \; & (c \vee w) \wedge (\neg w \vee \neg c) \\
\wedge \; & (v \vee x) \wedge (\neg x \vee \neg v) \\
\wedge \; & (\neg v \vee y) \wedge (\neg w \vee y) \wedge (\neg y \vee v \vee w) \\
\wedge \; & (y \vee z) \wedge (\neg z \vee \neg y) \\
\wedge \; & (\neg x \vee 1) \wedge (\neg z \vee 1) \wedge (\neg 1 \vee x \vee z)
\end{aligned}
$$

$$
\begin{aligned}
= \; & (\neg v \vee a) \wedge (\neg v \vee b) \wedge (\neg a \vee \neg b \vee v) \\
\wedge \; & (c \vee w) \wedge (\neg w \vee \neg c) \\
\wedge \; & (v \vee x) \wedge (\neg x \vee \neg v) \\
\wedge \; & (\neg v \vee y) \wedge (\neg w \vee y) \wedge (\neg y \vee v \vee w) \\
\wedge \; & (y \vee z) \wedge (\neg z \vee \neg y) \\
\wedge \; & 1 \wedge 1 \wedge (0 \vee x \vee z)
\end{aligned}
$$

$$
\begin{aligned}
= \; & (\neg v \vee a) \wedge (\neg v \vee b) \wedge (\neg a \vee \neg b \vee v) \\
\wedge \; & (c \vee w) \wedge (\neg w \vee \neg c) \\
\wedge \; & (v \vee x) \wedge (\neg x \vee \neg v) \\
\wedge \; & (\neg v \vee y) \wedge (\neg w \vee y) \wedge (\neg y \vee v \vee w) \\
\wedge \; & (y \vee z) \wedge (\neg z \vee \neg y) \\
\wedge \; & (x \vee z)
\end{aligned}
$$

Tseytin is pronounced "tsaytin". The Tseytin transformation is due to the Russian computer scientist and mathematician Grigory Samuilovich Tseytin (1936–), see ► https://www.math.spbu.ru/user/tseytin/

we get an expression in CNF that is satisfiable iff there are values of the inputs a, b, c that make the circuit produce 1. That is, this expression and the expression $\neg(a \wedge b) \vee \neg((a \wedge b) \vee \neg c)$ are **equisatisfiable**, with any solution to the transformed expression corresponding to a solution to the original expression and vice versa.

What we have just done is called the **Tseytin transformation**. The result it produces for this small example, which contains 13 clauses, is not impressive at all: applying Boolean algebra to $\neg(a \wedge b) \vee \neg((a \wedge b) \vee \neg c)$ gives the equivalent CNF expression

$$\neg a \vee \neg b$$

which contains only 1 clause.

The Tseytin transformation produces more useful results when it is applied to larger examples. Consider the following circuit:

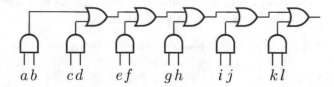

which is equivalent to the expression

$$(a \wedge b) \vee (c \wedge d) \vee (e \wedge f) \vee (g \wedge h) \vee (i \wedge j) \vee (k \wedge l)$$

Converting that expression to CNF using Boolean algebra gives an expression with 64 clauses, each containing 6 literals. Using the Tseytin transformation, we start by labelling the internal wires:

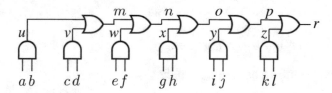

Writing down the list of equivalences and their corresponding CNF representations gives:

$$
\begin{array}{ll}
u \leftrightarrow a \wedge b : & (\neg u \vee a) \wedge (\neg u \vee b) \wedge (\neg a \vee \neg b \vee u) \\
v \leftrightarrow c \wedge d : & (\neg v \vee c) \wedge (\neg v \vee d) \wedge (\neg c \vee \neg d \vee v) \\
w \leftrightarrow e \wedge f : & (\neg w \vee e) \wedge (\neg w \vee f) \wedge (\neg e \vee \neg f \vee w) \\
x \leftrightarrow g \wedge h : & (\neg x \vee g) \wedge (\neg x \vee h) \wedge (\neg g \vee \neg h \vee x) \\
y \leftrightarrow i \wedge j : & (\neg y \vee i) \wedge (\neg y \vee j) \wedge (\neg i \vee \neg j \vee y) \\
z \leftrightarrow k \wedge l : & (\neg z \vee k) \wedge (\neg z \vee l) \wedge (\neg k \vee \neg l \vee z) \\
m \leftrightarrow u \vee v : & (\neg u \vee m) \wedge (\neg v \vee m) \wedge (\neg m \vee u \vee v) \\
n \leftrightarrow m \vee w : & (\neg m \vee n) \wedge (\neg w \vee n) \wedge (\neg n \vee m \vee w) \\
o \leftrightarrow n \vee x : & (\neg n \vee o) \wedge (\neg x \vee o) \wedge (\neg o \vee n \vee x) \\
p \leftrightarrow o \vee y : & (\neg o \vee p) \wedge (\neg y \vee p) \wedge (\neg p \vee o \vee y) \\
r \leftrightarrow p \vee z : & (\neg p \vee r) \wedge (\neg z \vee r) \wedge (\neg r \vee p \vee z)
\end{array}
$$

Taking the conjunction of these, setting r to 1 and simplifying gives the result:

$$(\neg u \lor a) \land (\neg u \lor b) \land (\neg a \lor \neg b \lor u)$$
$$\land (\neg v \lor c) \land (\neg v \lor d) \land (\neg c \lor \neg d \lor v)$$
$$\land (\neg w \lor e) \land (\neg w \lor f) \land (\neg e \lor \neg f \lor w)$$
$$\land (\neg x \lor g) \land (\neg x \lor h) \land (\neg g \lor \neg h \lor x)$$
$$\land (\neg y \lor i) \land (\neg y \lor j) \land (\neg i \lor \neg j \lor y)$$
$$\land (\neg z \lor k) \land (\neg z \lor l) \land (\neg k \lor \neg l \lor z)$$
$$\land (\neg u \lor m) \land (\neg v \lor m) \land (\neg m \lor u \lor v)$$
$$\land (\neg m \lor n) \land (\neg w \lor n) \land (\neg n \lor m \lor w)$$
$$\land (\neg n \lor o) \land (\neg x \lor o) \land (\neg o \lor n \lor x)$$
$$\land (\neg o \lor p) \land (\neg y \lor p) \land (\neg p \lor o \lor y)$$
$$\land (p \lor z)$$

which is 31 clauses, each containing 2 to 3 literals.

As we saw earlier, converting an expression of size n to CNF will produce an expression with 2^n clauses in the worst case. In contrast, the Tseytin transformation produces a CNF expression with at most $4n$ clauses, each containing 2 to 3 literals. Real-life circuits are built from thousands of gates, so the size difference between these CNF expressions can be enormous.

The limit is $3n$ clauses for circuits containing just AND, OR, and NOT gates, but each XOR gate requires 4 clauses.

Tseytin on Expressions

Applying the Tseytin transformation to logical expressions is very similar to applying it to circuits composed of logic gates. The first difference is that we need CNF equivalents for the remaining connectives, and—these are easy—for 0 and 1:

$$r \leftrightarrow (s \rightarrow t) : \quad (r \lor s) \land (r \lor \neg t) \land (\neg r \lor \neg s \lor t)$$
$$r \leftrightarrow (s \leftrightarrow t) : \quad (r \lor s \lor t) \land (r \lor \neg s \lor \neg t) \land (\neg r \lor s \lor \neg t) \land (\neg r \lor \neg s \lor t)$$
$$r \leftrightarrow 0 : \quad \neg r$$
$$r \leftrightarrow 1 : \quad r$$

Second, there are no wires in logical expressions so we can't label them. Instead, we label the sub-expressions. This is best understood by looking at an example.

Let's consider the expression

$$a \leftrightarrow ((b \lor c) \rightarrow (a \land d))$$

We start with the innermost sub-expressions, $b \lor c$ and $a \land d$. Taking first $b \lor c$, we introduce a new variable x_1, assert that it is equivalent to $b \lor c$:

$$x_1 \leftrightarrow (b \lor c)$$

and then replace $b \lor c$ in the original expression by x_1:

$$a \leftrightarrow (x_1 \rightarrow (a \land d))$$

Attacking $a \land d$ next, we introduce x_2, assert that it is equivalent to $a \land d$:

$$x_2 \leftrightarrow (a \land d)$$

and replace $a \land d$ by x_2:

$$a \leftrightarrow (x_1 \rightarrow x_2)$$

The innermost sub-expression is now $x_1 \rightarrow x_2$. We introduce x_3, assert that it is equivalent to $x_1 \rightarrow x_2$:

$$x_3 \leftrightarrow (x_1 \rightarrow x_2)$$

and replace $x_1 \rightarrow x_2$ by x_3:

$$a \leftrightarrow x_3$$

That leaves just one sub-expression, the expression $a \leftrightarrow x_3$ itself. We introduce x_4 with

$$x_4 \leftrightarrow (a \leftrightarrow x_3)$$

which reduces our expression to the variable

$$x_4$$

Now we convert each of the equivalences to CNF:

$$
\begin{array}{ll}
x_1 \leftrightarrow (b \vee c): & (\neg b \vee x_1) \wedge (\neg c \vee x_1) \wedge (\neg x_1 \vee b \vee c) \\
x_2 \leftrightarrow (a \wedge d): & (\neg x_2 \vee a) \wedge (\neg x_2 \vee d) \wedge (\neg a \vee \neg d \vee x_2) \\
x_3 \leftrightarrow (x_1 \rightarrow x_2): & (x_3 \vee x_1) \wedge (x_3 \vee \neg x_2) \wedge (\neg x_3 \vee \neg x_1 \vee x_2) \\
x_4 \leftrightarrow (a \leftrightarrow x_3): & (x_4 \vee a \vee x_3) \wedge (x_4 \vee \neg a \vee \neg x_3) \wedge (\neg x_4 \vee a \vee \neg x_3) \\
& \qquad\qquad\qquad\qquad\qquad\qquad \wedge (\neg x_4 \vee \neg a \vee x_3)
\end{array}
$$

Taking the conjunction of these, setting the final variable x_4 to 1 and simplifying gives the final result:

$$
\begin{aligned}
& (\neg b \vee x_1) \wedge (\neg c \vee x_1) \wedge (\neg x_1 \vee b \vee c) \\
& \wedge (\neg x_2 \vee a) \wedge (\neg x_2 \vee d) \wedge (\neg a \vee \neg d \vee x_2) \\
& \wedge (x_3 \vee x_1) \wedge (x_3 \vee \neg x_2) \wedge (\neg x_3 \vee \neg x_1 \vee x_2) \\
& \wedge (x_4 \vee a \vee x_3) \wedge (x_4 \vee \neg a \vee \neg x_3) \wedge (\neg x_4 \vee a \vee \neg x_3) \wedge (\neg x_4 \vee \neg a \vee x_3)
\end{aligned}
$$

$$
\begin{aligned}
= \; & (\neg b \vee x_1) \wedge (\neg c \vee x_1) \wedge (\neg x_1 \vee b \vee c) \\
& \wedge (\neg x_2 \vee a) \wedge (\neg x_2 \vee d) \wedge (\neg a \vee \neg d \vee x_2) \\
& \wedge (x_3 \vee x_1) \wedge (x_3 \vee \neg x_2) \wedge (\neg x_3 \vee \neg x_1 \vee x_2) \\
& \wedge (1 \vee a \vee x_3) \wedge (1 \vee \neg a \vee \neg x_3) \wedge (\neg 1 \vee a \vee \neg x_3) \wedge (\neg 1 \vee \neg a \vee x_3)
\end{aligned}
$$

$$
\begin{aligned}
= \; & (\neg b \vee x_1) \wedge (\neg c \vee x_1) \wedge (\neg x_1 \vee b \vee c) \\
& \wedge (\neg x_2 \vee a) \wedge (\neg x_2 \vee d) \wedge (\neg a \vee \neg d \vee x_2) \\
& \wedge (x_3 \vee x_1) \wedge (x_3 \vee \neg x_2) \wedge (\neg x_3 \vee \neg x_1 \vee x_2) \\
& \wedge (a \vee \neg x_3) \wedge (\neg a \vee x_3)
\end{aligned}
$$

$$
\frac{}{\Gamma, a \vDash a, \Delta} \; I
$$

$$
\frac{\Gamma \vDash a, \Delta}{\Gamma, \neg a \vDash \Delta} \; \neg L
\qquad
\frac{\Gamma, a \vDash \Delta}{\Gamma \vDash \neg a, \Delta} \; \neg R
$$

$$
\frac{\Gamma, a, b \vDash \Delta}{\Gamma, a \wedge b \vDash \Delta} \; \wedge L
\qquad
\frac{\Gamma \vDash a, \Delta \quad \Gamma \vDash b, \Delta}{\Gamma \vDash a \wedge b, \Delta} \; \wedge R
$$

$$
\frac{\Gamma, a \vDash \Delta \quad \Gamma, b \vDash \Delta}{\Gamma, a \vee b \vDash \Delta} \; \vee L
\qquad
\frac{\Gamma \vDash a, b, \Delta}{\Gamma \vDash a \vee b, \Delta} \; \vee R
$$

$$
\frac{\Gamma \vDash a, \Delta \quad \Gamma, b \vDash \Delta}{\Gamma, a \rightarrow b \vDash \Delta} \; \rightarrow L
\qquad
\frac{\Gamma, a \vDash b, \Delta}{\Gamma \vDash a \rightarrow b, \Delta} \; \rightarrow R
$$

$$
\frac{\Gamma, a \rightarrow b, b \rightarrow a \vDash \Delta}{\Gamma, a \leftrightarrow b \vDash \Delta} \; \leftrightarrow L
$$

$$
\frac{\Gamma \vDash a \rightarrow b, \Delta \quad \Gamma \vDash b \rightarrow a, \Delta}{\Gamma \vDash a \leftrightarrow b, \Delta} \; \leftrightarrow R
$$

The sequent calculus with $\rightarrow$ and $\leftrightarrow$

Exercises

1. Prove that the rules for implication ($\rightarrow L$, $\rightarrow R$) and bi-implication ($\leftrightarrow L$, $\leftrightarrow R$) are sound.
2. Prove that $a \rightarrow (b \vee c) \vDash ((b \rightarrow \neg a) \wedge \neg c) \rightarrow \neg a$ is universally valid.

22

3. Use the laws of Boolean algebra on page 222 to convert the follow expressions to CNF.

   (a) `isBlack` ∨ `isSmall` ↔ ¬ `isDisc`
   (b) $r \leftrightarrow (s \rightarrow t)$
   (c) $r \leftrightarrow (s \leftrightarrow t)$

   Check that the results for (b) and (c) correspond to their CNF equivalents given earlier:

   $r \leftrightarrow (s \rightarrow t):$ $(r \vee s) \wedge (r \vee \neg t) \wedge (\neg r \vee \neg s \vee t)$
   $r \leftrightarrow (s \leftrightarrow t):$ $(r \vee s \vee t) \wedge (r \vee \neg s \vee \neg t) \wedge (\neg r \vee s \vee \neg t) \wedge (\neg r \vee \neg s \vee t)$

4. Implement conversion to CNF using the laws of Boolean algebra as a Haskell function `toCNF :: Prop -> Form Name`.
   Hint: Phase 2 is `toNNF :: Prop -> Prop` in Exercise 16.5.

5. Define a version of `evalProp` on page 221 that interprets propositions as predicates rather than as Boolean values. Its type should be

   `evalProp :: PredValn u -> Prop -> Predicate u`

   where variables are given predicates as values using the following type of valuations:

   `type PredValn u = Name -> Predicate u`

   Hint: The right-hand sides of the equations defining `evalProp` on page 221 refer to functions on Boolean values like

   `(&&) :: Bool -> Bool -> Bool`

   You will need to replace these with functions on predicates like (see page 120)

   `(&:&) :: Predicate u -> Predicate u -> Predicate u`

6. Consider the following circuit:

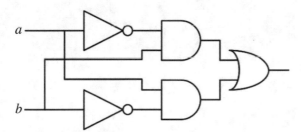

   (a) Give an equivalent logical expression.
   (b) Apply the Tseytin transformation to give an equisatisfiable CNF expression.

7. Apply the Tseytin transformation to the expression

   $$(\neg a \vee c) \wedge (b \rightarrow ((a \vee c) \leftrightarrow d))$$

   to give an equisatisfiable CNF expression.

Counting Satisfying Valuations

Contents

© The Author(s), under exclusive license to Springer Nature Switzerland AG 2021
D. Sannella et al., *Introduction to Computation*, Undergraduate Topics
in Computer Science, https://doi.org/10.1007/978-3-030-76908-6_23

2-SAT

We're now going to look into a special case of CNF expressions: those with clauses containing no more than two literals each. Surprisingly, there's a dramatic efficiency difference between satisfiability checking in this case and when the limit is increased, even to just three literals per clause.

Here's an example:

$$(\neg A \vee \neg C) \wedge (\neg B \vee C) \wedge (B \vee A) \wedge (\neg C \vee D) \wedge (\neg D \vee \neg B)$$

This corresponds to the following set of rules about allowed course combinations, where A stands for Archaeology and so on:

1. You may not take both Archaeology and Chemistry: $\neg(A \wedge C)$, which is equivalent to $\neg A \vee \neg C$.
2. If you take Biology you must take Chemistry: $B \rightarrow C$, which is equivalent to $\neg B \vee C$.
3. You must take Biology or Archaeology: $B \vee A$.
4. If you take Chemistry you must take Divinity: $C \rightarrow D$, which is equivalent to $\neg C \vee D$.
5. You may not take both Divinity and Biology: $\neg(D \wedge B)$, which is equivalent to $\neg D \vee \neg B$.

Confronted with this set of rules, you might want to know whether you are allowed to take both Archaeology and Divinity or not. You will be able to easily work out the answer (yes, that combination is allowed, provided you take neither Biology nor Chemistry) using a method called the **arrow rule**. And you'll be able to count the number of course combinations, out of the $2^4 = 16$ possible combinations of values of A, B, C, D, that the rules allow.

A permitted course combination is a valuation that satisfies the above CNF expression. So finding out whether there is a permitted course combination or not—that is, checking that the above set of rules isn't inconsistent—is the same as the satisfiability problem that we studied back in Chap. 19. The problem of checking the satisfiability of CNF expressions built from clauses containing no more than two literals each is called **2-SAT**. It's an interesting special case because it can always be solved in linear time, while 3-SAT is NP-complete, like the general case, with the best-known algorithms taking exponential time in the worst case.

The restriction of CNF to clauses containing no more than two literals is called **2-CNF**. See ▶ https://en.wikipedia.org/wiki/2-satisfiability for more on 2-CNF and 2-SAT.

Implication and Order

Because $a \rightarrow b$ is equivalent to $\neg a \vee b$, any clause of a CNF expression containing two literals

$$L_1 \vee L_2$$

is equivalent to an implication

$$\neg L_1 \rightarrow L_2$$

The same holds for a clause consisting of a single literal L: it's equivalent to the implication $1 \rightarrow L$.

The above CNF expression

$$(\neg A \vee \neg C) \wedge (\neg B \vee C) \wedge (B \vee A) \wedge (\neg C \vee D) \wedge (\neg D \vee \neg B)$$

is therefore equivalent to the following conjunction of implications:

$$(A \rightarrow \neg C) \wedge (B \rightarrow C) \wedge (\neg B \rightarrow A) \wedge (C \rightarrow D) \wedge (D \rightarrow \neg B)$$

23

Let's pick out three of those implications:

$$C \to D \qquad D \to \neg B \qquad \neg B \to A$$

We can add $0 \to C$ at the beginning, and $A \to 1$ at the end—both are true, no matter what the values of C and A are—and then write them as a chain of five implications:

$$0 \to C \to D \to \neg B \to A \to 1$$

Now, recalling that $a \to b$ is equivalent to $a \le b$, we get

$$0 \le C \le D \le \neg B \le A \le 1$$

This relationship between the values of C, D, $\neg B$ and A is very interesting. Because there are only two possible values for each literal, 0 and 1, it says that only the following five combinations of values are possible:

$$C = 0 \quad D = 0 \quad \neg B = 0 \quad A = 0$$
$$C = 0 \quad D = 0 \quad \neg B = 0 \quad A = 1$$
$$C = 0 \quad D = 0 \quad \neg B = 1 \quad A = 1$$
$$C = 0 \quad D = 1 \quad \neg B = 1 \quad A = 1$$
$$C = 1 \quad D = 1 \quad \neg B = 1 \quad A = 1$$

(Of course, $\neg B = 0$ means that $B = 1$, and $\neg B = 1$ means that $B = 0$.)

If we make the chain of implications into a diagram where all of the implications point upwards:

then each of these combinations corresponds to a horizontal "cut" through the chain, where all of the literals above the cut get value 1 and all of the literals below the cut get value 0:

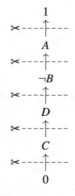

You can't cut above the 1, because it has value 1, obviously, and you can't cut below the 0.

There are two remaining implications in the list corresponding to our original CNF expression:

$$A \to \neg C \qquad B \to C$$

$A \to \neg C$ eliminates the last one in the list and $B \to C$ eliminates the first two.

Some of the above combinations of values don't satisfy these two implications, and it's easy to pick them out. We'll see later how to take them into account at the same time as the implications in the chain.

The Arrow Rule

What you've just seen in action is called the **arrow rule**. Given a chain of upward-pointing implications, starting with 0 and ending with 1, cuts through the chain divide literals that get value 1 (above) from literals that get value 0 (below). Each cut corresponds to one combination of values for the literals, so the number of cuts gives the number of combinations.

It's important for now that each literal only appears once in the chain, and that there are no **complementary literals** (the same atom both with and without negation). We'll be able to deal with both of these cases, but let's keep things simple to start with.

Let's look at an example that isn't just a simple chain of implications. We start with the CNF expression

$$(\neg R \vee Q) \wedge (\neg R \vee S)$$

which is equivalent to the implications

$$(R \to Q) \wedge (R \to S)$$

This gives the following diagram of upward-pointing implications, with implications to 1 added at the top and an implication from 0 added at the bottom. In this case, the implications don't form a linear chain:

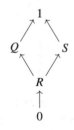

which corresponds to the inequalities:

$$0 \;\leq\; R \;\begin{smallmatrix} \leqslant \\ \leqslant \end{smallmatrix}\; \begin{smallmatrix} Q \\ S \end{smallmatrix} \;\begin{smallmatrix} \leqslant \\ \leqslant \end{smallmatrix}\; 1$$

Now there are five ways to cut through this diagram horizontally:

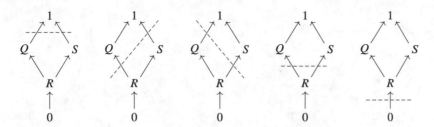

Each of these cuts corresponds to a combination of values for literals that respects the inequalities. They are:

top horizontal cut:	$R = 0$	$Q = 0$	$S = 0$
cut sloping upwards:	$R = 0$	$Q = 1$	$S = 0$
cut sloping downwards:	$R = 0$	$Q = 0$	$S = 1$
middle horizontal cut:	$R = 0$	$Q = 1$	$S = 1$
bottom horizontal cut:	$R = 1$	$Q = 1$	$S = 1$

If you just want to count the combinations, rather than list them, you could reason as follows:

- For the cuts across the diamond, there are two ways of starting the cut (above or below Q) and two ways of ending the cut (above or below S), giving $2 \times 2 = 4$ combinations.
- The cut below the diamond adds 1 combination.

This gives $4 + 1 = 5$ combinations.

Here's another example that's a little more complicated. We start with the CNF expression

$$(\neg A \vee B) \wedge (\neg B \vee C) \wedge (\neg C \vee D) \wedge (\neg A \vee E) \wedge (\neg E \vee D)$$

which is equivalent to the implications

$$(A \rightarrow B) \wedge (B \rightarrow C) \wedge (C \rightarrow D) \wedge (A \rightarrow E) \wedge (E \rightarrow D)$$

This gives the following diagram of upward-pointing implications. Again, the implications don't form a linear chain:

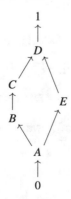

There are eight ways to cut through this diagram horizontally:

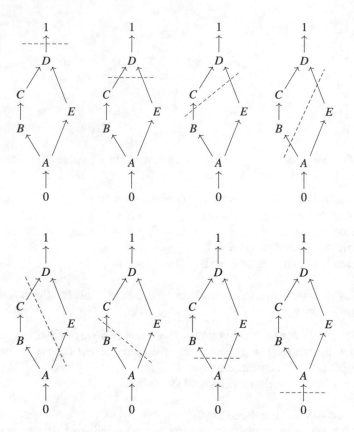

Again, we can count these without drawing them all by reasoning as follows:

- For the cuts across the pentagon, there are three ways of starting the cut (above C, above B, above A) and two ways of ending the cut (above or below E), giving $3 \times 2 = 6$ combinations.
- The cut above the pentagon adds 1 combination, and the cut below the pentagon adds another combination.

This gives $6 + 1 + 1 = 8$ cuts, i.e. 8 combinations of values for literals that satisfy the CNF expression.

Finally, here's an even more complicated example. We start with the CNF expression

$$(B \vee \neg A) \wedge (B \vee C) \wedge (\neg C \vee D) \wedge (A \vee E) \wedge (\neg E \vee D) \wedge (A \vee \neg F)$$

which is equivalent to the implications

$$(\neg B \to \neg A) \wedge (\neg B \to C) \wedge (C \to D) \wedge (\neg A \to E) \wedge (E \to D) \wedge (\neg A \to \neg F)$$

This gives the following diagram of upward-pointing implications:

23

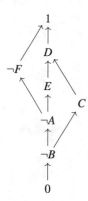

This is complicated enough that we need to make some notes on the diagram in order to work out the number of cuts, rather than trying to draw them. We start by indicating the number of ways of starting cuts on the left:

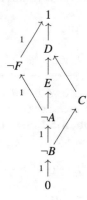

The next step is to indicate the number of ways of continuing cuts through the right-hand side of the upper triangle:

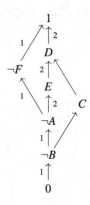

There are two ways of continuing the cut between D and 1: one from the cut that starts between $\neg F$ and 1, and one from the cut that starts between $\neg A$ and $\neg F$. Similarly, there are two ways of continuing the cut between E and D, and likewise between $\neg A$ and E.

Finally, we indicate the number of ways of finishing cuts through the right-hand side of the lower triangle:

$$
\begin{array}{c}
1 \\
\nearrow\uparrow 2 \\
{}^{1}\nearrow\quad D \\
\neg F \quad {}^{2}\uparrow\uparrow\quad\nwarrow {}^{5} \\
{}^{1}\nwarrow\quad E \\
\uparrow {}^{2}\quad C \\
\neg A\quad\nearrow {}^{5} \\
{}^{1}\uparrow \\
\neg B \\
{}^{1}\uparrow \\
0
\end{array}
$$

There are $2 + 2 + 1 = 5$ ways of finishing the cut between C and D, and $2 + 2 + 1 = 5$ ways of finishing the cut between $\neg B$ and C.

Doing the calculation from right to left gives the same result. (Check it!)

Adding up the numbers on the right-hand side of the diagram gives $2 + 5 + 5 + 1 = 13$ cuts, i.e. 13 combinations of values for literals that satisfy the CNF expression.

Complementary Literals

Now that you've had some practice with the arrow rule, it's time to consider what happens when the diagram includes complementary literals.

Here's an example that is a modification of one of the ones above. We start with the CNF expression

$$(\neg A \vee B) \wedge (\neg B \vee C) \wedge (\neg C \vee D) \wedge (\neg A \vee \neg B) \wedge (B \vee D)$$

which is equivalent to the implications

$$(A \rightarrow B) \wedge (B \rightarrow C) \wedge (C \rightarrow D) \wedge (A \rightarrow \neg B) \wedge (\neg B \rightarrow D)$$

This gives the following diagram of upward-pointing implications:

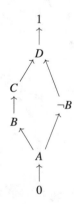

which includes the complementary literals B and $\neg B$.

In the earlier version of this example, which involved a diagram with the same shape, there were eight ways to cut through the diagram, giving rise to eight combinations of values for literals that satisfy the CNF expression. We can draw the same set of eight cuts through this diagram, but some of them are invalid. Why?

What makes some of the cuts invalid is the fact that complementary literals can't have the same value. Either $B = 0$ and $\neg B = 1$, or $B = 1$ and $\neg B = 0$. It

23

follows that any cut that doesn't separate B from $\neg B$ is invalid, since it would assign either 0 or 1 to both literals. There are five invalid cuts in this case:

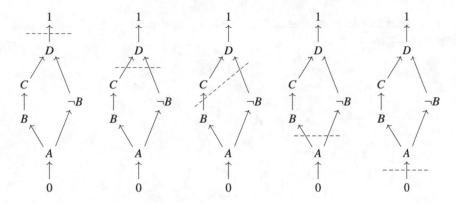

which leaves three valid ones:

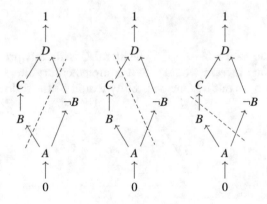

We can count the cuts by annotating the diagram:

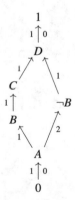

The numbers on the right-hand side of the diagram take into account the requirement for cuts to separate B and $\neg B$. For example, the annotation on the arrow between A and $\neg B$ is 2 because the cuts starting above and below C that cross this arrow will separate B and $\neg B$, but the one starting below B won't. Adding up the numbers on the right-hand side gives $0 + 1 + 2 + 0 = 3$ combinations of values for literals that satisfy the CNF expression.

When possible, you should eliminate complementary literals by considering the contrapositive of one or more of the implications. Here's a variation on another example above. Consider the CNF expression

$$(B \vee \neg A) \wedge (B \vee C) \wedge (\neg C \vee D) \wedge (A \vee E) \wedge (\neg E \vee D) \wedge (A \vee \neg F) \wedge (G \vee \neg C)$$

which is equivalent to the implications

$$(\neg B \to \neg A) \land (\neg B \to C) \land (C \to D) \land (\neg A \to E) \land (E \to D) \land (\neg A \to \neg F) \land (\neg G \to \neg C)$$

The first six implications give the diagram of upward-pointing implications that we had earlier:

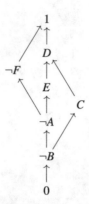

but what do we do with $\neg G \to \neg C$? The literal C is already present, so adding $\neg G \to \neg C$ to the diagram would add a complementary literal. But $\neg G$ isn't there, so where do we put it? One possibility would be the following:

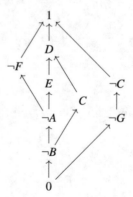

But a simpler solution, without complementary literals, is to instead use the contrapositive $C \to G$ of $\neg G \to \neg C$. This gives the following diagram:

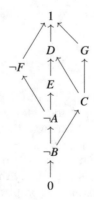

which can be annotated as follows:

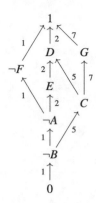

to work out that there are $7 + 7 + 5 + 1 = 20$ combinations of values for literals that satisfy the CNF expression.

Implication Diagrams with Cycles

Sometimes it is difficult or impossible to draw a diagram of upward-pointing implications that captures the entire CNF expression. An example is our first example of permitted course combinations, where the implications were

$$(A \to \neg C) \land (B \to C) \land (\neg B \to A) \land (C \to D) \land (D \to \neg B)$$

and our initial analysis considered just the three implications

$$(C \to D) \land (D \to \neg B) \land (\neg B \to A)$$

Adding in the implication $A \to \neg C$ gives the diagram

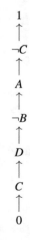

which contains the complementary literals C and $\neg C$, but we now know how to deal with those. But what about $B \to C$?

We have two choices. One is to add it to the bottom of the diagram, giving a linear chain with two pairs of complementary literals. Then the permissible cuts are the two that separate both of the pairs:

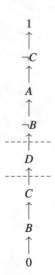

The other choice is to take its contrapositive, $\neg C \to \neg B$. But adding that implication to the diagram gives a cycle:

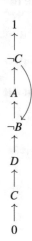

Now, the cycle corresponds to the inequalities

$$\neg B \leq A \leq \neg C \leq \neg B$$

for which the only solutions give all these literals the same value:

$$\neg B = 0 \quad A = 0 \quad \neg C = 0$$
$$\neg B = 1 \quad A = 1 \quad \neg C = 1$$

The same holds for any implication diagram containing a cycle: all literals in the cycle need to be given the same value. The permissible cuts are the two that separate the complementary literals C and $\neg C$, and that don't cut through the cycle:

23

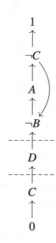

Let's look at another example that contains a cycle. We start with the CNF expression

$$(\neg A \vee B) \wedge (\neg B \vee C) \wedge (\neg C \vee \neg A) \wedge (A \vee D) \wedge (\neg D \vee A)$$

which is equivalent to the implications

$$(A \rightarrow B) \wedge (B \rightarrow C) \wedge (C \rightarrow \neg A) \wedge (\neg A \rightarrow D) \wedge (D \rightarrow A)$$

The closest we can get to a diagram of upward-pointing implications is the following:

Replacing one or more implications with their contrapositives doesn't help. (Try it!)

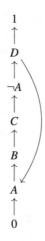

All of the literals in the cycle need to be given the same value. On the other hand, the cycle includes the complementary literals A and $\neg A$, which need to be given different values. So this example is **unsatisfiable**: there is no combination of values for literals that satisfies the CNF expression.

In some complex examples, it is difficult to draw a diagram of upward-pointing implications that capture all of the clauses. As a final example, consider the CNF expression

$$(\neg A \vee B) \wedge (\neg C \vee B) \wedge (\neg A \vee E) \wedge (\neg C \vee D) \wedge (\neg D \vee E)$$

which is equivalent to the implications

$$(A \rightarrow B) \wedge (C \rightarrow B) \wedge (A \rightarrow E) \wedge (C \rightarrow D) \wedge (D \rightarrow E)$$

This gives the following non-planar diagram of upward-pointing implications:

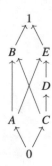

It's difficult to use such a diagram for counting cuts.

A trick that can be applied in such cases is to split the problem into two cases, treat them separately, and then combine the results.

Let's first consider the case where $A = 1$. Because $A \rightarrow B$ and $A \rightarrow E$, it follows that $B = 1$ and $E = 1$. This leaves the following simpler diagram, with the three possible cuts indicated:

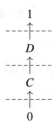

Now we consider the case where $A = 0$. We can't conclude anything about the values of the other literals from that, but nevertheless, it leaves a simpler diagram with $2 + 2 + 2 + 1 = 7$ possible cuts as indicated by the annotations:

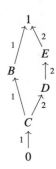

This gives a total of $3 + 7 = 10$ combinations of values for literals that satisfy the CNF expression.

23

Exercises

1. Use the Haskell implementation of DPLL in Chap. 19, and/or a modification to `satisfiable :: Prop -> Bool` from Chap. 16 that counts satisfying valuations instead of checking for their existence, to check that the solutions given for all of the examples in this chapter are correct.

2. Annotate the following implication diagram to work out how many cuts that separate the complementary literals C and $\neg C$ are possible. Compare with the result obtained above for the diagram that uses the contrapositive of $\neg G \rightarrow \neg C$.

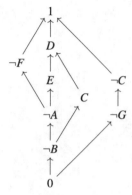

3. Use the arrow rule to count the number of combinations of values for literals that satisfy the following CNF expressions over the atoms A, B, C, D, E, F, G, H:

   (a) $E \vee F$
   (b) $(E \vee F) \wedge (\neg A \vee B) \wedge C$
   (c) $(\neg A \vee B) \wedge (\neg B \vee C) \wedge (\neg D \vee F) \wedge (\neg E \vee F) \wedge (\neg F \vee D)$
   (d) $(\neg A \vee B) \wedge (\neg C \vee B) \wedge (\neg D \vee E) \wedge (\neg E \vee F) \wedge (\neg F \vee C)$

   What do you need to do to the result of the calculation using the arrow rule to take account of atoms that aren't used in the expression?

4. Use the arrow rule to count the number of combinations of values for literals that satisfy the following CNF expressions:

   (a) $(\neg A \vee B) \wedge (\neg B \vee C) \wedge (\neg C \vee B)$
   (b) $(\neg A \vee B) \wedge (\neg B \vee C) \wedge (\neg D \vee \neg A) \wedge (\neg E \vee \neg A) \wedge (A \vee C)$
   (c) $(\neg A \vee B) \wedge (\neg B \vee C) \wedge (\neg C \vee D) \wedge (\neg D \vee \neg B) \wedge (\neg A \vee F) \wedge (\neg F \vee G) \wedge (\neg G \vee H) \wedge (\neg H \vee \neg B) \wedge (\neg B \vee H) \wedge (C \vee \neg D)$

5. Use the arrow rule to count the number of combinations of values for literals that satisfy the following CNF expression:

$$(A \vee B) \wedge (\neg B \vee \neg C) \wedge (\neg C \vee D) \wedge (\neg A \vee \neg E) \wedge (\neg E \vee D)$$

Type Classes

Contents

© The Author(s), under exclusive license to Springer Nature Switzerland AG 2021
D. Sannella et al., *Introduction to Computation*, Undergraduate Topics
in Computer Science, https://doi.org/10.1007/978-3-030-76908-6_24

Bundling Types with Functions

You've seen that there are some common functions, like == and show, that are available for many but not all types. For instance, neither == nor show will work on function types. The type of a polymorphic function can require that it be applied only on types that have such functions, and these requirements can be inferred automatically. An example is the definition of list membership:

```
x `elem` []     = False
x `elem` (y:ys) = x==y || x `elem` ys
```

Because of the way that == is used in the definition, the Haskell typechecker infers the type

```
elem :: Eq a => a -> [a] -> Bool
```

which says that elem will work on any type that supports equality testing.

You've also seen that == and show will be defined automatically for newly defined algebraic data types if the right magic words are added to the type definition. For example, once we define

```
data Season = Winter | Spring | Summer | Fall
  deriving (Eq,Show)
```

the functions == and show will work on values of type **Season**. But there are some subtleties. If we define:

```
data Maybe a = Nothing | Just a
  deriving (Eq,Show)
```

and T is a type, then == and show will work on values of type **Maybe** T, but only provided == and show work on type T.

Eq and **Show** are examples of **type classes**. Each type class provides one or more named functions over each type that belongs to the class. The type class **Show** provides show, while **Eq** provides the equality test == and its negation /=. A type can be declared as an **instance** of a type class by defining the functions that are required for types that belong to that class. This can be done explicitly, as you'll see below, or as part of an algebraic data type definition where the function definitions follow a general scheme. And as well as defining a type as an instance of an existing type class like **Show**, you can define your own type classes.

Type classes can be regarded as collections of types that are related in the sense that they all have the functions of the class. Alternatively, they can be regarded as a way of bundling functions together with a type, to make it more than a name for a collection of values: if a function of a type class is required for a type, and the type has been declared as an instance of that type class, then the required function is available.

Notice that there are two different situations in which a function might have more than one type in Haskell. The first is a function with a polymorphic type, which has a *single* definition that works for any instance of that type. An example is map :: (a -> b) -> [a] -> [b]. The second is a function belonging to a type class, where each type that is an instance of that type class defines the function for that type *separately*. An example is show :: T -> **String** for a type T, where converting a **Bool** to a **String** is quite different from converting an **Int** to a **String**.

24

These are often called **overloaded** functions, see ▶ https://en.wikipedia.org/wiki/Function_overloading. Another name for this is **ad hoc polymorphism**, in contrast with **parametric polymorphism** as in map.

Declaring Instances of Type Classes

Declaring that a type is an instance of a type class requires a definition for each of the functions in the type class. You've already seen some examples for the type class **Show**, and that declaring T as an instance of **Show** requires the definition of a function show :: T -> **String**. Here are some examples for the type class **Eq**, which is defined like so:

```
class Eq a where
  (==) :: a -> a -> Bool
```

This is a simplified version—see page 250 for the full definition.

This says that declaring T to be an instance of **Eq** requires the definition of a function (==) :: T -> T -> Bool.

Equality for **Char** is defined in terms of the function ord :: **Char** -> **Int** from the **Data.Char** library module, which converts a character to its numeric code:

```
instance Eq Char where
  x == y = ord x == ord y
```

This relies on the fact that **Int** is an instance of **Eq** to provide the equality test in ord x == ord y.

The definition of equality for pairs depends on equality being available for the types of both components. The instance declaration for (a,b) is therefore conditional on **Eq** a and **Eq** b:

```
instance (Eq a, Eq b) => Eq (a,b) where
  (u,v) == (x,y) = (u == x) && (v == y)
```

The declaration that [a] is an instance of **Eq**, which is conditional on **Eq** a, involves a recursive definition:

```
instance Eq a => Eq [a] where
  []   == []    = True
  []   == y:ys  = False
  x:xs == []    = False
  x:xs == y:ys  = (x == y) && (xs == ys)
```

Notice that the two occurrences of == on the right-hand side refer to two different instances of equality, on type a and (recursively) type [a] respectively.

In general, the definitions of functions in type class instances can be as complicated as you like.

A function == for testing equality should be **reflexive** ($x == x$), **symmetric** (if $x == y$ then $y == x$) and **transitive** (if $x == y$ and $y == z$ then $x == z$). Unfortunately, there is no way for Haskell to enforce such restrictions when you declare an instance of **Eq** or any other type class. Only the type matters. But provided each new instance of **Eq** is checked for these properties, users of == can assume that they hold.

You've seen examples of instances for **Eq** and **Show** being declared for newly defined algebraic data types using **deriving**. For polymorphic types like **Maybe** a above, the declaration is conditional on **Eq** a and **Show** a. The use of **deriving** is limited to those two built-in type classes, plus a few others. These include **Ord** and **Enum**, which will be covered later. The following example, based on the algebraic data type for arithmetic expressions in Chap. 16, demonstrates **deriving** for the type class **Read**:

```
data Exp = Lit Int
         | Add Exp Exp
         | Mul Exp Exp
  deriving Read
```

Including "deriving Read" declares **Exp** to be an instance of **Read**, with a function read for obtaining a value of type **Exp** from a **String**:

See page 144 for the definition of showExp and page 145 for evalExp.

```
> showExp (read "Add (Lit 3) (Mul (Lit 7) (Lit 2))")
"(3 + (7 * 2))"
> evalExp (read "Add (Lit 3) (Mul (Lit 7) (Lit 2))")
17
```

Of course, you can define an algebraic data type to be an instance of a type class using a function definition that is different from the one that would be generated automatically from the type definition. For instance, for

```
data Set = MkSet [Int]
```

in our abstract data type `AbstractSetAsList` (page 214), in which sets are represented as unordered lists, we could declare:

```
instance Eq Set where
  MkSet ms == MkSet ns = ms `subset` ns && ns `subset` ms
    where ms `subset` ns = and [ m `elem` ns | m <- ms ]
```

which defines == to be the correct notion of equality on sets:

This instance declaration would need to be inside the module `AbstractSetAsList` since it requires access to `MkSet`.

```
> MkSet [1,2] == MkSet [2,1]
True
```

This is in contrast to:

```
data Set = MkSet [Int] deriving Eq
```

which would define == on sets to inappropriately take account of the order of the items in the underlying list:

```
> MkSet [1,2] == MkSet [2,1]
False
```

Defining Type Classes

You don't have to stick to Haskell's built-in type classes: you can also define your own. To see how this works, let's look into how some of the built-in type classes are defined.

The type class `Show` just requires a function called `show` that converts a value to a string. Here is how it could have been defined, if it weren't already in Haskell:

```
class Show a where
  show :: a -> String
```

The type class `Eq` actually requires two functions, == and /=. But to declare a type to be an instance of `Eq`, you only need to define one of them, because the type class definition includes a **default definition** for /= in terms of ==, and for == in terms of /=:

```
class Eq a where
  (==) :: a -> a -> Bool
  (/=) :: a -> a -> Bool
  -- defaults
  x /= y = not (x == y)
  x == y = not (x /= y)
```

The `Ord` type class is for types whose values are ordered, with functions <, >, etc. It involves six functions, with default definitions for all but <=:

```
class Eq a => Ord a where
  (<)  :: a -> a -> Bool
  (<=) :: a -> a -> Bool
  (>)  :: a -> a -> Bool
  (>=) :: a -> a -> Bool
  min :: a -> a -> a
  max :: a -> a -> a
```

```
-- defaults
x < y            = x <= y && x /= y
x > y            = y < x
x >= y           = y <= x
min x y
  | x <= y       = x
  | otherwise = y
max x y
  | x <= y       = y
  | otherwise = x
```

The type class definition includes the requirement "Eq a =>", which means that in order for a type to be an instance of Ord, it needs to be an instance of Eq. Because of this, we can view the class Ord as **inheriting** the functions of Eq, with the default definitions making use of /=.

Algebraic data types can be declared as instances of Ord using deriving. The definition of <= that is generated is best illustrated through few examples. First is the built-in type Bool, which was introduced as an algebraic data type on page 132:

```
data Bool = False | True
  deriving (Eq,Show,Ord)
```

The order in which the constructors False and True appear in the definition of Bool determines the order that arises from "deriving (Eq,Show,Ord)", giving the following:

```
instance Ord Bool where
  False <= False = True
  False <= True  = True
  True  <= False = False
  True  <= True  = True
```

When the constructors take parameters, the parameter types are required to be instances of Ord. Then, after first taking the order in which the constructors appear in the definition into account, the definition of <= takes account of the orders on the parameter types. For example, taking the Maybe type from page 138:

```
data Maybe a = Nothing | Just a
  deriving (Eq,Show,Ord)
```

gives the following definition:

```
instance Ord a => Ord (Maybe a) where
  Nothing <= Nothing = True
  Nothing <= Just x  = True
  Just x  <= Nothing = False
  Just x  <= Just y  = x <= y
```

If a constructor has multiple parameters, the order in which the parameters appear is also taken into account in the definition of <=. This can be seen in the Pair type from page 136:

```
data Pair a b = Pair a b
  deriving (Eq,Show,Ord)
```

which gives:

```
instance (Ord a, Ord b) => Ord (Pair a b) where
  Pair x y <= Pair x' y' = x < x' || (x == x' && y <= y')
```

The terminology type *class*, *instance*, *inheriting*, etc. hints at a relationship to these concepts in **class-based programming languages** like Java, see ► https://en.wikipedia.org/wiki/Class-based_programming. But the differences in what those words mean in functional programming versus class-based programming make it important not to confuse them.

```
> Pair 1 35 <= Pair 2 7 -- 1st components determine result
True
> Pair 1 35 <= Pair 1 7 -- 1st components equal, compare 2nd components
False
```

In this case, an explicit instance declaration for `Ord` which overrides the default definition for < is more efficient because it avoids a repeated equality test:

```
instance (Ord a, Ord b) => Ord (Pair a b) where
  Pair x y <= Pair x' y' = x < x' || (x == x' && y <= y')
  Pair x y < Pair x' y'  = x < x' || (x == x' && y < y')
```

Finally, an algebraic data type involving recursion gives rise to a recursive definition of <=. For the `List` type from page 137:

```
data List a = Nil
            | Cons a (List a)
  deriving (Eq,Show,Ord)
```

we get:

```
instance Ord a => Ord (List a) where
  Nil        <= ys        = True
  Cons x xs <= Nil        = False
  Cons x xs <= Cons y ys = x < y || (x == y && xs <= ys)
```

For lists of length 2, this is the same as <= on pairs above:

```
> Cons 1 (Cons 35 Nil) <= Cons 2 (Cons 7 Nil)
True
> Cons 1 (Cons 35 Nil) <= Cons 1 (Cons 7 Nil)
False
```

The technical term for this is **lexicographic order**, see
► https://en.wikipedia.org/wiki/
Lexicographic_order.

For longer lists, it extends <= on pairs to give the familiar dictionary ordering of words. Using the usual notation for strings as lists of characters in place of `Nil` and `Cons`:

```
> "ashen" <= "asia"  -- 'a'=='a', 's'=='s', 'h' <= 'i'
True
> "ash" <= "as" -- 'a'=='a', 's'=='s', "h" <= "" is False
False
```

The `Enum` type class is for types whose values form a sequence. The functions it provides are the basis for Haskell's notations for enumeration expressions: `['a'..'z']`, `[0..]`, etc. It works by providing functions between the type and `Int`:

```
class Enum a where
  toEnum        :: Int -> a
  fromEnum      :: a -> Int
  succ, pred    :: a -> a
  enumFrom      :: a -> [a]            -- [x ..]
  enumFromTo    :: a -> a -> [a]       -- [x .. y]
  enumFromThen  :: a -> a -> [a]       -- [x, y ..]
  enumFromThenTo :: a -> a -> a -> [a]  -- [x, y .. z]
  -- defaults
  succ x = toEnum (fromEnum x + 1)
  pred x = toEnum (fromEnum x - 1)
  enumFrom x
    = map toEnum [fromEnum x ..]
  enumFromTo x y
```

24

```
  = map toEnum [fromEnum x .. fromEnum y]
enumFromThen x y
  = map toEnum [fromEnum x, fromEnum y ..]
enumFromThenTo x y z
  = map toEnum [fromEnum x, fromEnum y .. fromEnum z]
```

The functions `toEnum` and `fromEnum` are required, with the rest having default definitions. Any instance should satisfy `toEnum (fromEnum` x`) ==` x but there is no way to enforce this requirement.

The type `Season` on page 133 can be declared as an instance of `Enum`, using the functions `toInt :: Season -> Int` and `fromInt :: Int -> Season` defined on that page or repeating their definitions as follows:

```
instance Enum Season where
  toEnum 0 = Winter
  toEnum 1 = Spring
  toEnum 2 = Summer
  toEnum 3 = Fall

  fromEnum Winter = 0
  fromEnum Spring = 1
  fromEnum Summer = 2
  fromEnum Fall   = 3
```

The same can be done using `deriving` in the definition of `Season`, which only applies to algebraic data types like `Season` in which all constructors take no parameters:

```
data Season = Winter | Spring | Summer | Fall
  deriving (Eq,Show,Enum)
```

The n constructors are numbered left to right from 0 to $n - 1$, giving the same definitions of `toEnum` and `fromEnum` as above.

Either way, we get:

```
> [Spring .. Fall]
[Spring,Summer,Fall]
> [Fall, Summer ..]
[Fall,Summer,Spring,Winter]
```

and so on.

Numeric Type Classes

Haskell includes a rich collection of types for different kinds of numbers: fixed size integers `Int`, arbitrary precision integers `Integer`, single- and double-precision floating-point numbers `Float` and `Double`, rational numbers `Rational`, complex numbers `Complex`, etc. These are organised into a set of related type classes, with numeric types supporting different operations being instances of different type classes. The details can be looked up when you need them; this is just a brief overview of some of the main points.

See Sect. 6.4 of ► https://www.haskell.org/onlinereport/haskell2010/haskellch6.html for complete documentation of Haskell's numeric types and numeric type classes.

The simplest numeric type class, for all kinds of numbers, is `Num`:

```
class (Eq a, Show a) => Num a where
  (+), (-), (*) :: a -> a -> a
  negate        :: a -> a
  abs, signum   :: a -> a
  fromInteger   :: Integer -> a
  -- default
  negate x = fromInteger 0 - x
```

The functions abs and signum should satisfy abs $x *$ signum $x == x$.

All of the other numeric type classes—`Real`, `Fractional`, `Floating`, `Integral`, etc.—inherit from `Num`. For example, `Fractional` is for numbers that support division:

```
class Num a => Fractional a where
  (/)            :: a -> a -> a
  recip          :: a -> a
  fromRational :: Rational -> a
  -- default
  recip x = 1/x
```

while `Floating` supports logarithms, square roots, and trigonometry:

```
class Fractional a => Floating a where
  pi                     :: a
  exp, log, sqrt         :: a -> a
  (**), logBase          :: a -> a -> a
  sin, cos, tan          :: a -> a
  asin, acos, atan       :: a -> a
  sinh, cosh, tanh       :: a -> a
  asinh, acosh, atanh :: a -> a
  -- defaults
  x ** y       = exp (log x * y)
  logBase x y = log y / log x
  sqrt x       = x ** 0.5
  tan  x       = sin  x / cos  x
  tanh x       = sinh x / cosh x
```

Functors

The concept of functor comes from *category theory*, a branch of Mathematics devoted to the study of mathematical structure which has applications in programming language theory, see ▶ https://en.wikipedia.org/wiki/Category_theory.

Here's another one of Haskell's built-in type classes:

```
class Functor t where
  fmap :: (a -> b) -> t a -> t b
```

Before talking about what `Functor` is used for, look at its definition and notice that it's different from all of the other type classes you've seen so far. According to the definition, an instance of `Functor` won't be a type but rather a **type constructor**, like `Maybe`: it's applied to a type and yields a type.

`Functor` is used for types that can be *mapped over*, in a generalisation of what `map :: (a -> b) -> [a] -> [b]` does. As you've seen, given a function $f :: a \to b$, `map f :: [a] -> [b]` applies f to every element of a list, producing a list of the results:

```
> map even [1,2,3,4,5]
[False,True,False,True,False]
```

In this case, the type constructor `t` in the definition of `Functor` is the type constructor `[]` for lists, and `fmap` is `map`:

```
instance Functor [] where
  fmap = map
```

Here's another example, for the `Maybe` type:

```
instance Functor Maybe where
  fmap f (Just x) = Just (f x)
  fmap f Nothing  = Nothing
```

Notice that we didn't write "**instance Functor (Maybe a)**": only type constructors, like **Maybe**, can be instances of **Functor**, not types. The type of fmap wouldn't make sense otherwise. And, fmap only applies f :: a -> b to a value in the case (**Just** x) where there is a value of type a—namely x—to apply it to. In the other case, where there is nothing of type a—that is, **Nothing**—it does nothing.

Let's compare these two examples:

- fmap :: (a -> b) -> [a] -> [b] for the list type—that is, map—applies a function $f :: a \to b$ to each of the values of type a in something of type [a] to give something of type [b]. That is,

$$\text{fmap} f\ [a_1, a_2, \ldots, a_n] = [f\ a_1,\ f\ a_2,\ \ldots,\ f\ a_n]$$

- fmap :: (a -> b) -> Maybe a -> Maybe b for the Maybe type applies a function $f :: a \to b$ to each of the values of type a in something of type **Maybe** a to give something of type **Maybe** b. That is,

$$\text{fmap} f\ \text{Just}\ a\ = \text{Just}\ (f\ a)$$
$$\text{fmap} f\ \text{Nothing} = \text{Nothing}$$

Here's a function to square all of the integers in a data structure that is conditional on its type constructor being an instance of **Functor**:

```
squares :: Functor t => t Int -> t Int
squares = fmap (^2)
```

Then we get:

```
> squares [2..10]
[4,9,16,25,36,49,64,81,100]
> squares (Just 3)
Just 9
```

Instances of **Functor** should satisfy fmap id = id and fmap $(f\ .\ g)$ = fmap f . fmap g. Again, there's no way for Haskell to enforce these requirements. It's important to check it for each new instance declaration, so that users can rely on **Functor** instances behaving sensibly.

What about types that involve values of *two* types, like **Either** a b and **Pair** a b? For those you can use **Bifunctor** from the library module **Data.Bifunctor**, which is analogous to **Functor** but accommodates two types:

```
class Bifunctor t where
  bimap :: (a -> b) -> (c -> d) -> t a c -> t b d

instance Bifunctor Either where
  bimap f g (Left x)  = Left (f x)
  bimap f g (Right y) = Right (g y)

instance Bifunctor Pair where
  bimap f g (Pair x y) = Pair (f x) (g y)
```

Syntactic sugar is syntax in a programming language that makes things easier to express but adds no expressive power.

This is the same as the earlier definition but using prefix `elem` rather than infix `` `elem` `` for ease of comparison with `elem'` below.

Type Classes are Syntactic Sugar

An alternative to the use of type classes is to pass the functions that belong to the type class as extra parameters. For example, recall the definition of `elem`:

```
elem :: Eq a => a -> [a] -> Bool
elem x []     = False
elem x (y:ys) = x==y || elem x ys
```

Instead of requiring `Eq a`, we could supply the equality function as an explicit parameter:

```
elem' :: (a -> a -> Bool) -> a -> [a] -> Bool
elem' eq x []     = False
elem' eq x (y:ys) = x `eq` y || elem' eq x ys
```

Then, an application of `elem` such as:

```
> elem 5 [1..10]
True
```

would be replaced by:

```
> elem' eqInt 5 [1..10]
True
```

where `eqInt :: Int -> Int -> Bool` is equality on integers.

The same idea can be used to deal with conditional type class instances. For example, the declaration that `[a]` is an instance of `Eq`:

```
instance Eq a => Eq [a] where
  [] == []     = True
  [] == y:ys   = False
  x:xs == []   = False
  x:xs == y:ys = (x == y) && (xs == ys)
```

becomes

```
eqList :: (a -> a -> Bool) -> [a] -> [a] -> Bool
eqList eq [] []         = True
eqList eq [] (y:ys)     = False
eqList eq (x:xs) []     = False
eqList eq (x:xs) (y:ys) = (x `eq` y) && (eqList eq xs ys)
```

and then an application of `elem` to a list of strings:

```
> elem "of" ["list","of","strings"]
True
```

would be replaced by:

```
> elem' (eqList eqChar) "of" ["list","of","strings"]
True
```

where `eqChar :: Char -> Char -> Bool` is equality on characters.

The idea of replacing type class requirements in types by extra function parameters can be applied systematically to all type classes and type class instances in order to translate Haskell into a simplified version of Haskell without them; indeed, that's exactly how type classes are implemented in practice. In this sense, type classes don't actually extend Haskell's power; they just make things more convenient.

Type classes are introduced and the translation into Haskell without type classes is given in "How to make *ad-hoc* polymorphism less *ad hoc*" by Philip Wadler and Stephen Blott in *Proc. 16th ACM SIGPLAN-SIGACT Symp. on Principles of Programming Languages*, 60–76 (1989).

24

Exercises

1. The `Bounded` type class is for types that have a minimum and maximum value:

    ```
    class Bounded a where
      minBound, maxBound :: a
    ```

    with instances including `Int`:

    ```
    > minBound :: Int
    -9223372036854775808
    ```

    Algebraic data types can be declared as instances of `Bounded` using `deriving`, provided either all of their constructors take no parameters, or they have a single constructor. Explain how that works, for each of these two cases, without consulting the documentation. For the second case, use the type `Pair` from page 136 as an example.

2. In Exercise 21.6, you defined an abstract data type for natural numbers represented as non-negative integers. Define it to be an instance of `Num`.

3. The function

    ```
    evalProp :: (Name -> Bool) -> Prop -> Bool
    ```

    for evaluating propositions was extended to deal with implications and bi-implications on page 221. Later, in Exercise 22.5, you were asked to define a function

    ```
    evalProp :: (Name -> Predicate u) -> Prop -> Predicate u
    ```

    that interprets propositions as predicates rather than as Boolean values. This was relatively straightforward because both Boolean values and predicates, with corresponding definitions of the logical connectives, form Boolean algebras.

    Define a type class `BoolAlg` of Boolean algebras and a function

    ```
    evalProp :: BoolAlg a => (Name -> a) -> Prop -> a
    ```

    for evaluating propositions in an arbitrary Boolean algebra. Define instances of `BoolAlg` for Boolean values and for predicates.

4. The following algebraic data type gives a polymorphic version of binary trees, with labels of the given type at the nodes and leaves:

    ```
    data Tree a = Empty | Leaf a | Node (Tree a) a (Tree a)
    ```

    Define `Tree` as an instance of `Functor`. Check that it behaves as you would expect.

5. Define the partially applied type of functions, `(->) a`, as an instance of `Functor`. What type of functions does `fmap (replicate 3)` apply to, and what does it do when given a function of that type? (Try to figure it out before trying it in Haskell.)

6. `squares` (page 255) works on `[Int]`, on `Maybe Int`, and on t `Int` for other `Functor` instances t. But it doesn't work on `[Maybe Int]`, which involves a composition of two `Functor` instances.

    Write a function `squares'` that, given `Functor` instances t and s, will work on t (s `Int`).

Search in Trees

Contents

© The Author(s), under exclusive license to Springer Nature Switzerland AG 2021
D. Sannella et al., *Introduction to Computation*, Undergraduate Topics
in Computer Science, https://doi.org/10.1007/978-3-030-76908-6_25

Representing a Search Space

Trees are a common data structure, as you've seen. They were used to represent the syntax of languages like propositional logic in Chap. 16, and as the basis for an efficient way of representing sets of integers in Chap. 20.

More abstractly, trees can be used to represent **search spaces**. Nodes represent positions or situations, and a branch from a node to a sub-tree indicates a connection of some kind between the position represented by that node and the position given by the node at the root of that sub-tree.

An example would be a tree representing the game of chess or the game of Go. The nodes represent positions of the board, with the root of the tree representing the starting position, and the branches representing legal moves in the game from one board position to another. Another kind of example would be a tree representing a person's ancestry. The nodes represent people, and there is a branch from *a* to *b* if *b* is the mother or father of *a*. Or, a tree of airports, with a particular airport as the root node, where a branch from *a* to *b* represents the existence of a flight.

In a tree representing a game like chess or Go, you wouldn't have binary trees but trees with a *list* of sub-trees per node, to take account of the fact that the number of legal moves—almost always more than two—depends on the board position.

We're going to look into the problem of searching a tree for a node that satisfies a given property. For example, we could search for an ancestor who was born in Venice, or an airport in Samoa that is reachable by a sequence of flights from Pittsburgh. The trees in question won't satisfy the invariant on node labels that we saw in the representation of sets as ordered trees in Chaps. 20 and 21, so the efficient search algorithm used there doesn't apply.

Searching through a solution space is a matter of visiting a sequence of nodes to find a node that satisfies the property of interest. We're going to look into different ways of doing the search. The difference lies in the order in which the nodes are visited.

They also work for **graphs** where the connections between nodes aren't required to form a hierarchy, as they do in trees. And we're often interested in finding the best *route* to a node with a given property, not just the node itself. See for example *A* search*, ▶ https://en.wikipedia.org/wiki/A*_search_algorithm.

We're going to stick with binary trees because they are familiar, but the same ideas hold for trees with arbitrary branching. We'll use nodes with integer labels in examples, to keep things simple, and in order to concentrate on the structure of the search rather than the details of the data. The data at a node could obviously be much more complicated, and the search algorithms work for any type of data.

Trees, Again

We'll use simple binary trees, with labels at the nodes:

```
data Tree a = Nil | Node (Tree a) a (Tree a)
  deriving (Eq,Show)
```

Here's an example:

```
t :: Tree Int
t = Node (Node (Node Nil 4 Nil)
               2
               (Node Nil 5 Nil))
         1
         (Node (Node (Node Nil 8 Nil)
                     6
                     (Node Nil 9 Nil))
               3
               (Node Nil 7 Nil))
```

Drawing t as a tree gives:

25

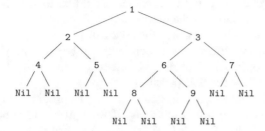

And you can build infinite trees, for instance:

```
inf :: Tree Int
inf = infFrom 0
  where
    infFrom x = Node (infFrom (x-1)) x (infFrom (x+1))
```

which defines `inf` using the recursively defined helper function `infFrom`. The infinite tree `infFrom n` has label n at its root, with its left sub-tree having a label of $n - 1$ at its root and its right sub-tree having a label of $n + 1$ at its root. Then `inf` corresponds to the diagram:

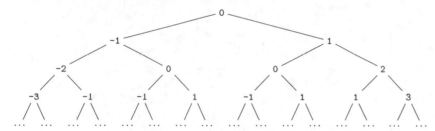

The way that the labels are arranged in `t` and `inf` will turn out to be useful later in the examples to come, but otherwise there's nothing special about them.

You can compute with infinite trees in Haskell, relying on lazy evaluation to avoid infinite computation. For example, here's a function that returns a list of the node labels at a given depth:

```
labelsAtDepth :: Int -> Tree a -> [a]
labelsAtDepth d Nil          = []
labelsAtDepth 0 (Node l x r) = [x]
labelsAtDepth n (Node l x r) =
  labelsAtDepth (n-1) l ++ labelsAtDepth (n-1) r
```

Applying it to `inf` will produce a finite list of node labels:

```
> labelsAtDepth 3 inf
[-3,-1,-1,1,-1,1,1,3]
```

When searching through such trees, we're going to be visiting their nodes in some particular order. A node is a tree of the form **Node** l x r, with sub-trees l and r. Sometimes we'll refer to this as a *tree*, especially when its sub-trees are important, and sometimes we'll refer to it as a *node*, especially when its label is the focus of attention.

Depth-First Search

Suppose we have a predicate—a function from node labels to **Bool**—that says whether a node label has the property we're looking for, or not. Our task is to search a given tree to find a node label that satisfies the predicate. (A different problem, that we're not going to consider, would be to search for *every* node label in the tree that satisfies the predicate.)

If we find a node label for which the predicate produces **True** then the search is successful, and we return it. But there may not be any node label in the tree that satisfies the predicate. The result of the search needs to be of type **Maybe a** rather than a to allow for that possibility.

If we're looking into something that satisfies the predicate in the empty tree, we're out of luck: the result is **Nothing**. If we're at a node and the predicate holds for the node label *x*, we're finished and return **Just** *x*. Otherwise, we have to look somewhere else. And this is where there's more than one choice.

The one that we consider first is called **depth-first search**. The idea is to exhaustively search the nodes in the left sub-tree before looking at the right sub-tree. Since we follow the same strategy if the node label at the root of the left sub-tree doesn't satisfy the predicate, the search proceeds down the leftmost path of the tree all the way to a leaf, before looking at the right sub-trees at any of the nodes that we have encountered along the way.

Here's a picture showing what happens if we search in the tree t above for a node label that is greater than 4:

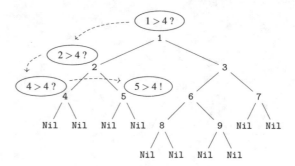

We potentially have to visit all of the nodes, because the node label we're looking for could be anywhere, but the *order* that we visit them is *depth-first* and *left-first*.

Here's a diagram that shows the order of visiting the nodes of t in case none of the node labels satisfies the predicate:

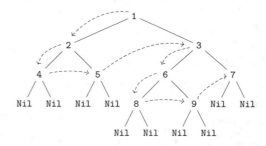

We can make the order of visiting the nodes explicit by defining the search in terms of a function that visits the nodes and records the node labels it sees. We call a procedure for visiting all of the nodes of the tree a **traversal** of the tree. In the case of a depth-first search, we do a **depth-first traversal**:

```
dfTraverse :: Tree a -> [a]
dfTraverse Nil          = []
dfTraverse (Node l x r) = x : (dfTraverse l) ++ (dfTraverse r)
```

Given a node, dfTraverse records the label of the node, then traverses the left sub-tree followed by the right sub-tree. For example:

```
> dfTraverse t
[1,2,4,5,3,6,8,9,7]
```

Then, given a list of the node labels in the tree, we can start with the first one and search down the list until we find one that satisfies the predicate:

```
depthFirst :: (a -> Bool) -> Tree a -> Maybe a
depthFirst p t =
  head( [ Just x | x <- dfTraverse t, p x ] ++ [Nothing] )
```

The fact that lazy evaluation is used means that the traversal doesn't need to be completed before the search begins. Only the portion of the traversal that is required, from the beginning to the point at which the search succeeds, is computed. For example:

```
> depthFirst (< -4) inf
Just (-5)
> dfTraverse inf
[0,-1,-2,-3,-4,-5,-6,-7,-8,-9,-10, ...
```

Of course, if the search fails or if the last node in the traversal is the one for which the predicate is satisfied, then the whole traversal does need to be completed.

Breadth-First Search

An alternative to depth-first search is **breadth-first search**, which visits the nodes of the tree in a different order.

When expressed in terms of a traversal, what is different in breadth-first search is that the traversal proceeds *across* the tree, in *layers*. After visiting the root, which is at depth 0, the nodes immediately below it, at depth 1, are visited. Then the nodes immediately below those, at depth 2, are visited, and so on. In each layer, the nodes at that depth are visited from left to right:

```
bfTraverse :: Tree a -> [a]
bfTraverse t = bft 0 t
  where
    bft :: Int -> Tree a -> [a]
    bft n t | null xs   = []
            | otherwise = xs ++ bft (n + 1) t
              where xs = labelsAtDepth n t
```

For example:

```
> bfTraverse t
[1,2,3,4,5,6,7,8,9]
```

The recursion in the definition of the helper function `bft` is a little unusual. The base case is not when n is 0 but rather when the layer of nodes at depth n is empty. In each recursive call, the depth of the nodes that are considered *increases*, from n to n + 1, rather than decreasing. But that's okay: except when t is an infinite tree, when the result of `bfTraverse` is an infinite list, the result of `labelsAtDepth n t` will always be empty for a large enough value of n.

A simpler way to compute breadth-first traversal, which makes no explicit reference to the depth of nodes in the tree, is to maintain a **queue** of nodes that remain to be visited. This queue is the parameter of the helper function `bft`:

See ▶ https://en.wikipedia.org/wiki/Queue_(abstract_data_type).

```
bfTraverse' :: Tree a -> [a]
bfTraverse' t = bft [t]
  where
    bft :: [Tree a] -> [a]
    bft []              = []
    bft (Nil : ts)      = bft ts
    bft (Node l x r : ts) = x : bft (ts ++ [l,r])
```

Here, the queue is represented as a list. See Exercise 2 for a version in which the representation of the queue is separated from the implementation of the traversal, as an abstract data type.

When a node is visited, its label is recorded and its sub-trees are added to the end of the queue. The nodes at depth d will be visited before any of the nodes at depth $d + 1$ since those will be added to the queue after all of the nodes at depth d are already there.

The code for breadth-first search is exactly the same as the code for depth-first search, except that it uses `bfTraverse` in place of `dfTraverse`:

```
breadthFirst :: (a -> Bool) -> Tree a -> Maybe a
breadthFirst p t =
  head( [ Just x | x <- bfTraverse t, p x ] ++ [Nothing] )
```

Using breadth-first search, we get the same results as we did for depth-first search in the examples above:

```
> breadthFirst (>4) t
Just 5
> breadthFirst (< -4) inf
Just (-5)
```

But for other examples, the result differs for these two search strategies:

```
> depthFirst (>2) t
Just 4
> breadthFirst (>2) t
Just 3
```

Both results are correct since both 4 and 3 are node labels in t and both satisfy the predicate (>2). The difference in results comes from the different order in which nodes are visited, since more than one node label in t satisfies the predicate.

The difference between depth-first search and breadth-first search becomes more important when searching in infinite trees.

Suppose we're searching for a node label that is greater than 0 in the infinite tree `inf`. Depth-first search will run forever, searching further and further down the infinite leftmost path through the tree:

```
> depthFirst (>0) inf
... runs forever ...
```

Here's a diagram:

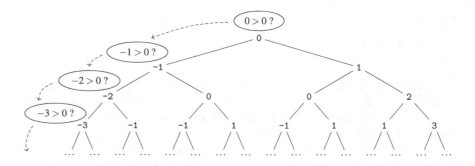

But breadth-first search succeeds:

```
> breadthFirst (>0) inf
Just 1
```

because it searches in layers:

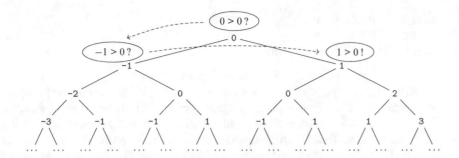

The most important advantage of breadth-first search is that it will *always* succeed if there is a node label somewhere in the tree that satisfies the predicate, while depth-first search will sometimes run forever on infinite trees even if a node label satisfying the predicate is present. For *finite* trees, both depth-first and breadth-first search will always terminate, of course.

The difference between depth-first search and breadth-first search is not merely of academic interest, since infinite search spaces are common. All of the motivating examples mentioned at the beginning of this chapter are infinite or nearly infinite. The tree representing the game of chess is infinite, or finite but huge. Similarly for the game of Go. The tree representing a person's ancestry is finite but very large, especially if ancestors preceding *Homo sapiens* are taken into account. The tree representing airports and flights is infinite for almost every choice of root airport, because of the presence of cyclical sequences of flights.

Whether or not these trees are infinite depends on whether rules that stop the game due to repetition of a position or lack of interesting action are disregarded or not. The number of legal positions in Go is about 2.1×10^{170}, which is greater than the number of atoms in the known universe, see ▶ https://en.wikipedia.org/wiki/Go_(game).

Best-First Search

In both depth-first search and breadth-first search, the order of nodes to be searched is fixed in advance. An alternative is **best-first search**, where the order depends on the nodes that are encountered during the search.

The idea behind best-first search is the observation that the node labels in typical search spaces aren't random. The information they contain often indicates how close the search is to success, and which direction of movement through the search space is most likely to be fruitful. In the example of airports and flights, a flight with a destination that reduces the distance to the goal airport is more promising than a flight that moves in the opposite direction. Or, big airports with many outgoing flights might be preferred to small airports with only a few flights.

A simple way to implement this idea is to use an **evaluation function** that estimates how close a position is to success, and choose the next node to visit based on its results. We'll use an evaluation function that takes a node as input and produces an integer score as output. The choice of evaluation function should depend on the search goal. For the airport/flight example, the geographical location of the airport in the node label and/or how many outgoing flights it has will be relevant to its score. In our final implementation of the DPLL algorithm on page 184, the lengths of the remaining clauses are used to decide which clause to choose next.

See ▶ https://en.wikipedia.org/wiki/Evaluation_function.

To implement best-first traversal—in terms of which best-first search will be expressed—we'll use a **priority queue** to keep track of nodes to be visited. A priority queue is a queue in which items—in this case trees—are ordered according to their priorities rather than when they joined the queue. Its API is:

See ▶ https://en.wikipedia.org/wiki/Priority_queue.

```
type PQ b
emptyPQ :: (b -> Int) -> PQ b
insertPQ :: b -> PQ b -> PQ b
```

```
topPQ :: PQ b -> b
popPQ :: PQ b -> PQ b
isemptyPQ :: PQ b -> Bool
```

where the type variable b will be instantiated to **Tree** a.

In these priority queues, priorities will be assigned by the evaluation function, and creation of an empty priority queue (emptyPQ) requires an evaluation function. The priority of each item added using insertPQ is calculated using that function, so that the highest-priority item is always at the front of the queue. That item is obtained using the function topPQ, and popPQ gives the remainder of the queue.

```
bestFirstTraverse :: (Tree a -> Int) -> Tree a -> [a]
bestFirstTraverse f t = bft (insertPQ t (emptyPQ f))
  where
    bft :: PQ (Tree a) -> [a]
    bft pq | isemptyPQ pq = []
           | otherwise     = x : bft (insertPQ' r (insertPQ' l pq'))
             where Node l x r = topPQ pq
                   pq'         = popPQ pq
```

The function insertPQ' inserts items into the priority queue using insertPQ. But it doesn't insert empty trees, which simplifies the definition of bft.

```
insertPQ' :: Tree a -> PQ (Tree a) -> PQ (Tree a)
insertPQ' Nil pq = pq
insertPQ' t pq   = insertPQ t pq

bestFirst :: (a -> Bool) -> (Tree a -> Int) -> Tree a -> Maybe a
bestFirst p f t =
  head( [ Just x | x <- bestFirstTraverse f t, p x ] ++ [Nothing] )
```

The priority queue is used to maintain a list of nodes that haven't been visited yet because they've been postponed in order to consider more promising nodes first. They are next in line for consideration in case the node currently being visited doesn't satisfy the predicate. Keeping them in a priority queue means that the next best choice is always at the front of the queue.

As an example, suppose we're searching inf for a node label that is greater than 19. There are many such node labels in inf, but the right sub-tree of the root looks more promising than its left sub-tree. Once we decide to look at that node, *its* right sub-tree looks more promising than its left sub-tree, and so on.

The type we are using for evaluation functions allows more information than the label of a node to be used to produce a score. The information in the sub-trees of that node could also be taken into account. But since we're looking for a node label that is larger than a given value, we'll just use the node label itself as its score, so that nodes with larger labels will have higher scores:

```
eval :: Tree Int -> Int
eval Nil = 0
eval (Node l x r) = x
```

That evaluation function works well in this case. Comparing it with breadth-first search, the same result is obtained:

```
> bestFirst (>19) eval inf
Just 20
> breadthFirst (>19) inf
Just 20
```

25

but best-first search is successful after visiting 21 nodes, while breadth-first search visited 2,097,151 nodes.

In fact, best-first search is optimal in this case, in terms of the number of nodes that are visited starting from the root. Suppose we try searching for a node label that is greater than 100:

```
> bestFirst (>100) eval inf
Just 101
```

Using best-first search visits 102 nodes. Using breadth-first search for the same problem visits 5,070,602,400,912,917,605,986,812,821,503 nodes. Wow, artificial intelligence!

The success of best-first search with respect to "blind" breadth-first search critically depends on the choice of evaluation function, together with the structure of the search space. The infinite tree `inf` has a very regular structure, which made it easy to get good results using a simple evaluation function.

In this sense, best-first search is related to the **hill climbing** optimisation technique. Hill climbing attempts to find an optimal solution to a problem by starting from an arbitrary solution and incrementally improving it. An evaluation function is used to decide which of the possible new solutions are improvements over the existing solution, and the best new solution is then used to continue climbing. If you want to get to the top of a hill, a good heuristic is to always go uphill, and the steeper the better if you want to get there fast.

Hill climbing will always find a *local* optimum but it is not guaranteed to find a *global* optimum if that would require temporarily going "downhill", as in the right-hand diagram below when starting from the left corner:

Best-first search may reach its goal in such cases, because it keeps track of unexplored sub-trees that look less promising than the one it has chosen, and considers them later. But it may waste time by first exploring regions of the search space that look promising but don't work out.

2,097,151= $2^{21} - 1$: the first node visited whose label satisfies the predicate is on the right-hand end of level 20 of the tree.

See ► https://en.wikipedia.org/wiki/Hill_climbing.

Exercises

1. In ► http://www.intro-to-computation.com/meadows.hs you will find the code in this chapter together with a tree

    ```
    meadowTreesTree :: Tree String
    ```

    containing data about each of the trees in The Meadows in Edinburgh, one per node. Use depth-first and/or breadth-first search to find out whether or not there is a *Sequoiadendron giganteum* (Giant redwood) there.
2. Re-write `bfTraverse'` so that it has exactly the same structure as `bestFirstTraverse`, but using a queue (implemented as an abstract data type) rather than a priority queue. The goal is to show that best-first search differs from breadth-first search only in the use of a priority queue in place of an ordinary queue.
3. Implement priority queues as an abstract data type, with the API given on page 265.

See ► https://en.wikipedia.org/wiki/The_Meadows,_Edinburgh. Data for all trees maintained by the City of Edinburgh Council can be found in ► https://data.edinburghopendata.info/dataset/edinburgh-council-trees-dataset.

See ▶ https://en.wikipedia.org/wiki/
Iterative_deepening_depth-
first_search.

Hint: One possible data representation is like the ordered list representation of sets on page 194, but the operations are different and the order is according to the result of applying the evaluation function to nodes.

4. **Iterative deepening search** is a hybrid between depth-first and breadth-first search. It visits nodes in the same order as depth-first search, but with a depth limit. The search is run repeatedly with increasing depth limits until the search is successful or the tree has been searched completely. Implement iterative deepening search in Haskell.

5. Define an evaluation function that would be appropriate for use with best-first search in a tree with integer labels when searching for a node label that satisfies the predicate (==7).

6. You've finished Chap. 25. Well done! It's high time for some real-life exercise. Go jogging, or cycling, or for a long walk in the fresh air.

Combinatorial Algorithms

Contents

© The Author(s), under exclusive license to Springer Nature Switzerland AG 2021
D. Sannella et al., *Introduction to Computation*, Undergraduate Topics
in Computer Science, https://doi.org/10.1007/978-3-030-76908-6_26

The Combinatorial Explosion

This is also known as *brute force search*, see ▶ https://en.wikipedia.org/wiki/Brute-force_search.

Some of the programs you have seen use clever algorithms or clever data representations to avoid unnecessary computation. But for some problems, there is no clever algorithm, or at least none that has been discovered yet. The best-known method might then be **generate and test**: simply enumerate all of the possible solutions and check them to find one that satisfies the given requirements. Or sometimes we need to find *all* the ones that satisfy the requirements, or the best ones according to some measure of quality, or to check if they all satisfy the requirements.

Whether or not this is tractable depends on the number of possible solutions and how easily they can be checked. A situation in which the number of possible solutions grows rapidly with increasing problem size, as in the case of finding all of the prime factors of a given n-digit integer, is referred to as a **combinatorial explosion**. Then only small problems can be handled in a reasonable period of time. The security of many cryptographic algorithms is based on the difficulty of solving large problems of this kind.

For example, the security of the widely-used RSA public-key cryptosystem (see ▶ https://en.wikipedia.org/wiki/RSA_(cryptosystem)) is based on the intractability of factorising large integers.

The tractability of such problems can sometimes be improved by exploiting symmetries or other structure of the problem space that makes it possible to reduce, often dramatically, the number of solutions that need to be checked. Sometimes heuristics can be used to improve the order in which potential solutions are checked to make it more likely that a solution will be found earlier.

In this chapter we're going to look at some programs for solving problems of this kind. We'll start with functions that can be used to enumerate potential solutions. All of the programs we'll look at are simple, but sometimes there are choices that make dramatic differences in run time.

Repetitions in a List

We'll be working mainly with lists containing distinct elements—i.e. without repetitions—because it makes things simpler for some of the problems below. To get started, here's the function nub from the **Data.List** library module which removes repetitions from a list:

```
nub :: Eq a => [a] -> [a]
nub []     = []
nub (x:xs) = x : nub [ y | y <- xs, x /= y ]
```

We'll need to test lists to see whether their elements are distinct or not. One way of doing that is to compare the list with the same list after nub has removed any repetitions. If those are the same, then the list's elements are distinct. This takes advantage of the fact that nub preserves the order of the elements in its input:

```
distinct :: Eq a => [a] -> Bool
distinct xs = xs == nub xs
```

This gives:

```
> distinct "avocado"
False
> distinct "peach"
True
```

Sublists

A list *xs* is a **sublist** of another list *ys* if every value in *xs* is also in *ys*. The elements in *xs* don't need to appear in the same order in *ys*. Note that *ys* is also a sublist of itself.

```
sub :: Eq a => [a] -> [a] -> Bool
xs `sub` ys = and [ x `elem` ys | x <- xs ]
```

For example:

```
> "pea" `sub` "apple"
True
> "peach" `sub` "apple"
False
```

This is the same as checking if one set is a subset of another set when sets are represented as unordered lists, see page 193.

Given a list *xs* containing distinct elements, let's consider the problem of generating all sublists of *xs*. How many are there? In any given sublist of *xs*, each item in *xs* is either in that sublist or not. That gives 2^n possible sublists, where $n = $ length *xs*. We can use that idea to generate all of the sublists of a list by recursion:

```
subs :: [a] -> [[a]]
subs []     = [[]]
subs (x:xs) = subs xs ++ map (x:) (subs xs)
```

The base case says that the only sublist of the empty list is the empty list. Notice that replacing this with subs [] = [] would be incorrect! It says that there are *no* sublists of the empty list, which is false, and which would lead to subs *xs* yielding [] for any list *xs*. For the recursive case x:xs, there are all the sublists of the tail xs (which don't contain x), plus all of those lists with x added (which do contain x, of course). For example:

```
> subs [0,1]
[[],[1],[0],[0,1]]
> subs "abc"
["","c","b","bc","a","ac","ab","abc"]
```

According to the recursive case, the length of subs (x:xs) is twice the length of subs xs, since it contains subs xs and another list produced from subs xs by adding x onto the front of each of its elements. That, together with length (subs []) = 1, imply that length (subs xs) = $2^{\text{length xs}}$.

The subs function should satisfy a number of properties. Restricting attention to lists *xs* containing distinct elements:

- everything in subs *xs* is indeed a sublist of *xs*;
- the elements of subs *xs* are distinct, so each sublist is only generated once;
- the elements of each of the sublists of *xs* are distinct; and
- the length is as we expect.

These properties can be expressed in QuickCheck as follows:

```
subs_prop :: [Int] -> Property
subs_prop xs =
  distinct xs ==>
    and [ ys `sub` xs | ys <- subs xs ]
    && distinct (subs xs)
    && all distinct (subs xs)
    && length (subs xs) == 2 ^ length xs
```

Because subs :: [a] -> [[a]] is polymorphic, it needs to be tested at an instance of its type, and so subs_prop tests it for lists of integers. The same goes for all of the tests of polymorphic functions in the rest of this chapter.

Unfortunately, it isn't feasible to check that this property holds using QuickCheck in the usual way because 100 random test cases are likely to include some long lists. We have seen that a list of length 100 has 2^{100} sublists. Since the run time of distinct is $O(n^2)$, checking that no sublist has been included twice will take a very long time.

Luckily, we can adapt QuickCheck by restricting test cases to be smaller than a given size, as follows:

See ► https://hackage.haskell.org/ package/QuickCheck-2.14.2/docs/ Test-QuickCheck.html for other ways of adjusting QuickCheck's behaviour.

```
sizeCheck n = quickCheckWith (stdArgs {maxSize = n})
```

Testing subs_prop for lists of length 10 should be okay, since $2^{10} = 1024$. And the test succeeds:

```
> sizeCheck 10 subs_prop
+++ OK, passed 100 tests; 22 discarded.
```

(Here, "22 discarded" refers to the lists that weren't tested because their elements aren't distinct and so they failed the pre-condition of subs_prop.)

Cartesian Product

Next, we'll look at the **Cartesian product**. Given two sets S and T, the Cartesian product $S \times T$ is the set of pairs (s, t) with $s \in S$ and $t \in T$. We're working with lists instead of sets, where the analogous function has type [a] -> [b] -> [(a,b)]:

```
cpair :: [a] -> [b] -> [(a,b)]
cpair xs ys = [ (x,y) | x <- xs, y <- ys ]
```

It's easy to see that the length of cpair xs ys is the product of the lengths of xs and ys.

Generalising this, the Cartesian product of a *list* of n lists yields a list containing lists of length n:

```
cp :: [[a]] -> [[a]]
cp []      = [[]]
cp (xs:xss) = [ y:ys | y <- xs, ys <- cp xss ]
```

This is analogous to the Haskell type () of 0-tuples, which has a single value ().

As with subs [], cp [] is [[]]: the Cartesian product of the empty list of lists contains just the empty list.

Again, the cp function should satisfy some properties, provided the elements in all of the lists in its input are distinct:

- the ith element of each list in cp *xss* is from the ith list in *xss*;
- the elements of cp *xss* are distinct;
- the elements of each of the lists in cp *xss* are distinct;
- each list in cp *xss* has the same length as *xss*; and
- the length of cp *xss* is the product of the lengths of the lists in *xss*.

The following function expresses these properties:

```
cp_prop :: [[Int]] -> Property
cp_prop xss =
  distinct (concat xss) ==>
    and [ and [ elem (ys !! i) (xss !! i)
                 | i <- [0..length xss-1] ]
          | ys <- cp xss ]
       && distinct (cp xss)
```

26

```
     && all distinct (cp xss)
     && all (\ys -> length ys == length xss) (cp xss)
     && length (cp xss) == product (map length xss)
```

This test will run using QuickCheck, but it gives up before finishing because most of the test cases it generates contain repetitions:

```
> quickCheck cp_prop
*** Gave up! Passed only 54 tests; 1000 discarded tests.
```

Smaller test cases are less likely to contain repetitions:

```
> sizeCheck 10 cp_prop
+++ OK, passed 100 tests; 130 discarded.
```

Permutations of a List

Now let's consider the problem of computing all of the **permutations** of a list: that is, all of the ways of rearranging its elements, including the rearrangement that keeps them in place. For example, there are six permutations of the list ["auld","lang","syne"]:

["auld","lang","syne"]	["auld","syne","lang"]
["lang","auld","syne"]	["lang","syne","auld"]
["syne","auld","lang"]	["syne","lang","auld"]

See ▶ https://en.wikipedia.org/wiki/Permutation. There are many applications of permutations in science and Mathematics. An application in a different area is to *change ringing* in church bell towers, where permutations of the bells are rung without repetitions following certain rules, see ▶ https://en.wikipedia.org/wiki/Change_ringing.

Again, we'll restrict attention to lists containing distinct elements. For a list of length *n*, there are *n*! permutations: *n* possible choices for the first element, then *n* − 1 remaining choices for the second element, etc.

We can compute all the permutations of a list xs of length *n* without repetitions using the Cartesian product function cp: just take the *n*-fold Cartesian product of xs with itself and remove lists whose elements are not distinct:

```
permscp :: Eq a => [a] -> [[a]]
permscp xs | distinct xs =
  [ ys | ys <- cp (replicate (length xs) xs), distinct ys ]
```

Properties of permscp *xs*, for lists *xs* of length *n* containing distinct elements, are:

- each of the permutations has the same elements as *xs*;
- the elements of permscp *xs* are distinct;
- the elements of each of the lists in permscp *xs* are distinct; and
- the length of permscp *xs* is *n*!

which are expressed as follows:

```
permscp_prop :: [Int] -> Property
permscp_prop xs =
  distinct xs ==>
    and [ sort ys == sort xs | ys <- permscp xs ]
    && distinct (permscp xs)
    && all distinct (permscp xs)
    && length (permscp xs) == fac (length xs)

fac :: Int -> Int
fac n | n >= 0 = product [1..n]
```

And these properties hold:

```
> sizeCheck 10 permscp_prop
+++ OK, passed 100 tests; 29 discarded.
```

This is a generate and test algorithm. The space of possible solutions is the n-fold Cartesian product of the input list with itself, and the result is obtained by testing all of those possibilities to find the ones that have no repetitions. Unfortunately, generating permutations this way is very inefficient. If xs has length n, then cp (replicate (length xs) xs) contains n^n lists. If n is 10, then $n^n = 10,000,000,000$, and all but 3,628,800 of those contain repetitions. The function $n!$ grows considerably more quickly than the exponential function 2^n, but the function n^n grows much faster even than $n!$, so the proportion not containing repetitions decreases rapidly with increasing n. Let's try to find something better.

We'll start by finding all of the possible ways of splitting a list, where one element is separated out from the rest. The following definition does this by progressing down the list, starting from the head, and separating out the selected element each time:

See Exercise 2 for an alternative definition of splits.

```
splits :: [a] -> [(a, [a])]
splits xs =
  [ (xs!!k, take k xs ++ drop (k+1) xs) | k <- [0..length xs-1] ]
```

Here's an example:

```
> splits "abc"
[('a',"bc"),('b',"ac"),('c',"ab")]
```

For a list of length n there will be n splits, one for each element in the list, and the length of the rest is always $n - 1$. If the elements of xs are distinct, then other properties of splits xs are:

- each of the splits has the same elements as xs;
- each of the lists of remaining elements is distinct;
- each of the separated-out elements is distinct; and
- all of the elements of each list of remaining elements are distinct.

Here they are as a QuickCheck property:

```
splits_prop :: [Int] -> Property
splits_prop xs =
  distinct xs ==>
    and [ sort (y:ys) == sort xs | (y,ys) <- splits xs ]
    && and [ 1 + length ys == length xs | (y,ys) <- splits xs ]
    && distinct (map snd (splits xs))
    && distinct (map fst (splits xs))
    && all distinct (map snd (splits xs))
    && length (splits xs) == length xs
```

There's no harm in doing this test without any restriction on the length of the test case:

```
> quickCheck splits_prop
+++ OK, passed 100 tests; 234 discarded.
```

Computing all the permutations of a list is easy, using the splits function. Here's a recursive definition:

```
perms :: [a] -> [[a]]
perms []     = [[]]
perms (x:xs) = [ y:zs | (y,ys) <- splits (x:xs), zs <- perms ys ]
```

The result for the empty list is, again, `[[]]`. For each split `(y,ys)` of `x:xs`, we add `y` to the beginning of each of the permutations of `ys`. This gives all of the permutations with `y` at the front, for each selected item `y`.

For example:

```
> perms "abc"
["abc","acb","bac","bca","cab","cba"]
```

The properties of `perms` are the same as before for `permscp`:

```
perms_prop :: [Int] -> Property
perms_prop xs =
  distinct xs ==>
    and [ sort ys == sort xs | ys <- perms xs ]
    && distinct (perms xs)
    && all distinct (perms xs)
    && length (perms xs) == fac (length xs)
```

And again, these properties hold:

```
> sizeCheck 10 perms_prop
+++ OK, passed 100 tests; 36 discarded.
```

We can also check that both versions give the same result:

```
perms_permscp_prop :: [Int] -> Property
perms_permscp_prop xs =
  distinct xs ==> perms xs == permscp xs
```

Then:

```
> sizeCheck 10 perms_permscp_prop
+++ OK, passed 100 tests; 26 discarded.
```

Choosing *k* Elements from a List

We'll now look at the problem of computing all of the ways of choosing *k* elements from a list of length *n*. The result will be a list of lists, where the order of the elements in the individual lists doesn't matter, in contrast to permutations.

In Mathematics, these are called *k*-**combinations**, see ▶ https://en.wikipedia.org/wiki/Combination.

This problem is related to the problem of finding all of the sublists of a list that we considered earlier: all of the lists that we choose *are* sublists, but they are required to have a particular length. The definition of `choose` is therefore similar to the definition of `subs` above:

See Exercise 4 for an alternative definition of `choose`.

```
choose :: Int -> [a] -> [[a]]
choose 0 xs           = [[]]
choose k []    | k > 0 = []
choose k (x:xs) | k > 0 =
  choose k xs ++ map (x:) (choose (k-1) xs)
```

Choosing 0 elements from any list is easy, and choosing *k* elements from the empty list, if *k* > 0, is impossible. Otherwise, we need to do some work. To choose *k* elements from `x:xs`, we can either choose them all from `xs`, not including x, or else choose *k* − 1 elements from `xs` and then add on x to make *k* elements.

For example:

```
> choose 3 "abcde"
["cde","bde","bce","bcd","ade","ace","acd","abe","abd","abc"]
```

The binomial coefficient $\binom{n}{k}$ is pronounced "n choose k".

For a list xs of length n whose elements are distinct and $0 \leq k \leq n$, the expected properties of choose k xs are:

- each of its elements is a sublist of xs that has the correct length;
- all of them are different;
- the elements of each of them are distinct; and
- there are $\binom{n}{k} = \frac{n!}{k!(n-k)!}$ of them.

which are expressed as follows:

```
choose_prop :: Int -> [Int] -> Property
choose_prop k xs =
  0 <= k && k <= n && distinct xs ==>
    and [ ys `sub` xs && length ys == k | ys <- choose k xs ]
    && distinct (choose k xs)
    && all distinct (choose k xs)
    && length (choose k xs) == fac n `div` (fac k * fac (n-k))
      where n = length xs
```

And altogether, taking all k such that $0 \leq k \leq n$, there are 2^n of them:

```
choose_length_prop :: [Int] -> Bool
choose_length_prop xs =
  sum [ length (choose k xs) | k <- [0..n] ] == 2^n
    where n = length xs
```

Finally, taking all of them together gives all of the sublists of xs:

```
choose_subs_prop :: [Int] -> Bool
choose_subs_prop xs =
  sort [ ys | k <- [0..n], ys <- choose k xs ] == sort (subs xs)
    where n = length xs
```

Then:

```
> sizeCheck 10 choose_prop
+++ OK, passed 100 tests; 286 discarded.
> sizeCheck 10 choose_length_prop
+++ OK, passed 100 tests.
> sizeCheck 10 choose_subs_prop
+++ OK, passed 100 tests.
```

Partitions of a Number

We'll now consider the problem of finding all of the ways of splitting a number n into a list of strictly positive numbers in ascending order that add up to n:

```
partitions :: Int -> [[Int]]
partitions 0 = [[]]
partitions n | n > 0
  = [ k : xs | k <- [1..n], xs <- partitions (n-k),
                            all (k <=) xs ]
```

There is no set of strictly positive numbers that sum up to 0. Otherwise, for each k between 1 and n, we add k to each of the partitions of $n-k$, provided it is no bigger than any of their elements. The latter property is what causes the resulting partitions to be arranged in ascending order.

Let's check that it works:

```
> partitions 5
[[1,1,1,1,1],[1,1,1,2],[1,1,3],[1,2,2],[1,4],[2,3],[5]]
```

26

The properties of partitions *n* are:

- adding up all of the numbers in each partition should give *n*; and
- sorting any list of strictly positive integers gives one of the partitions of its sum.

These are expressed as the following QuickCheck properties:

```
partitions_prop :: Int -> Property
partitions_prop n  =
  n >= 0 ==> all ((== n) . sum) (partitions n)

partitions_prop' :: [Int] -> Property
partitions_prop' xs =
  all (> 0) xs ==> sort xs `elem` partitions (sum xs)
```

and both of them hold:

```
> sizeCheck 10 partitions_prop
+++ OK, passed 100 tests; 70 discarded.
> sizeCheck 8 partitions_prop'
+++ OK, passed 100 tests; 131 discarded.
```

Making Change

Related to the partition problem is the problem of making change in a shop. Given an amount *n* of money to be returned to the customer, what are all the ways of doing that given a certain collection *xs* of coins? What we want are sublists of *xs* that add up to *n*, but *xs* and the sublists may contain repetitions.

The following definitions starts by sorting the list of coins into ascending order. Then, if n is 0, it's easy: you don't need any coins. Otherwise, we pick out a coin y using splits, leaving the remainder ys. If y <= n then including y is a candidate for making change for n, and we add it to all ways of making change for n−y, using coins that are greater than or equal to y:

```
type Coin = Int
type Total = Int

change :: Total -> [Coin] -> [[Coin]]
change n xs = change' n (sort xs)
  where change' 0 xs         = [[]]
        change' n xs | n > 0 =
          [ y : zs | (y, ys) <- nub (splits xs),
                     y <= n,
                     zs <- change' (n-y) (filter (y <=) ys) ]
```

See Exercise 5 for an alternative definition of change.

Applying nub to the result of splits is necessary to ensure that we get each result only once: otherwise we would get all the ways of making change involving the first 5p piece, then separately all the ways of making change involving the second 5p piece, etc.

So we get, for example:

```
> change 30 [5,5,10,10,20]
[[5,5,10,10],[5,5,20],[10,20]]
```

A property of change is that all of the ways of making change for *n* add up to *n*, provided *n* is positive and all of the coin denominations are greater than zero:

```
change_prop :: Total -> [Coin] -> Property
change_prop n xs =
```

```
    0 <= n && all (0 <) xs ==>
      all ((== n) . sum) (change n xs)
```

And it works:

```
> sizeCheck 10 change_prop
+++ OK, passed 100 tests; 486 discarded.
```

Eight Queens Problem

Queens are the most powerful pieces in chess: they can move any number of squares vertically, horizontally, or diagonally. Can you place eight queens on a standard 8 × 8 chessboard so that none of them is attacking any of the others? Here's a solution:

The eight queens problem is a famous combinatorial puzzle, see ► https://en.wikipedia.org/wiki/Eight_queens_puzzle. There are solutions to its generalisation to *n* queens and an *n* × *n* chessboard for all *n* except 2 and 3.

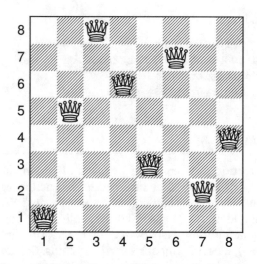

Our method for solving this problem will take into account the observation that there will always be one queen in each column: if there were two or more then they would be able to attack each other, and if there were none then some other column would have at least two.

We'll label rows and columns with integers, and positions on the board with their coordinates. The solutions are boards, which are lists giving the row numbers of the queens in each column, where the diagram above corresponds to [1,5,8,6,3,7,2,4].

Standard chess notation uses the letters a–h for columns, but numbers are a little easier to compute with.

```
type Row   = Int
type Col   = Int
type Coord = (Col, Row)
type Board = [Row]

queens :: Col -> [Board]
queens 0          = [[]]
queens n | n > 0 =
  [ q:qs | q <- [1..8],
           qs <- queens (n-1),
           and [ not (attack (1,q) (x,y))
               | (x,y) <- zip [2..n] qs ] ]
```

26

```
attack :: Coord -> Coord -> Bool
attack (x,y) (x',y') =
   x == x'        -- both in the same column
   || y == y'        -- both in the same row
   || x+y == x'+y'  -- both on the line y = -x + b
   || x-y == x'-y'  -- both on the line y = x + b
```

The first test is actually superfluous since no two queens will be placed in the same column.

We do this by recursion on the number of columns left to fill. The base case is 0: there is just one solution, with no queens. Every solution for n is built from a solution qs for n-1 by adding one more queen in the new column, in row q, provided it can't attack any of the queens in qs.

The first solution that this finds is the one shown above, and there are 92 solutions altogether:

```
> head (queens 8)
[1,5,8,6,3,7,2,4]
> length (queens 8)
92
```

An improved approach to this problem will be to use a generate and test algorithm. Some care is required, in view of the fact that there are $\binom{64}{8} =$ 4,426,165,368 ways to place 8 queens on an 8×8 chessboard. The function definition above takes into account the fact that there will always be one queen in each column, which reduces the number of possible solutions to $8^8 =$ 16,777,216. By also eliminating solutions that place two queens in the same row, we further reduce the problem to finding permutations of the row numbers [1..8]—of which there are 8! = 40,320—such that no queen attacks any other queen.

```
queens' :: [Board]
queens' = filter ok (perms [1..8])

ok :: Board -> Bool
ok qs = and [ not (attack' p p') | [p,p'] <- choose 2 (coords qs) ]

coords :: Board -> [Coord]
coords qs = zip [1..] qs

attack' :: Coord -> Coord -> Bool
attack' (x,y) (x',y') = abs (x-x') == abs (y-y')
```

In order to check the ok property, we just need to check the diagonals. And abs (x-x') == abs (y-y') gives **True** iff (x,y) and (x',y') are either both on the line $y = x + b$ or both on the line $y = -x + b$.

Comparing this with the previous solution, with Haskell set to display elapsed time and space usage, shows that it is much more efficient:

```
> :set +s
> head (queens 8)
[1,5,8,6,3,7,2,4]
(12.27 secs, 6,476,809,144 bytes)
> length (queens 8)
92
(159.74 secs, 83,589,840,216 bytes)
> head queens'
[1,5,8,6,3,7,2,4]
(0.11 secs, 41,654,024 bytes)
> length queens'
92
(1.25 secs, 575,798,296 bytes)
```

Exercises

1. Consider a version of `subs_prop` on page 271 in which the elements of `xs` are not required to be distinct. Which of the properties given would still hold? What about `cp_prop` on page 272? What about `splits_prop` on page 274?
2. Give a recursive definition of the `splits` function (page 274).
3. Define the world's worst sorting function, using a generate and test approach: it should generate all permutations of the input list and then select the one that is in ascending order.
4. Give an alternative definition of `choose` (page 275) that takes a generate and test approach, where `subs` is used to generate all of the possible solutions. How does its performance compare?
5. Give an alternative definition of `change` (page 277) that takes a generate and test approach, where `subs` is used to generate all of the possible solutions. How does its performance compare?

   Hint: The list of coins may contain repetitions—and multiple coins of the same denomination may be required to make change—so you will need to make sure (see Exercise 1) that `subs` behaves appropriately in that case. Then use `nub` on the result. (Why?)
6. A variation on the eight queens problem is the *fourteen bishops problem*, which requires placement of 14 bishops on a chessboard such that none of them can capture any other one. Bishops can move any number of squares diagonally but can't move vertically or horizontally. Write a Haskell program that finds a solution. How many are there altogether?

   (Hint: Instead of placing pieces in columns, place them in diagonals. Why is the maximum 14 instead of 15?)

See ▶ https://en.wikipedia.org/wiki/ Mathematical_chess_problem for similar chess problems.

The Fibonacci numbers (see ▶ https://en.wikipedia.org/wiki/ Fibonacci_number) are named after the Italian mathematician Leonardo of Pisa, also known as Fibonacci (c. 1170 –c. 1240–50), who introduced them in Europe after their first discovery by Indian mathematicians hundreds of years earlier.

See ▶ https://en.wikipedia.org/wiki/ Memoization. Memoisation was invented by Donald Michie (1923–2007), a British researcher in Artificial Intelligence.

7. The following function for computing the Fibonacci numbers

```
fib :: Int -> Int
fib 0 = 0
fib 1 = 1
fib n = fib (n-1) + fib (n-2)
```

takes exponential time, because the evaluation of `fib` n requires repeated evaluation of `fib` $(n-2), \ldots,$ `fib` 0. An alternative is the following definition, using **memoisation**, where computation is replaced by lookup in `fiblist`:

```
fiblist :: [Int]
fiblist = map fib' [0..]

fib' :: Int -> Int
fib' 0 = 0
fib' 1 = 1
fib' n = fiblist!!(n-1) + fiblist!!(n-2)
```

Compare the run times of these two definitions.

Finite Automata

Contents

Models of Computation

There are many other models of computation, including models based on quantum physics (▶ https://en.wikipedia.org/wiki/Quantum_computing) and even biology (▶ https://en.wikipedia.org/wiki/Membrane_computing), as well as models based on mechanisms that are more obviously computational, such as cellular automata (▶ https://en.wikipedia.org/wiki/Cellular_automaton). For more about models of computation, see ▶ https://en.wikipedia.org/wiki/Model_of_computation.

For more about finite-state transducers, see ▶ https://en.wikipedia.org/wiki/Finite-state_transducer.

As you saw in Chap. 13, everything in Haskell is ultimately based on lambda expressions, meaning that computation in Haskell is based on function application and substitution. That is Haskell's **model of computation**: the basic mechanisms that underlie its way of producing results of computational problems.

Starting in this chapter, you will learn about a dramatically different model of computation, embodied in **finite automata**. Computation in finite automata involves movement between states in response to symbols read from an input string. The response to the next input symbol, reading from left to right, depends on the state and (of course) what the input symbol is. The counterpart to writing a Haskell program as a sequence of definitions is drawing a finite automaton as a diagram.

There are different versions of finite automata. The kind you will learn about doesn't produce any output: it just signals acceptance or rejection of an input string, thereby defining the set containing all of the strings that it accepts. Finite automata that produce output are called **finite-state transducers**.

Finite automata have many applications. They are used in digital hardware design to model and design sequential circuits, whose output depends not just on the current input but on the history of the input. They are used as components in compilers, including the Haskell compiler, for decomposing the text of programs into units like variables, numeric literals, strings, etc. They are used to model and test communication protocols, for requirements modelling in software engineering, for phonological and morphological processing in computational linguistics, as models of development in multi-cellular organisms in biology, and to implement simple forms of artificial intelligence in computer games programming.

Haskell, and its underlying model of computation, is able to express any computable function. In contrast, while being very useful for certain kinds of tasks, finite automata have very limited computational power. Adding memory to a finite automaton gives a **Turing machine**, which turns out to have the same computational power as Haskell. Learning about finite automata will give you a solid basis for future study of computation using Turing machines.

States, Input and Transitions

For this reason, finite automata are often called **finite-state machines**, see ▶ https://en.wikipedia.org/wiki/Finite-state_machine.

The plural of "automaton" is "automata". Some authors of published papers on automata appear to think that "automata" is also the singular form, but it isn't.

A key component of a finite automaton is a finite collection of **states**. The fact that the collection of states is finite is important, and you'll see in Chap. 32 that it leads to a limitation in expressive power. But first, what's a state?

To understand that, let's look at an example. Here's a finite automaton with two states, shown as circles and labelled 0 and 1:

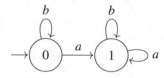

In some books, the start state is called the *initial state*.

During computation it will make a series of moves, from one state to another, in response to what it reads from its input. At each point it will either be in state 0 or in state 1. The arrow pointing to state 0 from the left says that its **start state** is state 0.

The arrows between states are called **transitions**, and they are labelled with the input symbol that causes the automaton to move from the state at the beginning of the arrow to the state at the end of the arrow. So if it is in state 0,

27

there are two possibilities: if the next symbol in the input is *a* then it will move to state 1; but if the next symbol is *b* then it will stay in state 0.

Let's see how that works, for the input string *ab*. We'll use the colour grey to indicate the current state, and a fat arrow for the last transition. The automaton starts in its start state:

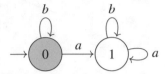

Reading *a*, the first input symbol, causes it to move to state 1:

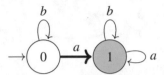

When it reads *b*, the second and final input symbol, it follows the looping transition labelled with *b* and stays in state 1:

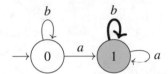

By looking at this computation and a few others, you can see that the automaton will be in state 1 iff it has read at least one *a*. That is, state 1 corresponds to it "remembering" that it has read at least one *a*, while state 0 corresponds to it "remembering" that it hasn't read *a* yet. This automaton doesn't keep track of how many times it has read *b*, or how many times it has read *a* after reading it once.

Let's look at a slightly more interesting example:

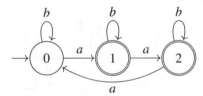

This automaton has two **accepting states**, state 1 and state 2, which are drawn using double circles. If computation ends in one of these states, the input string is **accepted**; otherwise it is **rejected**.

Doing a computation with the input *babaaba* leads to the following series of moves, starting in state 0:

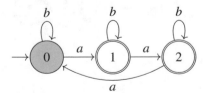

In some books, accepting states are called *final states*. That terminology is a little misleading because the computation can continue after a final (i.e. accepting) state is reached.

Read *b*, with remaining input *abaaba*:

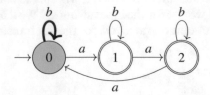

Read *a*, with remaining input *baaba*:

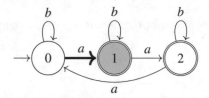

Read *b*, with remaining input *aaba*:

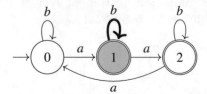

Read *a*, with remaining input *aba*:

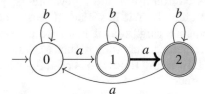

Read *a*, with remaining input *ba*:

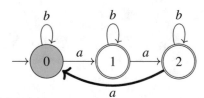

Read *b*, with remaining input *a*:

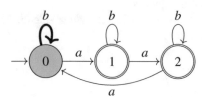

27

Read *a*, with no input remaining:

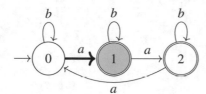

The input *babaaba* is accepted because state 1 is an accepting state.

Looking at this example input and others, and considering the structure of the automaton, you can see that:

- State 0 corresponds to it having read *a* some number of times that is divisible by 3;
- State 1 corresponds to it having read *a* some number of times that gives remainder 1 when divided by 3; and
- State 2 corresponds to it having read *a* some number of times that gives remainder 2 when divided by 3.

Again, this automaton doesn't keep track of how many times it has read *b*. Since states 1 and 2 are the accepting states, it will accept any string of *a*s and *b*s in which the number of *a*s is not divisible by 3.

These examples show that a finite automaton's states are its memory. But the memory isn't able to store numbers, or characters, or lists. It is limited to remembering certain things about the input that it has read. What those things are depends on the number of states and the arrangement of transitions.

A finite automaton is normally defined using a diagram showing the states and the transitions between states. The states are often labelled as in the examples above; the labels are just used to name the states, and the choice of names doesn't matter. Transitions are always labelled with the input symbol that triggers the transition. Moving along a transition is only possible if the next symbol in the input is the one labelling the arrow. Then moving to the target state consumes that symbol, leaving the remaining symbols to be read during later steps in the computation.

Each finite automaton partitions the set of all possible input strings into two subsets: the strings that it accepts and the strings that it rejects. The set of strings that a finite automaton M accepts is called the **language** accepted by M, written $L(M)$. The language accepted by a finite automaton is often an infinite set, even though its set of states and everything else about it is finite.

In some books, $L(M)$ is called the language *recognised* by M.

Some Examples

Here are some further simple examples, to help you develop your intuition.

The following finite automaton accepts strings composed of *a*s with no *b*s. That is, $L(M_1)$ is the set of all strings composed of *a*s and *b*s, including the empty string, that contain no *b*s.

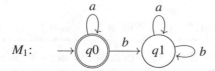

The following finite automaton accepts the string *abc* and nothing else. That is, $L(M_2) = \{abc\}$.

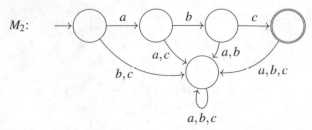

M_2:

Notice that some of the transitions in M_2 have labels like a, b. That's a convenient way of writing two transitions, one with label a and one with label b. The state at the bottom is a dead end, or **black hole**: once the automaton is in that state, there are no transitions leading to another state. The states in M_2 have no labels, and that's okay as long as you don't need to refer to them later.

The terminology "black hole" is by analogy with black holes in relativity theory, from which not even light can escape, see ▶ https://en.wikipedia.org/wiki/Black_hole.

This finite automaton accepts strings that start and end with an a separated by an even number (possibly 0) of bs:

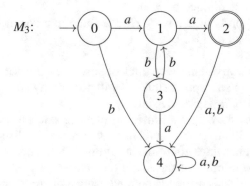

M_3:

This finite automaton accepts strings of binary digits that have *odd parity*, i.e. contain an odd number of 1s:

Parity checking is used to detect errors in data storage and transmission, see ▶ https://en.wikipedia.org/wiki/Parity_bit. When data is stored or transmitted, a *parity bit* is added and set to 0 or 1 in order to make the parity of the data plus parity bit odd or even, according to the parity scheme being used. If the parity is later found to be wrong, then one bit—actually, an odd number of bits—has been reversed.

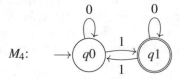

M_4:

This finite automaton accepts strings that alternate between a and b:

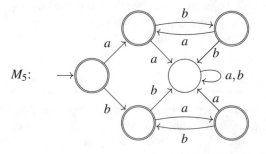

M_5:

Deterministic Finite Automata

All of the examples of finite automata above are **deterministic**, meaning that movement between states during computation is completely fixed ("determined") by the input string, with no choices between alternatives. A finite automaton is deterministic if:

- it has one start state;
- it has exactly one transition from each state for each input symbol; and
- it has no ε-transitions. (These are transitions that consume no input; they will be introduced in Chap. 29, when we move on to non-deterministic finite automata, so don't worry about them now.)

The Greek letter ε is pronounced "EPsilon" in American English and "epSEYElon" in British English. ε stands for the empty string.

Note that multiple accepting states *are* allowed.

A finite automaton that is deterministic is called a **deterministic finite automaton** or **DFA**.

Because of the requirement that there is exactly one transition from each state for each input symbol, we have to be clear about which input symbols are allowed by each DFA. Most of the DFAs above allow the input symbols a and b, with M_2 also allowing c, while the parity checking DFA M_4 allows the input symbols 0 and 1. The set of input symbols that are allowed by a DFA is called its **alphabet**, and it is required to be finite. If Σ is the alphabet of a DFA M, then $L(M)$ is a subset of Σ^*, the set of strings (including the empty string) over Σ.

The upper case Greek letter Σ is pronounced "Sigma". For any set Σ, Σ^* (pronounced "Sigma-star") is the set of strings built from the elements of Σ, including the empty string.

Some More Examples

The following examples of DFAs are a little more interesting than the ones above. Try drawing the diagram for each of them yourself after reading the description but before looking at the answer! If you need a hint, peek at the number of states and then try filling in the transitions yourself.

The following DFA, whose alphabet is $\Sigma = \{a, b\}$, accepts a string iff it contains aaa. That is, $L(M_6) = \{xaaay \in \Sigma^* \mid x, y \in \Sigma^*\}$:

The notation for strings in Mathematics is simpler than it is in Haskell: we write $xaaay$ for the string x followed by three occurrences of the symbol a, followed by the string y. In Haskell, this would be x ++ "aaa" ++ y. Later, we'll write a^n for n repetitions of a; in Haskell, this would be `replicate` n `'a'`.

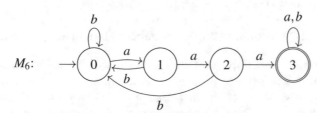

The following DFA, whose alphabet is $\Sigma = \{a, b\}$, accepts a string iff it ends with aba. That is, $L(M_7) = \{xaba \in \Sigma^* \mid x \in \Sigma^*\}$:

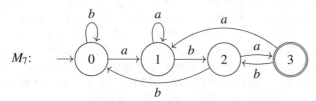

The following DFA, whose alphabet is $\Sigma = \{a, b\}$, accepts a string iff it either begins or ends with ab, so $L(M_8) = \{abx \in \Sigma^* \mid x \in \Sigma^*\} \cup \{xab \in \Sigma^* \mid x \in \Sigma^*\}$:

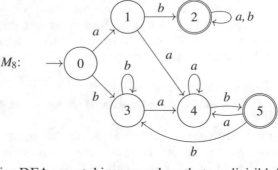

The following DFA accepts binary numbers that are divisible by 3, including the empty string (representing the number 0):

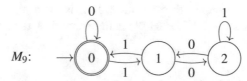

To understand M_9, think about what the states represent: if M_9 is in state 0, what is it "remembering" about the input so far? And what is the relationship between a binary number b and the binary numbers $b0$ and $b1$?

How to Build a DFA

Suppose you are asked to give a DFA that accepts a certain language. How should you proceed? We'll use the following example:

$L = \{x \in \{0, 1\}^* \mid x \text{ contains an even number of 0s and an odd number of 1s}\}$

The first and most important step is to think about what features of the input the DFA will need to remember. Being in state q after having read x versus being in state r after having read y is a DFA's way of remembering something that distinguishes x and y. What are the things that distinguish one string from another that are important?

In our example, there are two things that are important: the number of 0s in the input and the number of 1s. The order of the input symbols is unimportant; we're just interested in how many of each kind of symbol is present. In fact, the exact *numbers* of 0s and 1s don't matter, just whether they are even or odd.

To remember whether the DFA has read an even number or an odd number of 0s, it needs two states: one for even numbers of 0s, and the other for odd numbers of 0s. Likewise for even and odd numbers of 1s. To remember both of these features *simultaneously*, it needs four states:

- has read an even number of 0s and an even number of 1s;
- has read an even number of 0s and an odd number of 1s;
- has read an odd number of 0s and an even number of 1s; and
- has read an odd number of 0s and an odd number of 1s.

Those will be the states of the DFA. The start state is the first one since before reading any input the DFA has read an even number (namely 0) of 0s and 1s.

The second state will be the only accepting state, according to the definition of the language L. Let's start by writing down those states:

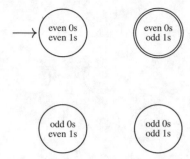

Now that you have the states, you can fill in the transitions, proceeding systematically. Suppose the DFA is in the start state, meaning that it has read an even number of 0s and an even number of 0s. What should happen if it reads another 0? Correct! It has now read an *odd* number of 0s and an even number of 1s, so it should move to that state. And if it instead reads another 1, that makes an even number of 0s and an odd number of *1s*. That gives:

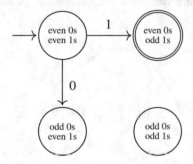

Continuing in the same way with the other three states gives the final result:

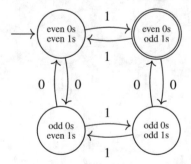

An alternative (BAD!) way to proceed is to start drawing immediately, without thinking about how many states you need or what they represent. Draw the start state, and the transitions from there to other states:

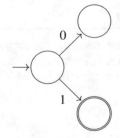

Then draw some more transitions to other states:

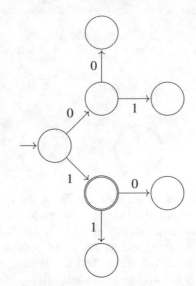

Stop when you're so confused that you can't think of any more transitions to add:

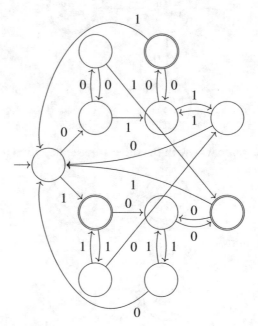

It's possible to produce a correct DFA by proceeding this way—the one above is an example—but it's error-prone and the result is likely to be larger than necessary and hard to understand.

Black Hole Convention

When building a DFA, it's often necessary to include non-accepting black hole states from which there is no way of reaching an accepting state. An example is the state at the bottom of the diagram for M_2, which accepts only the string *abc*:

M_2: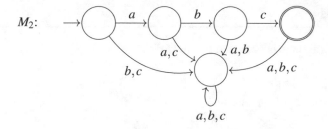

Under the **black hole convention**, such states can be omitted to simplify the diagram. This leads to a diagram in which there is *at most one* transition from each state for each symbol of the alphabet, rather than *exactly* one:

M_2:

In a DFA, one transition from every state is required for every symbol of the alphabet. The transitions that are absent when a black hole is omitted can easily be recreated by adding an additional state with all of the missing transitions leading to that state.

The DFAs M_3, which accepts strings that start and end with an a separated by an even number of bs, and M_5, which accepts strings that alternate between a and b, are also less cluttered and easier to understand when the black hole is omitted:

M_3: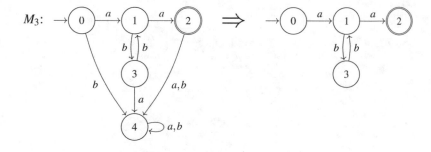

M_5: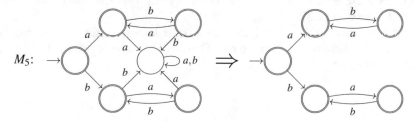

Except where otherwise specified, we're going to adopt the black hole convention when drawing DFA diagrams to make them simpler.

Exercises

1. What language is accepted by this DFA?

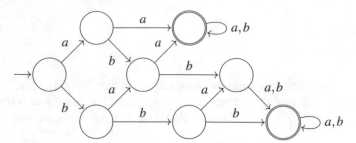

2. Draw a DFA that accepts the set of strings over the alphabet $\{a, b, c\}$ that start with a and end with either a or c.
3. Draw a DFA that accepts the set of binary numbers that are divisible by 4, including the empty string.
4. Draw a DFA that accepts the set of binary numbers that are divisible by 2, *not* including the empty string.
5. Draw a DFA with alphabet $\Sigma = \{0, 1\}$ that implements a combination lock with combination 1101. That is, it should accept the language $\{x1101 \in \Sigma^* \mid x \in \Sigma^*\}$.

There are several algorithms for transforming a DFA into a minimal DFA that accepts the same language, see ▶ https://en.wikipedia.org/wiki/DFA_minimization.

6. Give a DFA with fewer states that accepts the same language as the following DFA:

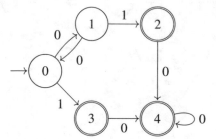

7. Draw a DFA that accepts the set of binary numbers that are divisible by 3 *and* have odd parity.
8. Explain how a DFA can accept an infinite language. Can you easily tell whether the language accepted by a DFA will be finite or infinite by looking at its structure?

Deterministic Finite Automata

Contents

© The Author(s), under exclusive license to Springer Nature Switzerland AG 2021
D. Sannella et al., *Introduction to Computation*, Undergraduate Topics
in Computer Science, https://doi.org/10.1007/978-3-030-76908-6_28

Diagrams and Greek Letters

The examples of deterministic finite automata in the last chapter were given in the form of simple diagrams, with states drawn as circles and transitions drawn as arrows between the circles. With a diagram, it's pretty easy to try out examples of input strings to see what happens—provided the diagram isn't so complicated that it resembles a bowl of spaghetti. We're now going to look at a mathematical definition of DFAs that is equivalent to giving a diagram, but phrased in terms of symbols including scary Greek letters. But why bother?

Diagrams, like the ones we use for DFAs, are excellent as a basis for intuition and understanding. However, they are a little awkward as a starting point for mathematical proofs. It's particularly hard if you want to prove that something is impossible: then you need to explain why a diagram of a certain kind can't exist.

On the other hand, a diagram is sometimes a good way to give *the intuition underlying* a proof, whether the proof is about DFAs or about something else.

In Chap. 32, we will prove that DFAs have limited expressibility, meaning that it's impossible to construct a DFA that will accept certain languages. Before that, we'll show how DFAs can be modified or combined to produce new DFAs that accept different languages. Explaining how these constructions work using diagrams is possible. But in order to be really clear, the actual constructions need to be defined in mathematical terms.

Deterministic Finite Automata, Formally

First, we'll say what a finite automaton is, and then we'll say what makes a finite automaton deterministic.

Definition. A **finite automaton** $M = (Q, \Sigma, \delta, S, F)$ has five components:

The Greek letter δ, which is the lower case version of Δ, is pronounced "delta".

- a finite set Q of **states**;
- a finite **alphabet** Σ of input symbols;
- a **transition relation** $\delta \subseteq Q \times \Sigma \times Q$;
- a set $S \subseteq Q$ of **start states**; and
- a set $F \subseteq Q$ of **accepting states**.

In most books and articles, a finite automaton is defined as having a single start state. Our definition is a useful generalisation.

The states $Q - F$ that are not accepting states are called **rejecting states**.

This definition fits perfectly with the way that we defined DFAs as diagrams. The states in Q were the circles in the diagram, and we can only draw diagrams with a finite number of circles. The symbols in the alphabet Σ were the labels on the arrows in the diagram. The start states S were the ones that had arrows coming into them with no label, and the accepting states F were the ones that were drawn using double circles. The sets S and F are always finite because they are subsets of Q, which is finite.

The transition relation $\delta \subseteq Q \times \Sigma \times Q$, which is a set of triples, captures the arrows in the DFA diagram and their labels. Each triple (q, a, q') in δ, also written $q \xrightarrow{a} q'$, corresponds to an arrow going from state q to state q' that is labelled with the symbol a. The relation δ is finite because Q and Σ are both finite.

Definition. A **deterministic finite automaton (DFA)** is a finite automaton $M = (Q, \Sigma, \delta, S, F)$ such that:

- S contains exactly one state, $S = \{q_0\}$; and

- δ is a total function $\delta : Q \times \Sigma \to Q$, which means that for each state $q \in Q$ and each input symbol $a \in \Sigma$, there is exactly one state $q' \in Q$ such that $q \xrightarrow{a} q' \in \delta$.

Again, this definition matches the limitations that were imposed on DFAs drawn as diagrams on page 287. The first requirement says that the diagram contains only one start state, and the second requirement says that there is one and only one transition from each state for each input symbol. It's convenient to give the transition function δ in the form of a table, with rows corresponding to states and columns corresponding to input symbols. Then the second requirement says that there will be exactly one state in each position of the table.

The following DFA from page 285, which accepts strings composed of *as* with no *bs*:

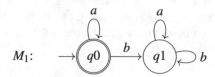

M_1:

corresponds to $(Q, \Sigma, \delta, S, F)$ where

$$Q = \{q0, q1\}$$
$$\Sigma = \{a, b\}$$
$$S = \{q_0\} \text{ where } q_0 = q0$$
$$F = \{q0\}$$

and δ is given by the following state transition table:

δ	a	b
$q0$	$q0$	$q1$
$q1$	$q1$	$q1$

Notice that any black hole states—in this example the state q1—whether they are omitted from the diagram according to the black hole convention or not, need to be included in Q and δ.

Here's a more complicated example from page 287:

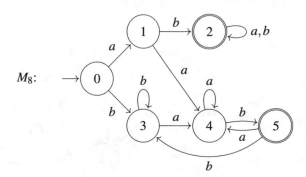

M_8:

This corresponds to $(Q, \Sigma, \delta, S, F)$ where

$$Q = \{0, 1, 2, 3, 4, 5\}$$
$$\Sigma = \{a, b\}$$
$$S = \{q_0\} \text{ where } q_0 = 0$$
$$F = \{2, 5\}$$

Recall from Chap. 1: a function $\delta : Q \times \Sigma \to Q$ is a relation $\delta \subseteq Q \times \Sigma \times Q$ such that for each $q \in Q$ and $a \in \Sigma$, if $(q, a, q'), (q, a, q'') \in \delta$ then $q' = q''$. It is a *total* function if for each $q \in Q$ and $a \in \Sigma$, there is some q' such that $(q, a, q') \in \delta$.

There is nothing about ε-transitions, which were excluded from DFAs on page 287, because they aren't allowed by the definition of finite automaton. See Chap. 29.

and δ is given by the following state transition table:

δ	a	b
0	1	3
1	4	2
2	2	2
3	4	3
4	4	5
5	4	3

The transition function δ says what happens when a symbol a is read: if we are in state q and $q \xrightarrow{a} q' \in \delta$, then the input symbol a is consumed and we move to state q'. For example, if M_8 is in state 1 and the next symbol is b, it moves to state 2 because $1 \xrightarrow{b} 2 \in \delta$, which corresponds to row 1 and column b of the state transition table above containing the entry 2. What happens when M reads a *string* of symbols, and what it means for M to accept a string, is explained by the next definition.

Definition. Let $M = (Q, \Sigma, \delta, \{q_0\}, F)$ be a DFA. The transition function $\delta : Q \times \Sigma \to Q$ is extended to a function $\delta^* : Q \times \Sigma^* \to Q$ on strings over Σ, including the empty string ε, as follows:

- for any state $q \in Q$, $\delta^*(q, \varepsilon) = q$;
- for any state $q \in Q$, symbol $x \in \Sigma$ and string $s \in \Sigma^*$, $\delta^*(q, xs) = \delta^*(\delta(q, x), s)$.

A string $s \in \Sigma^*$ is **accepted** by M iff $\delta^*(q_0, s) \in F$.

The definition of the extended transition function δ^* is by recursion. The first case says what it does for the empty string ε. The second case says what it does for a string xs consisting of a symbol x followed by a string s, in terms of what it does for s: first move to the state indicated by δ for the symbol x, and then proceed from there according to δ^* for s.

You're used to recursive definitions in Haskell, so this one should be easy to understand, but let's look at the example of M_8 accepting the string *aaab* anyway, one step at a time:

$\delta^*(0, aaab)$
$= \delta^*(\delta(0, a), aab)$
$= \delta^*(1, aab)$
$= \delta^*(\delta(1, a), ab)$
$= \delta^*(4, ab)$
$= \delta^*(\delta(4, a), b)$
$= \delta^*(4, b)$
$= \delta^*(\delta(4, b), \varepsilon)$
$= \delta^*(5, \varepsilon)$
$= 5$

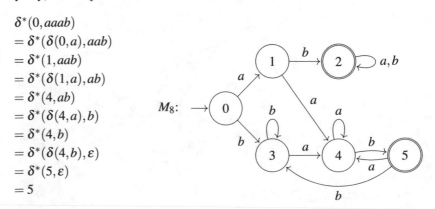

The string $aaab \in \Sigma^*$ is accepted because $\delta^*(0, aaab) = 5 \in F$.

It's interesting to look at the sequence of states that are visited by a DFA during its computation.

Definition. Let $M = (Q, \Sigma, \delta, \{q_0\}, F)$ be a DFA, and let $s = a_1 a_2 \cdots a_n$ be a string in Σ^*, with $a_i \in \Sigma$ for each $1 \le i \le n$. The **trace** of M on s is the sequence $q_0 \, q_1 \, \cdots \, q_n$ of states, where $q_0 \xrightarrow{a_1} q_1 \xrightarrow{a_2} \cdots \xrightarrow{a_n} q_n \in \delta$.

For example, the trace of M_8 on *aaab* is 0 1 4 4 5, as can be seen from the computation above. The first state in the trace is the start state q_0 and the last state in the trace is the state that is reached when the input string is exhausted. The string is accepted iff that state is an accepting state.

Finally, there is the notion of a language, and what it means for a language to be regular:

Definition. Let Σ be an alphabet. Any set of strings $L \subseteq \Sigma^*$ is called a **language**. The **language accepted by** a DFA M with alphabet Σ is the set $L(M) \subseteq \Sigma^*$ of strings that are accepted by M. A language L is **regular** if there is some DFA M such that $L = L(M)$.

All of the languages that appeared in Chap. 27 were regular. As you will see in Chap. 32, there are languages that are not regular.

Complement DFA

Suppose that you have a DFA M that accepts a certain language $L(M)$. One of the things you can do with M is to modify it or combine it with other DFAs to make it accept a different language. By doing a series of steps like this, you can build a complicated DFA out of very simple DFAs.

Let's start with a simple modification: changing M so that it accepts the **complement** of $L(M)$ with respect to Σ^*. That is, it should accept the strings in Σ^* that are *not* accepted by M. That's easy: just change the accepting states of M so that they reject the input, and change the rejecting states of M so that they accept the input.

Definition. The **complement** of a DFA $M = (Q, \Sigma, \delta, S, F)$ is the DFA $\overline{M} = (Q, \Sigma, \delta, S, Q - F)$.

The set $Q - F$ is the complement of F with respect to Q, the set of all states of M.

Then $L(\overline{M})$ will be the complement of $L(M)$ with respect to Σ^*. This shows that the complement of a regular language is a regular language. In other words, the set of regular languages is **closed under complement**: taking the complement of anything in that set gives us something that is also in that set.

Let's try that out on an example, the DFA on page 288 that accepts binary numbers that are divisible by 3:

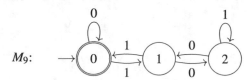

M_9:

The complement of M_9 is the following DFA, which accepts binary numbers that are *not* divisible by 3:

Don't forget any black hole states that are omitted from M's diagram because of the black hole convention: in $\overline{M}$, those states become accepting states.

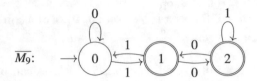

M_9 accepts the string 1100, which is the binary representation of 12. $\overline{M_9}$ rejects 1100, but it accepts 1011 (representing 11) and 1101 (representing 13).

Product DFA

Now suppose that you have two DFAs, M and M', accepting the languages $L(M)$ and $L(M')$ over the same alphabet Σ. How do you get a DFA that accepts their intersection, $L(M) \cap L(M')$?

The intuition behind the following **product construction** is simple: run M and M' *in parallel*, accepting a given string $s \in \Sigma^*$ iff it is accepted by *both M and M'*. If s is accepted by M then $s \in L(M)$, and if s is accepted by M' then $s \in L(M')$, so being accepted by both means that $s \in L(M) \cap L(M')$.

Turning that intuition into a DFA that accepts $L(M) \cap L(M')$ is a little more challenging. What you need is a single DFA that *simulates* the actions of M and M' running in parallel. The following definition does the job, showing that the set of regular languages is **closed under intersection**:

This construction can be extended to the case where the alphabets of M and M' are different, see Exercise 5.

Definition. The **product** of DFAs $M = (Q, \Sigma, \delta, \{q_0\}, F)$ and $M' = (Q', \Sigma, \delta', \{q_0'\}, F')$ is the DFA

$$M \times M' = (Q \times Q', \ \Sigma, \ \delta \times \delta', \ \{(q_0, q_0')\}, \ F \times F')$$

where the transition function $\delta \times \delta' : (Q \times Q') \times \Sigma \to (Q \times Q')$ is defined by

$$(\delta \times \delta')((q, q'), a) = (\delta(q, a), \delta'(q', a)).$$

That's why this is called the *product* construction, as in *Cartesian product*.

According to this definition, the states of $M \times M'$ are *pairs* consisting of a state from M and a state from M'. The product DFA being in state (q, q') corresponds to M being in state q and M' being in state q'. Then reading a symbol $a \in \Sigma$ causes a transition to $(\delta(q, a), \delta'(q', a))$, which corresponds to M moving to state $\delta(q, a)$ and M' moving to state $\delta'(q', a)$. Since the start state of $M \times M'$ is (q_0, q_0'), and its accepting states are pairs (q, q') such that $q \in F$ and $q' \in F'$, computations of $M \times M'$ correspond exactly to M and M' running in parallel, side by side, and accepting a string only when both M and M' would have accepted it.

Let's try that out on an example: take M to be the DFA M_4 on page 286 which accepts strings of binary digits that contain an odd number of 1s:

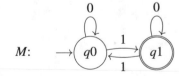

and take M' to be the following DFA which accepts strings that contain an even number of 0s:

28

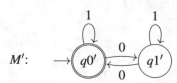

M':

The product $M \times M'$ will have four states: $(q0, q0')$, $(q0, q1')$, $(q1, q0')$, and $(q1, q1')$, with start state $(q0, q0')$ and one accepting state, $(q1, q0')$.

You can fill in the entries of the state transition table for $\delta \times \delta'$ using the definition $(\delta \times \delta')((q, q'), a) = (\delta(q, a), \delta'(q', a))$, giving

$\delta \times \delta'$	0	1
$(q0, q0')$	$(q0, q1')$	$(q1, q0')$
$(q0, q1')$	$(q0, q0')$	$(q1, q1')$
$(q1, q0')$	$(q1, q1')$	$(q0, q0')$
$(q1, q1')$	$(q1, q0')$	$(q0, q1')$

Alternatively, you can first concentrate on the part of each entry that comes from M:

$\delta \times \delta'$	0	1
$(q0, q0')$	$(q0,\ \ )$	$(q1,\ \ )$
$(q0, q1')$	$(q0,\ \ )$	$(q1,\ \ )$
$(q1, q0')$	$(q1,\ \ )$	$(q0,\ \ )$
$(q1, q1')$	$(q1,\ \ )$	$(q0,\ \ )$

and then fill in the part that comes from M':

$\delta \times \delta'$	0	1
$(q0, q0')$	$(q0, q1')$	$(q1, q0')$
$(q0, q1')$	$(q0, q0')$	$(q1, q1')$
$(q1, q0')$	$(q1, q1')$	$(q0, q0')$
$(q1, q1')$	$(q1, q0')$	$(q0, q1')$

Either way, this completes the construction of the DFA for $M \times M'$:

$M \times M'$:

Let's see what happens with the input string 1011. We'll look at what all three DFAs do with this input in order to show that $M \times M'$ is indeed simulating the actions of M and M' running in parallel. We start M in state $q0$, M' in state $q0'$, and $M \times M'$ in state $(q0, q0')$. The diagram for M' is rotated 90° to make it a little easier to see what's going on.

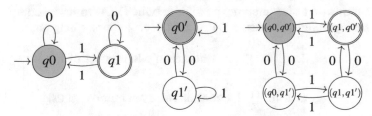

Read 1, with remaining input 011:

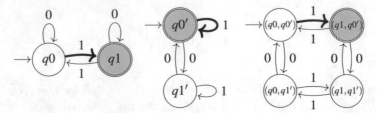

Read 0, with remaining input 11:

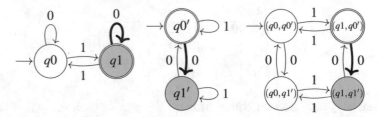

Read 1, with remaining input 1:

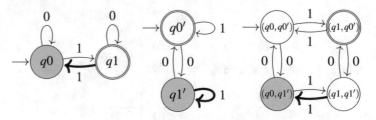

Read 1, with no input remaining:

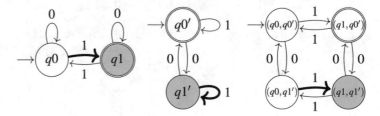

The input 1011 is accepted by M, because $q1 \in F$, but it is rejected by M', because $q1' \notin F'$, and by $M \times M'$, because $(q1, q1') \notin F \times F'$. That's correct, because $1011 \in L(M)$ but $1011 \notin L(M) \cap L(M')$.

But wait a minute, haven't we seen the DFA $M \times M'$ somewhere before? Yes! It's exactly the same as the DFA we built on page 289 to accept the language

$$\{x \in \{0, 1\}^* \mid x \text{ contains an even number of 0s and an odd number of 1s}\}$$

except that the names of the states are different.

This shows that it is sometimes possible to build DFAs to accept complicated languages by building separate DFAs that accept simpler languages—in this case,

$$\{x \in \{0, 1\}^* \mid x \text{ contains an odd number of 1s}\}$$

and

$$\{x \in \{0, 1\}^* \mid x \text{ contains an even number of 0s}\}$$

—and then using the product construction to combine them.

28

Sum DFA

Given DFAs M and M' accepting $L(M)$ and $L(M')$ over the same alphabet Σ, you now know how to produce a DFA that accepts their intersection $L(M) \cap L(M')$. But what if you want a DFA that accepts their *union* $L(M) \cup L(M')$?

The same basic idea as in the product construction—of producing a DFA that simulates M and M' running in parallel—works in this case as well. The only difference is in the set of accepting states. In the case of the intersection of languages, in order to accept a string we needed both M and M' to accept. Therefore, a state (q, q') was an accepting state of $M \times M'$ iff *both* q was an accepting state of M *and* q' was an accepting state of M'. In the case of the union, we need *either* q to be an accepting state of M *or* q' to be an accepting state of M'.

Here's the definition of the revised construction, which shows that the set of regular languages is **closed under union**:

Definition. The **sum** of two DFAs $M = (Q, \Sigma, \delta, \{q_0\}, F)$ and $M' = (Q', \Sigma, \delta', \{q_0'\}, F')$ is the DFA

The reason why this is called the sum will become clear in Chap. 31.

$$M + M' = (Q \times Q', \ \Sigma, \ \delta \times \delta', \ \{(q_0, q_0')\}, \ F + F')$$

where the transition function $\delta \times \delta' : (Q \times Q') \times \Sigma \to (Q \times Q')$ is defined by

$$(\delta \times \delta')((q, q'), a) = (\delta(q, a), \delta'(q', a))$$

and the set of accepting states is

$$F + F' = \{(q, q') \in Q \times Q' \mid q \in F \text{ or } q' \in F'\}.$$

Let's try that out on the same example as for the product construction, where M accepts strings of binary digits that contain an odd number of 1s:

$$M: \quad \to q0 \ \underset{1}{\overset{1}{\rightleftarrows}} \ q1 \quad (0 \text{ loops})$$

and M' accepts strings that contain an even number of 0s:

$$M': \quad \to q0' \ \underset{0}{\overset{0}{\rightleftarrows}} \ q1' \quad (1 \text{ loops})$$

The construction of $M + M'$ is exactly the same as for the product $M \times M'$, except that now $(q0, q1')$ is the only state that is not an accepting state:

$M + M':$
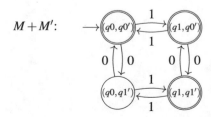

When given the input string 1011, this DFA will follow the same sequence of steps as $M \times M'$ did above, ending in the state $(q1, q1')$. But this time it will accept, since $(q1, q1') \in F + F'$. And that is the correct result, because 1011 contains an odd number of 1s.

Exercises

1. Give the DFA in the middle of page 290 as a 5-tuple.
2. Prove that for any string st, $\delta^*(q, st) = \delta^*(\delta^*(q, s), t)$. Then show that $\delta^*(q, sx) = \delta(\delta^*(q, s), x)$.
3. Prove that $L(\overline{M}) = \overline{L(M)}$ for any DFA M.
4. Construct the product of M_1 on page 285 (which accepts strings composed of as with no bs) and M_6 on page 287 (which accepts strings that contain aaa).
5. Extend the product construction to work with DFAs having different alphabets.
6. Construct the sum of M_4 on page 286 (which accepts strings of binary digits that have odd parity) and M_9 on page 288 (which accepts binary numbers that are divisible by 3).
7. Prove that $L(M \times M') = L(M) \cap L(M')$ and $L(M + M') = L(M) \cup L(M')$ for any DFAs M and M'.
8. Starting with M_4 on page 286 and M_9 on page 288, construct the DFA $\overline{\overline{M_4} \times \overline{M_9}}$ and compare it with the DFA $M_4 + M_9$ constructed in Exercise 6.
9. Prove that the set of regular languages is closed under set difference.

Non-deterministic Finite Automata

Contents

Choices, Choices

Should you take the lemon cake or the crème brûlée? Or, consider a more important decision: you're lost in the mountains in a blizzard without a GPS, and you're wondering whether to take the path to the left or to the right. It's sometimes hard to decide. What if you could decide not to decide: try *all* of the possible choices, and see what happens in each case?

The restrictions built into the definition of DFAs meant that no decisions are required: there is one state to start from, and there is always one transition available from every state for each input symbol. Relaxing these restrictions gives a **non-deterministic finite automaton (NFA)**, which can have more than one start state and where there can be any number of transitions—including zero—from any state for any input symbol.

One way of thinking of the operation of an NFA is that it offers many choices, and a series of decisions is required during computation. This leads to different possible computations, having potentially different outcomes: accept or reject. The NFA accepts an input if at least one series of decisions leads it to accept.

Equivalently, you can think of it trying all of the possible choices *simultaneously*, and accepting an input if *at least one* of them results in it finishing in an accepting state. As you will see, in such a computation it will always be in a *set* of states, with the available transitions from those states and the next input symbol dictating which set of states comes next.

Creating an NFA that accepts a given language is often much easier than creating a DFA that does the same thing. That makes sense: it's like being able to decide not to decide, trying all of the possible choices and taking the one that works out best. But surprisingly, it turns out that non-determinism actually gives no additional power: every NFA can be converted to a DFA that accepts the same language! In fact, you'll see that even adding a bit more non-determinism to NFAs in the form of ε-transitions, which give the ability to move freely between states without consuming input symbols, gives no additional power.

NFAs and the proof that every NFA can be converted to an equivalent DFA—and the Pumping Lemma in Chap. 32—are due to 1976 Turing Award winners Michael Rabin (1931–), see ▶ https://en.wikipedia.org/wiki/Michael_O._Rabin, and Dana Scott (1932–), see ▶ https://en.wikipedia.org/wiki/Dana_Scott.

Comparing a DFA with an NFA

Recall the following DFA from page 287 which accepts a string iff it contains *aaa*:

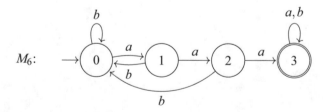

Here's an NFA that accepts the same language:

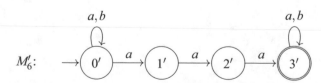

29

You can immediately see that M_6' is simpler than M_6, since it lacks the backwards transitions. And it's not a DFA because there are two transitions from state $0'$ for the symbol a.

Let's see what M_6 and M_6' do when they read the input *abaaaba*. We'll compare the two computations, side by side, to get a feel for how an NFA works, how it compares with a DFA that accepts the same language, and the idea behind the design of M_6'.

Each one starts in its start state:

There are no transitions in M_6' from states $1'$ and $2'$ for the symbol b. If M_6' were a DFA then they would be filled in according to the black hole convention, but for NFAs the same effect is achieved by just leaving them out.

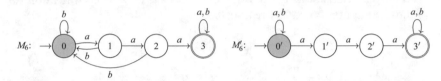

Read a, with remaining input *baaaba*. M_6 has no choice, moving to state 1. M_6' can stay in state $0'$ or move to $1'$; it tries both options at the same time:

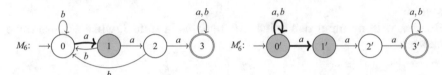

Read b, with remaining input *aaaba*. Again, M_6 has no choice, moving to state 0. From state $0'$, M_6' can stay in state $0'$. But from state $1'$ there is no transition for b so this computation path is aborted, shown below as the (non-existent) transition on b running into a brick wall. That leaves state $0'$ as the only option:

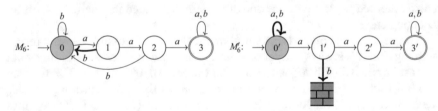

Read a, with remaining input *aaba*. M_6 moves to state 1 again. M_6' has two options, and again it tries both:

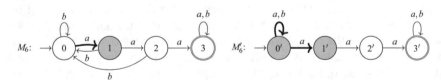

Read a, with remaining input *aba*. M_6 moves to state 2. M_6' now has three options—from state $0'$ it can either remain in $0'$ or move to $1'$, and from state $1'$ it can move to $2'$—so it tries all three:

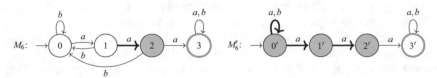

Read a, with remaining input *ba*. M_6 moves to state 3, and M_6' has four options:

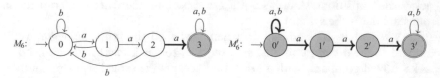

Read b, with remaining input a. M_6 remains in state 3. From states $0'$ and $3'$, M_6' has looping transitions for b. But from $1'$ and $2'$ there is no transition for b, so those computation paths are aborted:

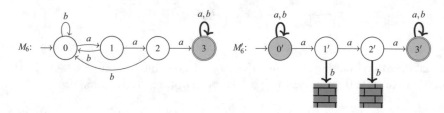

Read a, with no input remaining. M_6 remains in state 3, while M_6' has three choices:

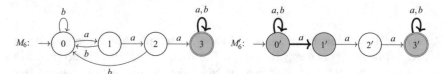

The input *abaaaba* is accepted by M_6 because it ended in an accepting state. It is also accepted by M_6', because one of the states it ended in—namely $3'$—is an accepting state.

You've just seen how an NFA simultaneously tries all possible computations, meaning that it is always in a set of states. Another way of understanding what has just happened with M_6' is to focus on the trace $0'$ $0'$ $0'$ $1'$ $2'$ $3'$ $3'$ $3'$ that led it to finish in state $3'$ and so to accept *abaaaba*. In this computation path, M_6' began by following the looping transition from $0'$ twice on the first two input symbols *ab* before following transitions on the subsequent input symbols *aaa* to $3'$, where it remained on the final two input symbols *ba*.

Focusing just on this trace, it feels like M_6' waited patiently in state $0'$ until the next three input symbols were *aaa*, at which point it cleverly took the transition to $1'$ in order to eventually reach $3'$ and acceptance. But no lookahead to see what's coming in the rest of the input was involved, and of course M_6' is not being patient or clever. By pursuing *all* possible choices in parallel, M_6' will accept if *any* series of choices leads to acceptance, and this trace is just the one that worked.

Nevertheless, there's no harm in thinking this way, especially when designing NFAs. In fact, that's the thinking behind the design of M_6': state $0'$ and its looping transition represents "wait patiently for *aaa*", with the series of *a*-transitions from $0'$ to $3'$ corresponding to "found *aaa*". Finally, state $3'$ and its looping transition represents "having found *aaa*, hang around in an accepting state until the input is finished". It doesn't matter *how* the NFA decides to make the right series of transitions, since it tries all of them. What is important is only whether the right series of transitions is *possible*, and—just as important—that there is no series of transitions that will lead it to accept an incorrect input.

29

Some More Examples

Here are some more DFAs from Chap. 27, along with a simpler NFA that accepts the same language.

The following DFA from page 287 accepts a string iff it ends with *aba*:

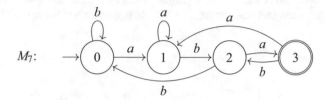

Here's an NFA that accepts the same language:

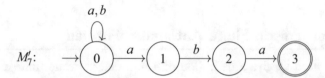

The backward transitions and one of the looping transitions in M_7 are required to keep track of how much of the string *aba* was contained in the last three input symbols read. In M_7', these are replaced by a looping transition on the start state for both input symbols to allow it to "wait patiently" until it's time to read the last three symbols of the input, accepting if they are *aba*.

The following DFA from page 287 accepts a string iff it either begins or ends with *ab*:

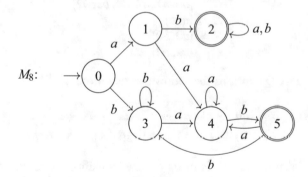

Here's an NFA that accepts the same language:

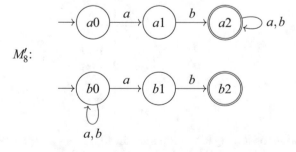

This is our first example of an NFA with more than one start state. In this case, there are no transitions between the "top half" and the "bottom half" of M_8', but not all NFAs with multiple start states have that property. The top half is responsible for accepting if the input starts with *ab*, and the bottom half is responsible for accepting if it ends with *ab*. If either one accepts, then the input is accepted.

Exercise 27.5 asked you to draw a DFA with alphabet $\Sigma = \{0, 1\}$ that implements a combination lock with combination 1101, accepting the language $\{x1101 \in \Sigma^* \mid x \in \Sigma^*\}$. The solution is a little complicated because of the need to keep track of how much of the combination has already been read, partially backtracking when it turned out that some of the input was not actually part of the combination after all. Here's a simple NFA that does the job, waiting until it's time to read the last four symbols of the input and then accepting if they are 1101:

Non-deterministic Finite Automata, Formally

There's no need to give a definition of a non-deterministic finite automaton because the definition of a finite automaton on page 294 as a five-tuple $(Q, \Sigma, \delta, S, F)$ is exactly what is required. Some NFAs are DFAs because they obey the restrictions in the definition of DFAs on page 294, having a single start state and a transition relation $\delta \subseteq Q \times \Sigma \times Q$ that is a total function $\delta : Q \times \Sigma \to Q$. In general—and in all of the examples of NFAs in this chapter—these restrictions are not satisfied.

That is, non-deterministic finite automata includes both finite automata that are deterministic and those that are not deterministic. Make sure that you understand this point! The terminology is unfortunately not very helpful.

For example, the NFA M_8' on page 307 corresponds to $(Q, \Sigma, \delta, S, F)$ where

$$Q = \{a0, a1, a2, b0, b1, b2\}$$
$$\Sigma = \{a, b\}$$
$$S = \{a0, b0\}$$
$$F = \{a2, b2\}$$

and $\delta \subseteq Q \times \Sigma \times Q$ contains the following transitions:

$$\delta = \{a0 \xrightarrow{a} a1,\ a1 \xrightarrow{b} a2,\ a2 \xrightarrow{a} a2,\ a2 \xrightarrow{b} a2,$$
$$b0 \xrightarrow{a} b0,\ b0 \xrightarrow{b} b0,\ b0 \xrightarrow{a} b1,\ b1 \xrightarrow{b} b2)\}$$

Another way of presenting δ is using a state transition table but with entries that are *sets* of states—namely, all of the ones to which there is a transition from the given state on the given symbol—rather than individual states:

δ	a	b
$a0$	$\{a1\}$	$\varnothing$
$a1$	$\varnothing$	$\{a2\}$
$a2$	$\{a2\}$	$\{a2\}$
$b0$	$\{b0, b1\}$	$\{b0\}$
$b1$	$\varnothing$	$\{b2\}$
$b2$	$\varnothing$	$\varnothing$

This corresponds to viewing it as a function $\delta : Q \times \Sigma \to \wp(Q)$, so $\delta(q, x) \subseteq Q$. If all of the entries in the table are singleton sets and there is a single start state, then the NFA is a DFA.

You saw above how computation works in NFAs for the case of M_6'. Here's the formal definition of what happens when an NFA reads a string of symbols, and what's required for the string to be accepted:

29

Definition. Let $M = (Q, \Sigma, \delta, S, F)$ be a (non-deterministic) finite automaton. The transition relation $\delta \subseteq Q \times \Sigma \times Q$ is extended to a function $\delta^* : Q \times \Sigma^* \to \wp(Q)$ on strings over Σ, including the empty string ε, as follows:

- for any state $q \in Q$, $\delta^*(q, \varepsilon) = \{q\}$;
- for any state $q \in Q$, symbol $x \in \Sigma$ and string $s \in \Sigma^*$,

$$\delta^*(q, xs) = \bigcup \{\delta^*(q', s) \mid q \xrightarrow{x} q' \in \delta\}.$$

A string $s \in \Sigma^*$ is **accepted** by M iff

$$\bigcup \{\delta^*(q, s) \mid q \in S\} \cap F \neq \varnothing.$$

The **language accepted by** M is the set $L(M) \subseteq \Sigma^*$ of strings that are accepted by M.

As already mentioned, any DFA M is also an NFA. For such an NFA, this definition of δ^* and acceptance is consistent with the definitions for DFAs on pages 296–297 in the sense that $\delta^*_{\text{NFA}}(q, s) = \{\delta^*_{\text{DFA}}(q, s)\}$ and $s \in L_{\text{NFA}}(M)$ iff $s \in L_{\text{DFA}}(M)$, where the subscripts NFA/DFA refer to the respective definitions of δ^* and $L(M)$.

The notation makes some parts of this definition a little hard to read, but the ideas are simple. First, $\delta^*(q, s)$ is the set of all states that can be reached from state q by following a sequence of transitions via the consecutive symbols in the string s. That is, if $s = a_1 a_2 \cdots a_n$ then $q_n \in \delta^*(q, s)$ iff $q \xrightarrow{a_1} q_1 \xrightarrow{a_2} \cdots \xrightarrow{a_n} q_n \in \delta$. The definition of δ^* is recursive, like the one for the case of DFAs on page 296. Then s is accepted if at least one of the states in $\delta^*(q, s)$ is an accepting state, for some start state $q \in S$.

Let's see how that works for M_8' when it reads the input string aba, starting from state $b0$:

$$
\begin{aligned}
& \delta^*(b0, aba) \\
={} & \bigcup\{\delta^*(q', ba) \mid b0 \xrightarrow{a} q' \in \delta\} \\
={} & \delta^*(b0, ba) && \cup\; \delta^*(b1, ba) \\
={} & \bigcup\{\delta^*(q', a) \mid b0 \xrightarrow{b} q' \in \delta\} && \cup \bigcup\{\delta^*(q', a) \mid b1 \xrightarrow{b} q' \in \delta\} \\
={} & \delta^*(b0, a) && \cup\; \delta^*(b2, a) \\
={} & \bigcup\{\delta^*(q', \varepsilon) \mid b0 \xrightarrow{a} q' \in \delta\} && \cup \bigcup\{\delta^*(q', \varepsilon) \mid b2 \xrightarrow{a} q' \in \delta\} \\
={} & (\delta^*(b0, \varepsilon) \cup \delta^*(b1, \varepsilon)) && \cup\; \varnothing \\
={} & \{b0, b1\}
\end{aligned}
$$

because $b0 \xrightarrow{a} b0$ and $b0 \xrightarrow{a} b1$

because $b0 \xrightarrow{b} b0$ and $b1 \xrightarrow{b} b2$

because $b0 \xrightarrow{a} b0$ and $b0 \xrightarrow{a} b1$ but $b2 \xrightarrow{a} \blacksquare$

A similar calculation for the other start state, $a0$, gives $\delta^*(a0, aba) = \{a2\}$. Then M_8' accepts $aba \in \Sigma^*$, meaning that $aba \in L(M_8')$, because

$$\bigcup\{\delta^*(q, aba) \mid q \in S\} \cap F = \{a2, b0, b1\} \cap \{a2, b2\} = \{a2\}$$

and $\{a2\} \neq \varnothing$.

NFAs in Haskell

It's straightforward to translate the formal definitions above into Haskell, using lists to represent sets. We'll assume that alphabets contain symbols that are values of Haskell's **Char** type, so input strings are values of type **String**. On the other hand, we'll define the type of transitions and other types that involve states to be polymorphic on the type of states, in order to accommodate different types of states. This will make it easier to define certain constructions on DFAs and NFAs in Haskell.

```haskell
type Sym = Char
type Trans q = (q, Sym, q)
data FA q = FA [q] [Sym] [Trans q] [q] [q] deriving Show
```

A value (q,x,q') of type `Trans` q represents a transition $q \xrightarrow{x} q'$. Then a value of type `FA` q represents an NFA, which is a DFA if it has a single start state and its transition relation is a total function:

```
isDFA :: Eq q => FA q -> Bool
isDFA (FA qs sigma delta ss fs) =
  length ss == 1
  && and [ or [ (q,x,q') `elem` delta | q' <- qs]
             | q <- qs, x <- sigma ]
  && and [ q' == q'' | (q,x,q') <- delta, q'' <- qs,
                          (q,x,q'') `elem` delta ]
```

The type of `isDFA` requires `Eq` q because testing that the transition relation is a function involves testing equality of states.

For example, the NFA M_8' on page 307 is represented as follows:

```
data State = A0 | A1 | A2 | B0 | B1 | B2 deriving (Eq,Show)
qs8' = [A0,A1,A2,B0,B1,B2]
sigma8' = ['a','b']
delta8' = [(A0,'a',A1), (A1,'b',A2), (A2,'a',A2), (A2,'b',A2),
             (B0,'a',B0), (B0,'b',B0), (B0,'a',B1), (B1,'b',B2)]
ss8' = [A0,B0]
fs8' = [A2,B2]

m8' :: FA State
m8' = FA qs8' sigma8' delta8' ss8' fs8'
```

and it isn't a DFA:

```
> isDFA m8'
False
```

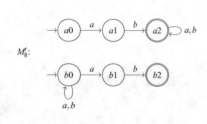

M_8':

The heart of the formalisation of NFAs is the extension of the transition relation δ to a function $\delta^* : Q \times \Sigma^* \to \wp(Q)$ on strings, which is defined by recursion. Here's the same definition in Haskell, where δ^* is written `star` δ:

```
star :: Eq q => [Trans q] -> q -> [Sym] -> [q]
star delta q "" = [q]
star delta q (x:xs) =
  nub (concat [ star delta q' xs
                  | (r,y,q') <- delta, r == q, x == y ])
```

In a DFA, the result of `star` will always be a singleton.

Based on δ^* is the definition of when a string is accepted by an NFA:

```
accept :: Eq q => FA q -> [Sym] -> Bool
accept (FA qs sigma delta ss fs) xs =
  concat [ star delta q xs | q <- ss ] `intersect` fs /= []
```

as `intersect` *bs* /= []
gives the same result as
(nub *as*) `intersect` *bs* /= [].

We can use these to check the computation for M_8' on page 309:

```
> star delta8' B0 "aba"
[B0,B1]
> accept m8' "aba"
True
```

The following function is similar to `star` but it also records the states that are encountered during computation.

```
traces :: Eq q => [Trans q] -> q -> [Sym] -> [[q]]
traces delta q "" = [[q]]
```

```
traces delta q (x:xs) =
  nub (concat [ map (q:) (traces delta q' xs)
              | (r,y,q') <- delta, r == q, x == y ])
```

With `traces`, you can see which computation traces led to which of the states in the result of δ^*:

```
> traces delta8' B0 "aba"
[[B0,B0,B0,B0],[B0,B0,B0,B1]]
```

which says that the traces were B0 B0 B0 B0 and B0 B0 B0 B1.

Constructions like the product of two DFAs (page 298) can be defined in Haskell:

```
productDFA :: (Eq a,Eq b) => FA a -> FA b -> FA (a,b)
productDFA fa@(FA qs sigma delta ss fs)
           fa'@(FA qs' sigma' delta' ss' fs')
  | not (isDFA fa) || not (isDFA fa')
    = error "not DFAs"
  | sigma/=sigma'
    = error "alphabets are different"
  | otherwise
    = FA (cpair qs qs') sigma dd [(q0,q0')] (cpair fs fs')
      where dd    = [ ((q1,q1'), x, (q2,q2'))
                    | (q1,x,q2) <- delta,
                      (q1',x',q2') <- delta', x==x' ]
            [q0]  = ss
            [q0'] = ss'
```

This definition uses an **as-pattern** `fa@(FA qs sigma delta ss fs)`, which is pronounced "fa as FA qs sigma delta ss fs". If the input matches `FA qs sigma delta ss fs` then the variables qs, sigma, delta, ss and fs are bound to the respective components of the input, while the variable fa is bound to the whole input.

where `cpair :: [a] -> [b] -> [(a,b)]` is the Cartesian product of two lists, defined on page 272. Notice the type of the result: the product of DFAs having states of type a and b has states of type (a,b).

Converting an NFA to a DFA

You've seen that—at least in some cases—it's easier to design an NFA than a DFA that accepts a given language. It may therefore come as quite a surprise to discover that every NFA M can be converted into a DFA $\widehat{M}$ that accepts the same language. This means that non-determinism doesn't actually add any expressive power: for any NFA M, $L(M) = L(\widehat{M})$ where $\widehat{M}$ is a DFA, so $L(M)$ is a regular language.

The reason why NFAs seem to be more expressive than DFAs is that converting an NFA with n states to an equivalent DFA yields one with 2^n states. Designing a DFA with 16, 32 or 64 states is a complicated and error-prone affair, while keeping track of 4, 5 or 6 states in the equivalent NFA is much easier.

The basic idea of the conversion is simple. First, the states of the DFA $\widehat{M}$ are *sets* of the states of the NFA M. Each state $\{q_1, \ldots, q_m\}$ of $\widehat{M}$ corresponds to M being in states q_1 and $\cdots$ and q_m simultaneously. You've seen how that works for the computation of M_6' on pages 305–306. Then the transitions of $\widehat{M}$ record how M moves from being in one set of states to being in another set of states when reading a symbol. For instance, the last move of M_6' in response to an input of a took it from being simultaneously in $0'$ and $3'$ to being simultaneously in $0'$, $1'$ and $3'$. It follows that there is a transition $\{0', 3'\} \xrightarrow{a} \{0', 1', 3'\}$ in the DFA $\widehat{M_6'}$.

That's why there are 2^n of them: remember that $|\wp(Q)| = 2^{|Q|}$.

The powerset construction is often called the **subset construction**, see ▶ https://en.wikipedia.org/wiki/Powerset_construction.

$\mathscr{Q} \in \hat{Q}$, which is Q in a script font, is a state of $\widehat{M}$. We'll call $\mathscr{Q}$ a **superstate** to emphasize that it's a set of states of M, that is, $\mathscr{Q} \subseteq Q$.

See Exercise 7 for a proof that $L(M) = L(\widehat{M})$.

Definition. Let $M = (Q, \Sigma, \delta, S, F)$ be a (non-deterministic) finite automaton. The **powerset construction** produces the DFA $\widehat{M} = (\hat{Q}, \Sigma, \hat{\delta}, \hat{S}, \hat{F})$ where the set of states is

$$\hat{Q} = \wp(Q),$$

the transition function $\hat{\delta} : \hat{Q} \times \Sigma \to \hat{Q}$ is defined by

$$\hat{\delta}(\mathscr{Q}, a) = \{q' \in Q \mid q \in \mathscr{Q}, q \xrightarrow{a} q' \in \delta\},$$

the set of start states is

$$\hat{S} = \{S\} \subseteq \hat{Q},$$

and the set of accepting states is

$$\hat{F} = \{\mathscr{Q} \in \hat{Q} \mid \mathscr{Q} \cap F \neq \varnothing\} \subseteq \hat{Q}.$$

$\widehat{M}$ will often contain superstates that are unused in the sense that they can't be reached by any series of transitions from its start superstate. Because the number of superstates is potentially very large, it's therefore convenient to construct $\widehat{M}$ "lazily", starting with its start superstate and adding new superstates only when they are required as targets of transitions. This process omits the unreachable superstates.

Let's see how the powerset construction works for the NFA M'_6 on page 304:

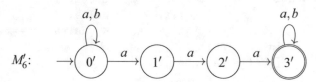

We'll build $\hat{\delta}$ one transition at a time, starting with the start superstate $\{0'\}$ which is the set of all of the start states of M'_6:

$\hat{\delta}$	a	b
$\{0'\}$		

There are transitions in M'_6 via a from $0'$ to both $0'$ and $1'$, so we add an entry for $\{0'\} \xrightarrow{a} \{0', 1'\}$ to the table. The superstate $\{0', 1'\}$ is new, so we add a new row for recording its transitions:

$\hat{\delta}$	a	b
$\{0'\}$	$\{0', 1'\}$	
$\{0', 1'\}$		

The only transition via b from $0'$ is back to $0'$, so we add an entry for $\{0'\} \xrightarrow{b} \{0'\}$ to the table. The destination superstate $\{0'\}$ is already there, so we don't need to add it:

$\hat{\delta}$	a	b
$\{0'\}$	$\{0', 1'\}$	$\{0'\}$
$\{0', 1'\}$		

29

Now we have to fill in the second row of the table. From $0'$, M'_6 can either remain in $0'$ or move to $1'$ when reading the symbol a, while from state $1'$ it can move to $2'$. So we add an entry for $\{0', 1'\} \xrightarrow{a} \{0', 1', 2'\}$ to the table, and a new row for $\{0', 1', 2'\}$:

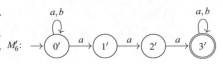

M'_6:

$\hat{\delta}$	a	b
$\{0'\}$	$\{0', 1'\}$	$\{0'\}$
$\{0', 1'\}$	$\{0', 1', 2'\}$	
$\{0', 1', 2'\}$		

Filling in the second entry in the second row: there is a looping b-transition from $0'$ but no b-transition from $1'$ (this corresponds to a "brick wall"). So we add an entry for $\{0', 1'\} \xrightarrow{b} \{0'\}$ to the table, and no new row because the superstate $\{0'\}$ is already there:

$\hat{\delta}$	a	b
$\{0'\}$	$\{0', 1'\}$	$\{0'\}$
$\{0', 1'\}$	$\{0', 1', 2'\}$	$\{0'\}$
$\{0', 1', 2'\}$		

Much of the reasoning involved in filling in this state transition table repeats the explanation of the computation of M'_6 on pages 305–306. For instance, the reasoning behind this entry repeats the second computation step there. Exceptions are for superstates and transitions that didn't arise during that computation.

There are a-transitions from $0'$ to $0'$ and $1'$, from $1'$ to $2'$, and from $2'$ to $3'$, which means that we need to add an entry for $\{0', 1', 2'\} \xrightarrow{a} \{0', 1', 2', 3'\}$, and a new row for $\{0', 1', 2', 3'\}$. There is a b-transition from $0'$ to $0'$ but no b-transitions from $1'$ or $2'$, so we add $\{0', 1', 2'\} \xrightarrow{b} \{0'\}$:

$\hat{\delta}$	a	b
$\{0'\}$	$\{0', 1'\}$	$\{0'\}$
$\{0', 1'\}$	$\{0', 1', 2'\}$	$\{0'\}$
$\{0', 1', 2'\}$	$\{0', 1', 2', 3'\}$	$\{0'\}$
$\{0', 1', 2', 3'\}$		

There are a-transitions from $0'$ to $0'$ and $1'$, from $1'$ to $2'$, from $2'$ to $3'$ and from $3'$ to $3'$, so we add $\{0', 1', 2', 3'\} \xrightarrow{a} \{0', 1', 2', 3'\}$. There is a b-transition from $0'$ to $0'$ and from $3'$ to $3'$ but no b-transitions from $1'$ or $2'$, so we add $\{0', 1', 2', 3'\} \xrightarrow{b} \{0', 3'\}$ and a new row for the superstate $\{0', 3'\}$:

$\hat{\delta}$	a	b
$\{0'\}$	$\{0', 1'\}$	$\{0'\}$
$\{0', 1'\}$	$\{0', 1', 2'\}$	$\{0'\}$
$\{0', 1', 2'\}$	$\{0', 1', 2', 3'\}$	$\{0'\}$
$\{0', 1', 2', 3'\}$	$\{0', 1', 2', 3'\}$	$\{0', 3'\}$
$\{0', 3'\}$		

There are a-transitions from $0'$ to $0'$ and $1'$ and from $3'$ to $3'$, so we add $\{0', 3'\} \xrightarrow{a} \{0', 1', 3'\}$ and a new row for the superstate $\{0', 1', 3'\}$. There is a b-transition from $0'$ to $0'$ and from $3'$ to $3'$, so we add $\{0', 3'\} \xrightarrow{b} \{0', 3'\}$:

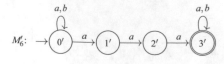

$\hat{\delta}$	a	b
$\{0'\}$	$\{0', 1'\}$	$\{0'\}$
$\{0', 1'\}$	$\{0', 1', 2'\}$	$\{0'\}$
$\{0', 1', 2'\}$	$\{0', 1', 2', 3'\}$	$\{0'\}$
$\{0', 1', 2', 3'\}$	$\{0', 1', 2', 3'\}$	$\{0', 3'\}$
$\{0', 3'\}$	$\{0', 1', 3'\}$	$\{0', 3'\}$
$\{0', 1', 3'\}$		

There are a-transitions from $0'$ to $0'$ and $1'$, from $1'$ to $2'$, and from $3'$ to $3'$, so we add $\{0', 1', 3'\} \overset{a}{\to} \{0', 1', 2', 3'\}$. There is a b-transition from $0'$ to $0'$ and from $3'$ to $3'$, but no b-transition from $1'$ so we add $\{0', 1', 3'\} \overset{b}{\to} \{0', 3'\}$. No new superstates have been added, so that completes the table:

$\hat{\delta}$	a	b
$\{0'\}$	$\{0', 1'\}$	$\{0'\}$
$\{0', 1'\}$	$\{0', 1', 2'\}$	$\{0'\}$
$\{0', 1', 2'\}$	$\{0', 1', 2', 3'\}$	$\{0'\}$
$\{0', 1', 2', 3'\}$	$\{0', 1', 2', 3'\}$	$\{0', 3'\}$
$\{0', 3'\}$	$\{0', 1', 3'\}$	$\{0', 3'\}$
$\{0', 1', 3'\}$	$\{0', 1', 2', 3'\}$	$\{0', 3'\}$

M_6' has 4 states, so there are $2^4 - 6 = 10$ superstates missing.

$\widehat{M_6'}$ will have the 6 superstates listed in the left-hand column of the table. The 10 remaining superstates can be omitted because they aren't reachable from the start superstate. The accepting superstates are the ones that have a non-empty intersection with the set $\{3'\}$ of accepting states in M_6', namely $\{0', 1', 2', 3'\}$, $\{0', 3'\}$ and $\{0', 1', 3'\}$. This gives the result:

To avoid clutter, the curly brackets are omitted from the superstate labels.

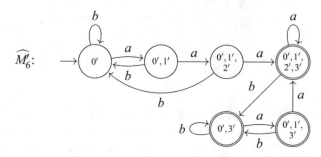

Now let's compare what M_6' and $\widehat{M_6'}$ do when they read the input $abaaaba$. For M_6', the computation is the one shown on pages 305–306. While watching the progress of the computations, notice that the set of states that M_6' is in always matches the superstate that $\widehat{M_6'}$ is in. This begins with their start states:

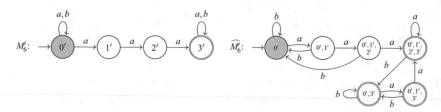

Read a, with remaining input $baaaba$. M_6' can stay in $0'$ or move to $1'$, while $\widehat{M_6'}$ moves to $\{0', 1'\}$:

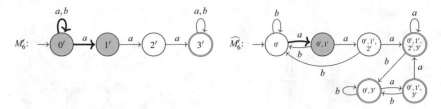

Read b, with remaining input *aaaba*. From $0'$, M_6' can stay in $0'$, but there is no b-transition from $1'$, so this computation path is aborted, leaving $0'$ as the only option. Meanwhile, $\widehat{M_6'}$ moves back to $\{0'\}$:

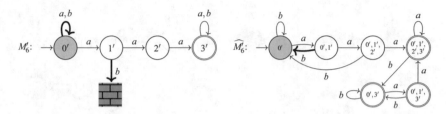

Read a, with remaining input *aaba*. M_6' has two options, and again it tries both. $\widehat{M_6'}$ moves to $\{0', 1'\}$ again:

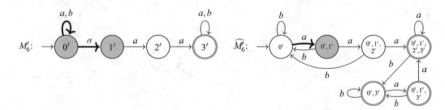

Read a, with remaining input *aba*. M_6' has three options: from $0'$ it can either remain in $0'$ or move to $1'$, and from $1'$ it can move to $2'$. $\widehat{M_6'}$ moves to $\{0', 1', 2'\}$:

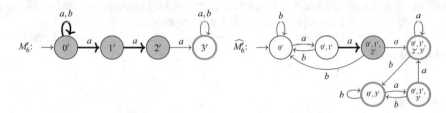

Read a, with remaining input *ba*. M_6' has four options, and $\widehat{M_6'}$ moves to $\{0', 1', 2', 3'\}$:

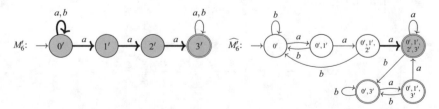

Read b, with remaining input *a*. M_6' has looping b-transitions from $0'$ and $3'$, but there are no b-transitions from $1'$ or $2'$. And $\widehat{M_6'}$ moves to $\{0', 3'\}$:

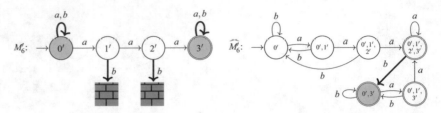

Read a, with no input remaining. M'_6 has three choices, while $\widehat{M'_6}$ moves to $\{0', 1', 3'\}$:

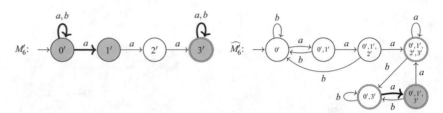

The input *abaaaba* is accepted by M'_6, because one of the states it ended in is an accepting state, and it's also accepted by $\widehat{M'_6}$.

It's interesting to compare $\widehat{M'_6}$ with the original DFA M_6:

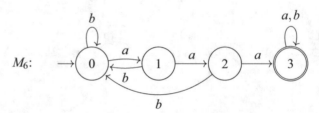

$\widehat{M'_6}$ is similar to M_6, and its computation above almost matches the computation of M_6 for the same input string on pages 305–306. The difference is that M_6 has one accepting state while $\widehat{M'_6}$ has three. Since all of the transitions from the accepting superstates in $\widehat{M'_6}$ go to accepting superstates, collapsing them into a single accepting superstate as in M_6 gives an equivalent DFA. They are distinguished in $\widehat{M'_6}$ because each one corresponds to a different combination of states of M'_6.

ε-NFAs

A useful extension of NFAs allows transitions between selected states "for free", without consuming input symbols. The extra transitions are labelled with ε, the empty string. An example is the following ε-NFA that accepts the language $\{a^m ab^n a^p \mid m, n, p \geq 0\}$:

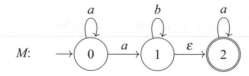

Computation with an ε-NFA is as with a normal NFA, except that when an ε-transition is encountered, there is a choice between making the transition or not. Let's look at what M does when it reads the input *aba*.

M starts in its start state 0:

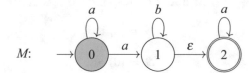

Read *a*, with remaining input *ba*. There is a looping *a*-transition on 0, and an *a*-transition from 0 to 1. From 1, M can then follow the ε-transition to 2 or stay in 1:

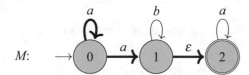

Read *b*, with remaining input *a*. There is a looping *b*-transition on 1, but no *b*-transitions from 0 or 2. However, there is an ε-transition from 1 to 2 that M can take (or not) after the looping *b*-transition:

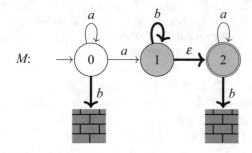

Read *a*, with no input remaining. There is a looping *a*-transition on 2, but no *a*-transitions from 1:

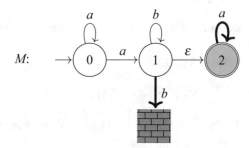

The input *aba* is accepted, because one of the states it ended in—there is only one, in this example—is an accepting state.

Formally, an ε-NFA is just like a normal NFA except that the transition relation allows transitions labelled with ε:

Definition. A non-deterministic finite automaton with ε-transitions (ε-NFA) $M = (Q, \Sigma, \delta, S, F)$ has five components:

- a finite set Q of **states**;
- a finite **alphabet** Σ of input symbols, such that $\varepsilon \notin \Sigma$;

The requirement that $\varepsilon \notin \Sigma$ and the "type" of δ are the only changes.

- a **transition relation** $\delta \subseteq Q \times (\Sigma \cup \{\varepsilon\}) \times Q$;
- a set $S \subseteq Q$ of **start states**; and
- a set $F \subseteq Q$ of **accepting states**.

For example, the ε-NFA M above corresponds to $(Q, \Sigma, \delta, S, F)$ where

It's important to understand that, although ε can be used as a transition label, it's not a symbol in Σ!

$$Q = \{0, 1, 2\}$$
$$\Sigma = \{a, b\}$$
$$S = \{0\}$$
$$F = \{2\}$$

and $\delta \subseteq Q \times (\Sigma \cup \{\varepsilon\}) \times Q$ contains the following transitions:

$$\delta = \{0 \xrightarrow{a} 0,\ 0 \xrightarrow{a} 1,\ 1 \xrightarrow{b} 1,\ 1 \xrightarrow{\varepsilon} 2,\ 2 \xrightarrow{a} 2)\}$$

When presented as a state transition table, δ needs an extra column for ε-transitions:

δ	a	b	ε
0	$\{0, 1\}$	$\varnothing$	$\varnothing$
1	$\varnothing$	$\{1\}$	$\{2\}$
2	$\{2\}$	$\varnothing$	$\varnothing$

Taking ε-transitions into account in the definition of computation of an ε-NFA uses the ε-closure $\langle\!\langle R \rangle\!\rangle$ of a set R of states, consisting of R together with all states that can be reached from R by following only ε-transitions. Then, as in the example of M above, a computation step via an input symbol x yields not just the set of states that are destinations of x-transitions from current states but the ε-closure of that set. In symbols:

Definition. Let $M = (Q, \Sigma, \delta, S, F)$ be an ε-NFA.

Note that this covers chains of ε-transitions, not just individual ε-transitions.

The ε-**closure** of a set $R \subseteq Q$ of states in M is the smallest set $\langle\!\langle R \rangle\!\rangle \supseteq R$ of states such that if $q \in \langle\!\langle R \rangle\!\rangle$ and $q \xrightarrow{\varepsilon} q'$ then $q' \in \langle\!\langle R \rangle\!\rangle$.

The transition relation $\delta \subseteq Q \times (\Sigma \cup \{\varepsilon\}) \times Q$ is extended to a function $\delta^* : Q \times \Sigma^* \to \wp(Q)$ on strings over Σ, including the empty string ε, as follows:

- for any state $q \in Q$, $\delta^*(q, \varepsilon) = \langle\!\langle \{q\} \rangle\!\rangle$;
- for any state $q \in Q$, symbol $x \in \Sigma$ and string $s \in \Sigma^*$,

According to this definition, $\delta^*(q, s) = \langle\!\langle \delta^*(q, s) \rangle\!\rangle$. (Why?)

$$\delta^*(q, xs) = \bigcup \{\delta^*(q'', s) \mid q \xrightarrow{x} q' \in \delta \text{ and } q'' \in \langle\!\langle \{q'\} \rangle\!\rangle\}.$$

A string $s \in \Sigma^*$ is **accepted** by M iff

$$\bigcup \{\delta^*(q, s) \mid q \in \langle\!\langle S \rangle\!\rangle\} \cap F \neq \varnothing.$$

The **language accepted by** M is the set $L(M) \subseteq \Sigma^*$ of strings that are accepted by M.

For example, applying the definition of δ^* to M on page 316 for the input string ab starting from state 0 repeats the first two steps of the computation above:

$\delta^*(0, ab)$

$= \bigcup\{\delta^*(q'', b) \mid 0 \xrightarrow{a} q' \in \delta \text{ and } q'' \in \langle\!\langle\{q'\}\rangle\!\rangle\}$

$= \bigcup\{\delta^*(q'', b) \mid q' \in \{0, 1\} \text{ and } q'' \in \langle\!\langle\{q'\}\rangle\!\rangle\}$ because $0 \xrightarrow{a} 0$ and $0 \xrightarrow{a} 1$

$= \bigcup\{\delta^*(q'', b) \mid q'' \in \langle\!\langle\{0\}\rangle\!\rangle\}$
$\qquad \cup \bigcup\{\delta^*(q'', b) \mid q'' \in \langle\!\langle\{1\}\rangle\!\rangle\}$

$= \delta^*(0, b) \;\cup \bigcup\{\delta^*(q'', b) \mid q'' \in \{1, 2\}\}$ because $\langle\!\langle\{0\}\rangle\!\rangle = \{0\}$ and

$= \delta^*(0, b) \;\cup (\delta^*(1, b) \;\cup\; \delta^*(2, b))$ $\langle\!\langle\{1\}\rangle\!\rangle = \{1, 2\}$

$= \bigcup\{\delta^*(q'', \varepsilon) \mid 0 \xrightarrow{b} q' \in \delta \text{ and } q'' \in \langle\!\langle\{q'\}\rangle\!\rangle\}$
$\qquad \cup (\bigcup\{\delta^*(q'', \varepsilon) \mid 1 \xrightarrow{b} q' \in \delta \text{ and } q'' \in \langle\!\langle\{q'\}\rangle\!\rangle\}$
$\qquad\qquad \cup\; \bigcup\{\delta^*(q'', \varepsilon) \mid 2 \xrightarrow{b} q' \in \delta \text{ and } q'' \in \langle\!\langle\{q'\}\rangle\!\rangle\})$

$= \varnothing \qquad \cup (\bigcup\{\delta^*(q'', \varepsilon) \mid q'' \in \langle\!\langle\{1\}\rangle\!\rangle\}$ because $0 \xrightarrow{b} \blacksquare$, $1 \xrightarrow{b} 1$ and $2 \xrightarrow{b} \blacksquare$
$\qquad\qquad\qquad \cup\;\; \varnothing)$

$= \qquad\qquad \bigcup\{\delta^*(q'', \varepsilon) \mid q'' \in \{1, 2\}\}$ because $\langle\!\langle\{1\}\rangle\!\rangle = \{1, 2\}$

$= \qquad\qquad \delta^*(1, \varepsilon) \cup \delta^*(2, \varepsilon)$

$= \qquad\qquad \langle\!\langle 1 \rangle\!\rangle \cup \langle\!\langle 2 \rangle\!\rangle$

$= \qquad\qquad \{1, 2\} \cup \{2\}$ because $\langle\!\langle\{1\}\rangle\!\rangle = \{1, 2\}$ and

$= \qquad\qquad \{1, 2\}$ $\langle\!\langle\{2\}\rangle\!\rangle = \{2\}$

M accepts ab because

$$\bigcup\{\delta^*(q, s) \mid q \in \langle\!\langle S \rangle\!\rangle\} \cap F = \delta^*(0, s) \cap F = \{1, 2\} \cap \{2\} = \{2\}$$

     because $\langle\!\langle S \rangle\!\rangle = \langle\!\langle\{0\}\rangle\!\rangle = \{0\}$

and $\{2\} \neq \varnothing$.

As with ordinary NFAs, there is a powerset construction for converting any ε-NFA M to an equivalent DFA $\widehat{M}$, meaning that ϵ-transitions provide no additional expressive power. The construction is very similar to the one for ordinary NFAs, with exactly the same intuition and the same potentially exponential increase in the number of states. The only changes that are required are the addition of ε-closure operations in a few places to take ε-transitions into account.

Definition. Let $M = (Q, \Sigma, \delta, S, F)$ be an ε-NFA. The **powerset construction** produces the DFA $\widehat{M} = (\hat{Q}, \Sigma, \hat{\delta}, \hat{S}, \hat{F})$ where the set of states is

$$\hat{Q} = \wp(Q),$$

the transition function $\hat{\delta} : \hat{Q} \times \Sigma \to \hat{Q}$ is defined by

$$\hat{\delta}(\mathscr{Q}, a) = \{q'' \in \langle\!\langle\{q'\}\rangle\!\rangle \mid q \in \mathscr{Q}, q \xrightarrow{a} q' \in \delta\},$$

the set of start states is

$$\hat{S} = \langle\!\langle\{S\}\rangle\!\rangle \subseteq \hat{Q},$$

and the set of accepting states is

$$\hat{F} = \{\mathscr{Q} \in \hat{Q} \mid \mathscr{Q} \cap F \neq \varnothing\} \subseteq \hat{Q}.$$

Obviously, any NFA M is an ε-NFA that happens to have no ε-transitions. The result of the powerset construction for M viewed as an NFA is the same as the result for M viewed as an ε-NFA, since $\langle\!\langle R \rangle\!\rangle = R$ for any $R \subseteq Q$ in the absence of ε-transitions.

Applying the powerset construction to M above—proceeding "lazily" from the start superstate $\langle\!\langle\{0\}\rangle\!\rangle = \{0\}$ to avoid generating unreachable superstates—yields the following state transition table:

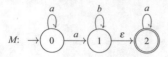

$\hat{\delta}$	a	b
$\{0\}$	$\{0, 1, 2\}$	$\varnothing$
$\{0, 1, 2\}$	$\{0, 1, 2\}$	$\{1, 2\}$
$\varnothing$	$\varnothing$	$\varnothing$
$\{1, 2\}$	$\{2\}$	$\{1, 2\}$
$\{2\}$	$\{2\}$	$\varnothing$

The accepting superstates are the ones that have a non-empty intersection with the set $\{2\}$ of accepting states in M, namely $\{0, 1, 2\}$, $\{1, 2\}$ and $\{2\}$. This gives the result:

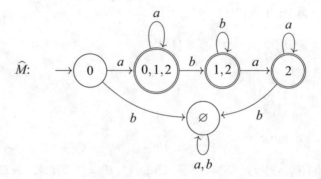

To avoid clutter, the curly brackets are omitted from the superstate labels. Applying the black hole convention to omit superstate $\varnothing$ would further simplify the diagram.

Concatenation of ε-NFAs

Suppose that you have two ϵ-NFAs, M and M', accepting languages $L(M)$ and $L(M')$. The availability of ε-transitions makes it easy to construct an ε-NFA MM' that accepts strings composed of a string from $L(M)$ followed by a string from $L(M')$:

$$L(MM') = \{xy \mid x \in L(M) \text{ and } y \in L(M')\}.$$

MM' is built from M and M' by connecting all of the accepting states of M via ε-transitions to all of the start states of M':

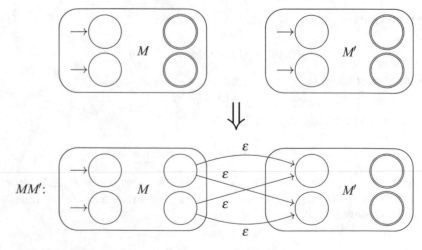

The definition of the construction is simpler under the assumption that the state sets of M and M' don't overlap.

29

Definition. Let $M = (Q, \Sigma, \delta, S, F)$ and $M' = (Q', \Sigma', \delta', S', F')$ be ε-NFAs such that $Q \cap Q' = \varnothing$. The **concatenation** of M and M' is the ε-NFA

$$MM' = (Q \cup Q', \ \Sigma \cup \Sigma', \ \delta\delta', \ S, \ F')$$

where the transition relation $\delta\delta' : (Q \cup Q') \times \Sigma \times (Q \cup Q')$ is defined by

$$\delta\delta' = \delta \cup \delta' \cup \{q \xrightarrow{\varepsilon} q' \mid q \in F \text{ and } q' \in S'\}.$$

For example, the concatenation of the ε-NFA M on page 316

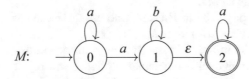

M:

and the NFA M_8' (which is an ε-NFA without ε-transitions) on page 307

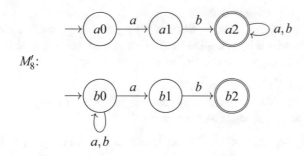

M_8':

is the ε-NFA MM_8':

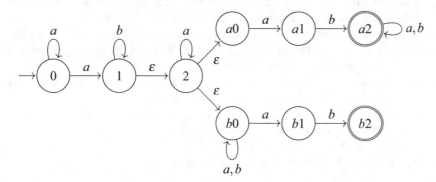

Concatenation of ε-NFAs, taken together with the powerset construction, shows that the set of regular languages is **closed under concatenation**.

Exercises

1. Draw an NFA with alphabet $\Sigma = \{0, 1, 2\}$ that implements a multi-user combination lock, with combinations 1101, 1220 and 2111 for different users. That is, it should accept the language

$$\{x1101 \in \Sigma^* \mid x \in \Sigma^*\} \cup \{x1220 \in \Sigma^* \mid x \in \Sigma^*\} \cup \{x2111 \in \Sigma^* \mid x \in \Sigma^*\}.$$

2. Draw an NFA with alphabet $\Sigma = \{a, b\}$ that accepts strings which contain aa followed later by either bbb or another aa. That is, it should accept the language

$$\{x\,aa\,y\,bbb\,z \in \Sigma^* \mid x, y, z \in \Sigma^*\} \cup \{x\,aa\,y\,aa\,z \in \Sigma^* \mid x, y, z \in \Sigma^*\}.$$

3. Let $M = (Q, \Sigma, \delta, S, F)$ be an NFA. Applying the complement construction for DFAs gives an NFA $\overline{M} = (Q, \Sigma, \delta, S, Q{-}F)$. Show that $L(\overline{M}) \neq \overline{L(M)}$ by giving a counterexample.

4. Define M_8 on page 295 in Haskell, and use `QuickCheck` to test that it accepts the same strings as M_8'.
   Define M_6 on page 287 in Haskell, and use `QuickCheck` with `productDFA` to test that $L(M_6 \times M_8) = L(M_6) \cap L(M_8)$.
   Hint: Using test inputs of type `String` with a conditional property that restricts to strings over a particular alphabet will reject almost all inputs. Instead, restrict to test inputs over the alphabet `['a','b']` by applying `QuickCheck` to a property of the form

   ```
   forAll (listOf (elements ['a','b'])) prop
   ```

   where `prop :: [Sym] -> Bool`.

5. The reverse of an NFA is obtained by reversing the direction of each transition and exchanging the sets of start and accepting states. Replace `undefined` to complete this definition:

   ```
   reverseNFA :: FA q -> FA q
   reverseNFA (FA qs sigma delta ss fs) =
     FA qs sigma undefined fs ss
   ```

   Define M_7' on page 307 in Haskell, and use `QuickCheck` to test that $s \in L(M_7')$ iff `reverse s` $\in$ L(`reverseNFA` M_7'). (See the hint in Exercise 4.) If m is a DFA, is `reverseNFA` m a DFA?

6. The black hole convention allows non-accepting black hole states and transitions to them to be omitted from DFA diagrams. Such diagrams are not DFAs—there are states for which transitions for some symbols of the alphabet are missing—until the omitted states and transitions are added. But they are NFAs. How can that fact be used to add the missing states and transitions, turning them into DFAs?

7. Let $M = (Q, \Sigma, \delta, S, F)$ be an NFA, with $\widehat{M} = (\hat{Q}, \Sigma, \hat{\delta}, \hat{S}, \hat{F})$ being the DFA that is produced by the powerset construction.
   Prove that for any $R \subseteq Q$ and any input string $s \in \Sigma^*$, $\hat{\delta}^*(R, s) = \bigcup\{\delta^*(q, s) \mid q \in R\}$. **Hint:** Use induction on s.
   Use that fact to prove that $L(M) = L(\widehat{M})$.

8. Give ε-NFAs M and M' that accept the languages

   $$L = \{a^n b^m \mid n, m \geq 0\} \quad \text{and} \quad L' = \{b^n a^m \mid n, m \geq 0\}$$

   Give their concatenation MM', and then apply the powerset construction to produce $\widehat{MM'}$.

9. Give an example of what can go wrong with concatenation MM' if the state sets for M and M' aren't disjoint. Give an improved definition.

29

Input/Output and Monads

Contents

Interacting with the Real World

All of the Haskell programs you've seen so far have been pure mathematical functions that consume input values and produce output values. Your interaction with these functions has been by applying them to inputs during an interactive Haskell session, and looking at the outputs that appear on your screen once computation is finished. But your past experience with computers and software, including phone apps, websites and computer games, has led you to expect that Haskell programs must have some way of interacting with the real world: reacting to keyboard or touchscreen input, displaying images, fetching information from the internet, playing sounds, accessing the current GPS position, etc. How does all of that fit in?

Now you're finally going to learn how to do these things. We'll concentrate on input from the keyboard and printing text on the screen, to keep things simple, but the same ideas apply to other modes of interaction. If you're familiar with other programming languages, you'll probably find that Haskell's approach is dramatically different from what you've seen before.

The problem with real-world interaction is that it seriously complicates the business of understanding and reasoning about programs. If an expression's value can depend on the results of interaction, it can change if the expression is evaluated at different times, or in different places, or when data in a file or on the internet changes. Then simple methods for figuring out what programs do, starting with basic rules like $exp - exp = 0$, are no longer valid. And even if running a test delivered a correct result this afternoon, who knows what it will do next Wednesday?

For this reason, Haskell carefully separates the domain of pure functional computation from the domain of real-world interaction. As you'll see, the vast majority of Haskell programming—including most programming related to interaction—is done using the programming concepts that you've become familiar with: recursion, higher-order functions, lists, etc., with types keeping everything organised. There is a narrow channel through which interaction with the real world takes place, in a way that avoids complicating the domain of pure functional computation.

When an computation changes something in the world, or reacts to a change in the world, it is said to have an **effect**—sometimes known as a *side effect*, because it happens "off to the side"—in addition to its normal result value, see ▶ https://en.wikipedia.org/wiki/Side_effect_(computer_science). Pure functional computations are effect-free.

Commands

We'll start with a new built-in type, `IO ()`, which you should think of as the type of **commands**. So why isn't it called `Command`? Partly because `IO ()` is a special case of something else, as you'll see shortly: that's the reason why it's `IO ()` and not just `IO`. And partly because `IO` is itself an instance of something more general, where the name says that these are commands that can do *input/output*.

One of the simplest commands is to print a character on the screen. This can be done using the Prelude function

```
putChar :: Char -> IO ()
```

So, `putChar 'a'` is a command to print the character `'a'`. Here's the first surprise: applying `putChar` to `'a'` doesn't actually print anything! It produces a command that—*when it is performed*—will print `'a'` on the screen. You'll learn how to cause commands to be performed soon.

Even more exciting than printing one character is printing two characters! You can produce a command to do that using the Prelude function

`>>` is pronounced "then".

```
(>>) :: IO () -> IO () -> IO ()
```

For instance,

```
putChar 'a' >> putChar 'b'
```

produces the command that, when it is performed, prints `'a'` followed by `'b'`.

Obviously, `>>` isn't commutative: printing `'a'` followed by `'b'` isn't the same as printing `'b'` followed by `'a'`. However, it is associative: printing `'a'` followed by `'b'`, and then printing `'c'`, is the same as printing `'a'`, followed by printing `'b'` and then `'c'`. Its identity is the command that, when it is performed, does nothing:

```
done :: IO ()
```

done isn't in the Haskell Prelude, but it can be defined as done = return (), using the function return below (page 327).

For instance, printing `'a'` and then doing nothing is the same as just printing `'a'`. So we have:

$$(m \gg n) \gg o = m \gg (n \gg o)$$
$$m \gg \text{done} = m = \text{done} \gg m$$

We can put these together to give a recursive definition of the Prelude function `putStr`. Given a string, `putStr` produces a command that, when it is performed, will print the string:

```
putStr :: String -> IO ()
putStr []     = done
putStr (x:xs) = putChar x >> putStr xs
```

So `putStr "ab"` produces `putChar 'a' >> (putChar 'b' >> done)`.

A variation on `putStr` is the Prelude function `putStrLn` which adds a newline character at the end:

```
putStrLn :: String -> IO ()
putStrLn cs = putStr cs >> putChar '\n'
```

`'\n'` is the newline character, pronounced "backslash-n" or "newline", not the beginning of a lambda expression.

Performing Commands

Now you know how to put together simple commands to make more complicated commands. But how is a command performed?

When you use Haskell interactively, typing an expression at Haskell's prompt causes the expression to be evaluated to produce a value. When that value is a command of type `IO ()`, the command is performed. What you see on the screen is then the output produced by the command:

```
> putStrLn "Hello World!"
Hello World!
```

An alternative is to create a file `Main.hs` containing a module `Main` that defines a variable `main` of type `IO ()`:

```
module Main where
  main :: IO ()
  main = putStrLn "Hello World!"
```

Running this using `runghc` performs the command that is bound to `main`:

```
$ runghc Main.hs
Hello World!
```

You should type `runghc Main.hs` into a command-line terminal window or "shell", not into a Haskell interactive session!

This way of performing commands may appear to be too limited to be useful. But the fact that you can put arbitrarily complex combinations of actions together to make a single command—which can include input as well as output commands, see below—means that it's all you need.

Why does Haskell make a distinction between *producing* a command and *performing* a command? One reason is that only the latter takes us outside the domain of pure functional computation. When defining functions like putStr and putStrLn, you don't have to bother about interaction with the messy real world. And this distinction is what allows the same simple reasoning methods to be applied in this case as everywhere else in Haskell.

Here's an example. As a general principle, repeated sub-expressions can always be factored out in Haskell, in the sense that the expression

```
(1 + 2) * (1 + 2)
```

is equivalent to the expression

```
x * x
  where x = 1 + 2
```

with both versions producing 9 :: Int. This principle also applies when the sub-expression in question is a command, so the expression

```
putStr "Boing" >> putStr "Boing"
```

is equivalent to the expressiom

```
m >> m
  where m = putStr "Boing"
```

with both versions producing a command that, when performed, prints "BoingBoing" on the screen.

Now, if applying putStr to a string would cause the string to be printed immediately, rather than producing a command that can be performed later, then the second version would only print "Boing" when **where** m = putStr "Boing" is evaluated, with m >> m combining the value produced by putStr, whatever it is, with itself. That would violate the general factorisability principle. Separating producing commands from performing them is the key to keeping things sane and simple.

Commands That Return a Value

Now you know how to do output to the screen. What about input from the keyboard? And how can input affect what is produced as output?

The first step is to look more carefully at the type of commands. The type IO () is actually for commands that might do some input/output *and then return a value of type* (). Recall that () is the type of 0-tuples, for which there is just one value, also written (). This value isn't very interesting because it carries no information, but that's just fine in this case because there's no need to return information from a command that just prints to the screen.

For a command that receives input from the keyboard, there *is* useful information to be returned: namely, the character(s) typed. So, input commands have types like IO Char. For example:

```
getChar :: IO Char
```

is a Prelude command that, when it is performed, reads a character from the keyboard. Performing the command getChar when the characters "zip" are typed yields the value 'z', leaving the remaining input "ip" to be read by subsequent commands. In general, IO *a* is the type of commands that return a value of type *a*. As you'll see, *a* isn't always Char or String, despite the fact that you can only type characters on the keyboard, because commands can process what is typed before returning a value.

The Prelude function

```
return :: a -> IO a
```

is like done :: IO () was for output commands: given a value, it produces the command that does nothing, and then returns the given value. For example, performing the command

```
return [] :: IO String
```

when the characters "zip" are typed yields the value [], leaving "zip" to be read by subsequent commands. The function return isn't very useful on its own but—like done—it will turn out to be an essential ingredient in functions that produce input commands.

The analogue of >> for commands that return values is the Prelude function

```
(>>=) :: IO a -> (a -> IO b) -> IO b
```

>>= is pronounced "bind".

which is used to put two commands into sequence, passing on the value produced by the first command for use by the second command.

The type of >>= takes some getting used to. Let's start with a simple example, the command

```
getChar >>= \x -> putChar (toUpper x)
```

which has type IO (). Performing this command when the characters "zip" are typed produces the output "Z" on the screen, and leaves "ip" to be read by subsequent commands. Here it is in action:

```
> getChar >>= \x -> putChar (toUpper x)
zip
Z>
```

The first parameter of >>= is getChar, which has type IO Char. Its second parameter is the function \x -> putChar (toUpper x). This has type Char -> IO () because toUpper has type Char -> Char and putChar has type Char -> IO (). So, according to the type of >>=, the whole command has type IO ().

When that command is performed and the characters "zip" are typed:

1. the character 'z' is read by getChar;
2. the function \x -> putChar (toUpper x) is applied to its result 'z', leading to putChar (toUpper 'z') printing Z;
3. finally, the command yields () as its result.

The characters "ip" haven't been read so they're still available to be read later.

In general: let m :: IO a be a command yielding a value of type a, and let k :: a -> IO b be a function taking a value of type a to a command yielding a value of type b. Then m >>= k :: IO b is the command that, when it is performed, does the following:

1. first, perform command m, yielding a value x of type a;
2. then, perform command k x, yielding a value y of type b;
3. finally, yield the value y.

A more interesting example of the use of >>= is the Prelude command getLine which reads a whole line of input from the keyboard. When getLine is performed, it reads the input using getChar until a newline character is encountered, returning a list of the characters read up to that point.

```
getLine :: IO String
getLine =
  getChar                        -- read a character
    >>= \x ->                    -- and call it x
          if x == '\n'           -- if it's newline, we're done
            then return ""       -- so return the empty string
          else                   -- otherwise
            getLine              -- read the rest of the line
              >>= \xs ->         -- and call it xs
                    return (x:xs) -- and then return x:xs
```

Here it is in action:

```
> getLine
zip
"zip"
```

Note the use of recursion in the definition of getLine. But there seems to be something wrong: since getLine isn't a function, there's no parameter that decreases in size each time. So what's going on?

The answer is that each time getLine is performed, it consumes keyboard input, and that causes the list of those characters typed on the keyboard that are yet to be read to decrease in size. Since that list isn't explicitly represented anywhere in the code, neither as a parameter of getLine nor otherwise, the decrease in size is also not explicit.

Here's another example: a command echo :: IO () that reads lines of input, echoing them back in upper case, until an empty line is entered.

```
echo :: IO ()
echo =
  getLine
    >>= \line ->
          if line == "" then
            return ()
          else
            putStrLn (map toUpper line)
              >> echo
```

For example:

```
> echo
Haskell is fun
HASKELL IS FUN
oh yes it is
OH YES IT IS

>
```

The command done that you saw earlier is a special case of return:

```
done :: IO ()
done = return ()
```

and the function >> is a special case of >>=:

```
(>>) :: IO () -> IO () -> IO ()
m >> n = m >>= \_ -> n
```

You saw earlier that >> is associative and done is its identity. Later, you'll see that >>= and return have properties that are like associativity and identity, but different. (Why? Because their types make things a little more complicated!)

do **Notation**

Commands written using >>= can be made easier to read and write by a change in notation. With the change, it becomes clearer that what we are doing is actually fairly simple!

Here's the Prelude command getLine again:

```
getLine :: IO String
getLine =
  getChar                      -- read a character
    >>= \x ->                  -- and call it x
        if x == '\n'           -- if it's newline, we're done
          then return ""       -- so return the empty string
        else                   -- otherwise
          getLine              -- read the rest of the line
            >>= \xs ->         -- and call it xs
                  return (x:xs) -- and then return x:xs
```

And now here's exactly the same thing, written in **do notation**:

```
getLine :: IO String
getLine = do {
          x <- getChar;
          if x == '\n' then
            return ""
          else do {
            xs <- getLine;
            return (x:xs)
              }
        }
```

This notation replaces the function >>= followed by an explicit lambda expression \x -> ... with something that looks a lot like a binding of x to the result of performing the command before the >>=, with semicolons used for sequencing.

Partly because of this, >>= is sometimes referred to as the "programmable semicolon".

All we've done here is to replace

cmd >>= \x -> *exp*

by

```
do {
  x <- cmd;
  exp
}
```

x <- *cmd*; ... is pronounced "*x* is drawn from *cmd* in ...". Or alternatively, "let *x* be the result of doing *cmd* in ...".

The braces and semicolon makes the version in do notation look a lot like programs for performing sequences of commands in more traditional languages.

The same idea works for commands that involve >> instead of >>=, but no value passing is involved so it's simpler. For instance, here's the Prelude function putStr again:

```
putStr :: String -> IO ()
putStr []     = done
putStr (x:xs) = putChar x >> putStr xs
```

The same thing, written in do notation, is

```
putStr :: String -> IO ()
putStr []     = done
putStr (x:xs) = do {
                putChar x;
                putStr xs
              }
```

Here, we've replaced

cmd1 >> *cmd2*

by

```
do {
  cmd1 ;
  cmd2
}
```

These notations can be mixed, with repetitions of `do` and nested braces omitted. For example,

```
do {
  x1 <- cmd1 ;
  x2 <- cmd2 ;
  cmd3 ;
  x4 <- cmd4 ;
  cmd5 ;
  cmd6
}
```

is equivalent to

```
cmd1 >>= \x1 ->
cmd2 >>= \x2 ->
cmd3 >>
cmd4 >>= \x4 ->
cmd5 >>
cmd6
```

Monads

The term "monad" was first used in philosophy, see ▶ https://en.wikipedia.org/wiki/Monad_(philosophy). That inspired its use in category theory, see ▶ https://en.wikipedia.org/wiki/Monad_(category_theory). The Haskell use of the term is an application of that concept, via its use in the theory of programming languages, see ▶ https://en.wikipedia.org/wiki/Monad_(functional_programming).

`IO` is an example of a **monad**. In order to understand monads, it's best to first understand **monoids**, a mathematical concept that is related but simpler.

A monoid is just a name for the situation where you have an associative binary operator $\odot$ and a value e that is the identity for $\odot$. That is:

$$(x \odot y) \odot z = x \odot (y \odot z)$$
$$x \odot e = x$$
$$e \odot x = x$$

You already know lots of examples of monoids: addition with identity 0; multiplication with identity 1; Haskell's disjunction (`||`) with identity `False`; conjunction (`&&`) with identity `True`; list append (`++`) with identity `[]`; and finally, `>>` with identity `done`.

A monad like `IO` has two functions, `>>=` and `return`. They satisfy properties that are generalised versions of the associative and identity laws:

$$(m \text{ >>= } \backslash x \text{ -> } n) \text{ >>= } \backslash y \text{ -> } o = m \text{ >>= } \backslash x \text{ -> } (n \text{ >>= } \backslash y \text{ -> } o)$$
$$\text{return } v \text{ >>= } \backslash x \text{ -> } m = m[x := v]$$
$$m \text{ >>= } \backslash x \text{ -> return } x = m$$

For instance, `(x * x)[x := 3]` is `3 * 3`.

where $m[x := v]$ is m with v substituted for all occurrences of x.

Remembering that >> is a special case of >>= and done is a special case of return, you can get a feeling for the relationship between these properties and the monoid laws for >> and done by putting them next to each other, like so:

$$(m \; \texttt{>>=} \; \backslash x \; \texttt{->} \; n) \; \texttt{>>=} \; \backslash y \; \texttt{->} \; o = m \; \texttt{>>=} \; \backslash x \; \texttt{->} \; (n \; \texttt{>>=} \; \backslash y \; \texttt{->} \; o)$$
$$(m \; \texttt{>>} \qquad n) \; \texttt{>>} \qquad o = m \; \texttt{>>} \qquad (n \; \texttt{>>} \qquad o)$$

$$\texttt{return} \; v \; \texttt{>>=} \; \backslash x \; \texttt{->} \; m = m[x := v]$$
$$\texttt{done} \qquad \texttt{>>} \qquad m = m$$

$$m \; \texttt{>>=} \; \backslash x \; \texttt{->} \; \texttt{return} \; x = m$$
$$m \; \texttt{>>} \qquad \texttt{done} \quad = m$$

What you can see from this is that the relationship is close, once you take into account the way that >>= handles value-passing between one command and the next. Note that in the first identity law, substitution needs to be used to take account of the way that the value passed by return v might affect m.

Haskell provides a built-in type class Monad, of which the type constructor IO is an instance:

```
class Monad m where
  return :: a -> m a
  (>>=)  :: m a -> (a -> m b) -> m b
  -- default
  (>>)   :: m a -> m b -> m b
  x >> y = x >>= \_ -> y
```

As with the **Functor** type class (page 254), instances of Monad are *type constructors*, not types. And as usual, there is no way for Haskell to enforce the monad laws, so it is important to check that they hold for each instance.

Using the Monad type class, you can write functions that work in an arbitrary monad. Such functions will have types of the form Monad m => .... Probably more useful for you at this stage is the fact that Haskell provides do notation for *any* instance of Monad, not just for IO, as a more convenient way of handling sequencing and value-passing than direct use of >>=.

As of Haskell version 7.10, the type class Monad is an extension of a type class called Applicative which is an extension of Functor, see ▶ https://wiki.haskell.org/Functor-Applicative-Monad_Proposal. Since IO and the other instances of Monad that you'll see here are also already defined in the Haskell Prelude as instances of those other type classes, the details are omitted.

Lists as a Monad

Since you've probably never heard of monads before, you may be astonished to learn that lots of things turn out to have the structure of a monad. One important example is lists:

```
instance Monad [] where
  return :: a -> [a]
  return x = [ x ]

  (>>=)  :: [a] -> (a -> [b]) -> [b]
  m >>= k = [ y | x <- m, y <- k x ]
```

Type signatures in **instance** declarations aren't allowed in standard Haskell, so this won't be accepted. To allow them, add {-# LANGUAGE InstanceSigs #-} at the top of your file.

To understand these definitions, it will help to forget about commands consuming input and producing output, and what return and >>= meant in that context. Think instead of lists as modelling **non-deterministic values**: values for which there are different possibilities. For example, the result of tossing a coin might be a value **Heads**, or a value **Tails**, but if the coin hasn't yet been tossed and you're interested in keeping track of the possible outcomes of future events then you might represent the result as a list of possible outcomes,

You've learned about non-determinism already in the context of NFAs. For more about non-determinism in programming, see ▶ https://en.wikipedia.org/wiki/Nondeterministic_algorithm.

[Heads,Tails]. And then look at the types of return and >>= for lists to figure out what they mean for non-deterministic computations.

The type of return is a -> [a]. It converts the value it is given into a non-deterministic value, listing that value as its only possibility.

The type of >>= is [a] -> (a -> [b]) -> [b]. This says that it takes as input a non-deterministic value m of type a, together with a non-deterministic function k from a to b, and produces a non-deterministic value of type b. The result collects all of the possible outcomes that can be obtained from k when applied to possible values of m. The definition above uses list comprehension but the same function can be defined using recursion:

```
[] >>= k     = []
(x:xs) >>= k = (k x) ++ (xs >>= k)
```

or using map to produce a list of lists of possible results, which is flattened to a list using concat:

```
m >>= k = concat (map k m)
```

Now you can write definitions using do notation that model non-deterministic computations. Here's an example:

```
pairs :: Int -> [(Int, Int)]
pairs n = do {
            i <- [1..n];
            j <- [(i+1)..n];
            return (i,j)
          }
```

Given an integer *n*, this produces the list of all of the possible pairs that can be obtained by first picking an integer *i* between 1 and *n*, then picking an integer *j* between *i* + 1 and *n*, and then returning the pair (*i*,*j*). Let's see if it works:

```
> pairs 4
[(1,2),(1,3),(1,4),(2,3),(2,4),(3,4)]
```

This example can also be written easily using list comprehension:

```
pairs :: Int -> [(Int, Int)]
pairs n = [ (i,j) | i <- [1..n], j <- [(i+1)..n] ]
```

In fact, do notation has a lot in common with list comprehension notation, with the "drawn from" arrows <-, the dependency between multiple generators as in this example reflecting value passing between subsequent lines in do notation, and return used for what is the "result part" of a list comprehension. In a way, the list monad and do notation explain list comprehension: if list comprehension were not built into Haskell, then it could be added as an notational alternative to do notation in the list monad.

What's missing in do notation, compared with list comprehension, is guards. But all is not lost! Some monads can be given extra structure that makes it possible to define something that works like a guard, and the list monad is one of those.

The extra structure we need is another associative binary operator together with a value that is the identity for that operator. Extending the Monad type class with these additional components gives the built-in type class MonadPlus, with lists as an instance:

```
class Monad m => MonadPlus m where
  mzero :: m a
  mplus :: m a -> m a -> m a

instance MonadPlus [] where
  mzero :: [a]
  mzero = []

  mplus :: [a] -> [a] -> [a]
  mplus = (++)
```

As of Haskell version 7.10, MonadPlus is an extension of a type class called **Alternative** which is an extension of **Monad**, see ▶ https://wiki.haskell.org/Functor-Applicative-Monad_Proposal, and guard below is defined in any instance of **Alternative**. Lists are defined as an instance of **Alternative**, so the details are omitted.

For lists as an instance of MonadPlus, mzero represents a non-deterministic value that lists no possible outcomes: the result of a failed computation. The function mplus combines two non-deterministic values into one, appending the two lists of alternatives.

Given an instance of MonadPlus, we can define the following function:

```
guard :: MonadPlus m => Bool -> m ()
guard False = mzero
guard True  = return ()
```

For lists as an instance of MonadPlus, guard False gives a failed computation, while guard True produces the list [()]:

```
> guard (1 > 3) :: [()]
[]
> guard (3 > 1) :: [()]
[()]
```

You need to add type annotations to these so that Haskell knows how to output the result. All Haskell can infer by itself is that
guard (1 > 3) :: m ()
for some instance m of MonadPlus.

This is useful in combination with >>. It either causes the rest of the computation to be aborted or to continue.

```
> guard (1 > 3) >> return 1 :: [Int]
[]
> guard (3 > 1) >> return 1 :: [Int]
[1]
```

The use of () in the definition of guard is unimportant: any type, and any value of that type, would do just as well.

What guard provides in connection with **do** notation is the real point:

```
pairs' :: Int -> [(Int, Int)]
pairs' n = do {
            i <- [1..n];
            j <- [1..n];
            guard (i < j);
            return (i,j)
          }
```

This gives:

```
> pairs' 4
[(1,2),(1,3),(1,4),(2,3),(2,4),(3,4)]
```

and corresponds to the following definition using list comprehension, which includes a guard:

```
pairs' :: Int -> [(Int, Int)]
pairs' n = [ (i,j) | i <- [1..n], j <- [1..n], i < j ]
```

Parsers as a Monad

You'll have already discovered that Haskell includes a parser, if you've ever encountered an error messages of the form "`parse error on input ...`" after making a mistake with Haskell's syntax! In a compiler, like the Haskell compiler, a parser has two phases, the first being a **lexical analyser**—typically implemented using a DFA—which turns the input string into a list of **tokens**, where tokens are variable names, literals, symbols, etc. The parser here combines lexical analysis and parsing in a single phase.

A **parser** converts a string of characters into a tree-structured representation of the syntactic entity that the string represents. An example is a parser for the algebraic data type of arithmetic expressions `Exp` defined on page 144:

```
data Exp = Lit Int
         | Add Exp Exp
         | Mul Exp Exp
    deriving Eq
```

for which the `showExp` function was defined on page 144 to produce familiar parenthesized notation for such expressions:

```
> showExp (Add (Lit 1) (Mul (Lit 2) (Lit 3)))
"(1 + (2 * 3))"
> showExp (Mul (Add (Lit 1) (Lit 2)) (Lit 3))
"((1 + 2) * 3)"
```

A parser for `Exp` is a function `readExp :: String -> Exp` that turns a string into the value of type `Exp` that the string represents—essentially, computing the reverse of `showExp :: Exp -> String`—while rejecting strings that are syntactically ill-formed:

```
> readExp "(1 + (2 * 3))"
Add (Lit 1) (Mul (Lit 2) (Lit 3))
> readExp "((1 + 2) * 3)"
Mul (Add (Lit 1) (Lit 2)) (Lit 3)
> readExp "(1 + 2))(* 3)"
*** Exception: no parse
```

However, that type turns out to be too restrictive when we try to build complex parsers in stages by combining simpler parsers.

First, because the definition of `Exp` refers to the type `Int`, a parser for `Exp` will need to use a parser for `Int` to deal with parts of the input string that represent integers. And parsing a string like "`((1 + 2) * (3 * 4))`" that contains substrings representing smaller values of type `Exp`—in this case, "`(1 + 2)`" and "`(3 * 4)`" representing `Add (Lit 1) (Lit 2)` and `Mul (Lit 3) (Lit 4)`, plus "`1`" representing `(Lit 1)`, etc.—will require recursive calls to deal with these substrings. That is, a parser will generally be able to deal with *only part of* the input string, leaving the remainder of the string to be parsed by other parsers or by other calls of the same parser.

Second, parsing a string is not guaranteed to yield a single value of type `Exp`. The string might be syntactically ill-formed, like "`(1 + 2))(* 3)`", so that parsing produces *no* value of type `Exp`. Or the input string might be **ambiguous**, requiring the parser to return *more than one* result of type `Exp`. The entire input string being ambiguous is normally regarded as an error, but the surrounding context can often be used to disambiguate candidate parses of substrings.

Taking all of these considerations into account, a parser for `Exp` will be a function of type `String -> [(Exp,String)]`. Applied to a string s, it will produce a list of possible parses of initial substrings of s, paired with the part of s that remains after that initial substring is removed. For the string $s =$ "`(1 + 2))(* 3)`" the result will be `[(Add (Lit 1) (Lit 2), ")(* 3)")]`, signifying that there is one way that an initial substring of s—namely "`(1 + 2)`"—can be parsed to give an `Exp`—namely `Add (Lit 1) (Lit 2)`—without touching the remainder of s—namely, "`)(* 3)`". For the string $s' =$ "`((1 + 2) * (3 * 4))`", the result will be

```
[(Mul (Add (Lit 1) (Lit 2)) (Mul (Lit 3) (Lit 4)), "")]
```

with the empty string indicating that all of s' is consumed to produce `Mul (Add (Lit 1) (Lit 2)) (Mul (Lit 3) (Lit 4))`.

Parsers in this form can be given the structure of a monad, and **do**-notation can be used to combine parsers for simple syntax to give parsers for more complicated syntax. We prepare the basis for that by giving the type of parsers as an algebraic data type, together with functions for applying a parser to a string:

```
data Parser a = Parser (String -> [(a, String)])

apply :: Parser a -> String -> [(a, String)]
apply (Parser f) s = f s

parse :: Parser a -> String -> a
parse p s = one [ x | (x,"") <- apply p s ]
  where one []                  = error "no parse"
        one [x]                 = x
        one xs | length xs > 1 = error "ambiguous parse"
```

There's an implicit invariant on values of type **Parser** a, namely that each of the strings produced is a suffix of the input string. The function parse, which parses the entire input string, requires a result from apply p s that completely consumes s and is non-ambiguous.

The following simple examples of parsers, for single characters, don't require the use of monads. The function char parses a single character from the input, if there is one; spot parses a character provided it satisfies a given predicate, for example isDigit; and token c recognises the character c, if it is the next character in the input string:

```
char :: Parser Char
char = Parser f
  where f []    = []
        f (c:s) = [(c,s)]

spot :: (Char -> Bool) -> Parser Char
spot p = Parser f
  where f []                 = []
        f (c:s) | p c        = [(c, s)]
                | otherwise = []

token :: Char -> Parser Char
token c = spot (== c)
```

For example:

```
> apply (spot isDigit) "123"
[('1',"23")]
> apply (spot isDigit) "(1+2)"
[]
> apply (token '(') "(1+2)"
[('(',"1+2)")]
```

As of Haskell 7.10, the following
additional definitions are required:

```
instance Functor Parser
   where fmap = liftM
```

```
instance Applicative Parser
   where pure = return
         (<*>) = ap
```

To go much further, we need to define **Parser** as an instance of **Monad**:

```
instance Monad Parser where
  return x = Parser (\s -> [(x,s)])
  m >>= k  = Parser (\s ->
                [ (y, u) |
                  (x, t) <- apply m s,
                  (y, u) <- apply (k x) t ])
```

The definition of return converts a value into a parser that produces that value as result, without consuming any input. The definition of >>= chains together two consecutive parsers m : : Parser a and k : : a -> Parser b. This deals with **sequencing**: situations where a string can be broken into consecutive substrings that are parsed separately, with the results combined to give the parse of the combined substrings. Given a string s, m >>= k:

- first applies the parser m to s, producing parses (x, t) where t is the substring of s that remains after parsing x;
- then applies the parser k x to t, producing parses (y, u) where u is the final unparsed substring that remains after parsing y; and
- the result is all of the possible parses (y, u).

Note that the parsing done by k, and its result, can depend on the result x of the first parser; thus the second parser is k x rather than k.

These can be used to define a function, using recursion and **do**-notation, that parses a given string appearing at the beginning of the input:

```
match :: String -> Parser String
match []     = return []
match (x:xs) = do {
                  y <- token x;
                  ys <- match xs;
                  return (y:ys)
               }
```

For example:

```
> apply (match "abc") "abcdef"
[("abc","def")]
> :{
| apply (do {
|           x <- match "abc";
|           y <- match "de";
|           return (x++y)
|        }) "abcdef"
| :}
[("abcde","f")]
```

Parser can also be defined as an instance of **MonadPlus**:

As of Haskell 7.10, the following
additional definition is required:

```
instance Alternative Parser
   where
     empty = mzero
     (<|>) = mplus
```

```
instance MonadPlus Parser where
  mzero     = Parser (\s -> [])
  mplus m n = Parser (\s -> apply m s ++ apply n s)
```

Here, mzero is a parser that always fails, while mplus m n combines the results of parsers m and n. The **MonadPlus** structure enables parsing of **alternatives**, including the use of guard. For example, here is another definition of the function spot above, using **do**-notation and guard:

30

```
spot :: (Char -> Bool) -> Parser Char
spot p = do {
            c <- char;
            guard (p c);
            return c
        }
```

The functions `star` and `plus` are mutually recursive, with `star` applying a parser to parse zero or more occurrences of a fragment of syntax, and `plus` applying a parser to parse one or more occurrences, using **do**-notation:

```
star :: Parser a -> Parser [a]
star p = plus p `mplus` return []

plus :: Parser a -> Parser [a]
plus p = do {
            x <- p;
            xs <- star p;
            return (x:xs)
        }
```

The use of `star`/`plus` for repetition of a fragment of syntax and the way that `mplus` is used to combine alternatives come from regular expressions, see Chap. 31.

These can be applied to parse positive and negative integers:

```
parseInt :: Parser Int
parseInt = parseNat `mplus` parseNeg
  where parseNat = do {
                        s <- plus (spot isDigit);
                        return (read s)
                    }
        parseNeg = do {
                        token '-';
                        n <- parseNat;
                        return (-n)
                    }
```

For example:

```
> apply parseInt "-123+4"
[(-123,"+4"),(-12,"3+4"),(-1,"23+4")]
```

This is an example of ambiguity in parsing. A negative integer is "–" followed by a natural number, which is a sequence of one or more digits. There is a choice of how many digits to parse, so all three options are produced, from longest to shortest, each paired with the substring that remains.

A parser for `Exp` can now be built by combining parsers for its three alternative forms:

```
parseExp :: Parser Exp
parseExp = parseLit `mplus` parseAdd `mplus` parseMul
  where parseLit = do {
                        n <- parseInt;
                        return (Lit n)
                    }
```

```
parseAdd = do {
                token '(';
                d <- parseExp;
                token '+';
                e <- parseExp;
                token ')';
                return (Add d e)
              }
parseMul = do {
                token '(';
                d <- parseExp;
                token '*';
                e <- parseExp;
                token ')';
                return (Mul d e)
              }
```

For example:

To get this output, you need to define Exp to use the default show function, rather than showExp on page 144:

```
data Exp = Lit Int
         | Add Exp Exp
         | Mul Exp Exp
  deriving (Eq,Show)
```

```
> parse parseExp "(-142+(26*3))"
Add (Lit (-142)) (Mul (Lit 26) (Lit 3))
> parse parseExp "((-142+26)*3)"
Mul (Add (Lit (-142)) (Lit 26)) (Lit 3)
> parse parseExp "(-142+26)*3"
*** Exception: no parse
```

As the last example shows, this parser requires expressions to be fully parenthesized. It would be easy to add other operators, and refinements to make the notation less restrictive.

Exercises

1. Give an alternative definition of putStr :: `String -> IO ()` (page 325) using foldr and map.
2. Write the definition of echo :: `IO ()` (page 328) using do notation.
3. Define a command getInt :: `IO Int` that reads a string from the keyboard, like getLine, and converts it to an `Int`. (**Hint: Int** is an instance of **Read**.)
4. Define the Prelude function sequence :: `Monad m => [m a] -> m [a]` which, for an arbitrary monad, takes as input a list of actions and returns a combined action that performs each of those actions, one after the other, and returns a list of their results.
5. Translate the function pairs (page 332) from do notation to a definition in terms of >>= in order to understand how the value of i is passed from i <- [1..n], which produces it, to j <- [(i+1)..n] and return (i,j), which depend on it.
6. Define the functions star :: `Eq q => [Trans q] -> q -> [Sym] -> [q]` (page 310) and accept :: `Eq q => FA q -> [Sym] -> Bool` (page 310) using do notation in the list monad.
7. Check that return and >>= in the list monad satisfy the monad laws.
8. Define **Maybe** as an instance of **Monad**. (**Hint:** If it helps, think of functions of type a -> Maybe b as modelling possibly-failing computations.)
9. The parser for **Exp** on page 337 doesn't allow spaces to appear anywhere in expressions. Modify it to allow zero or more spaces to precede and follow arithmetic operators.

Regular Expressions

Contents

© The Author(s), under exclusive license to Springer Nature Switzerland AG 2021
D. Sannella et al., *Introduction to Computation*, Undergraduate Topics
in Computer Science, https://doi.org/10.1007/978-3-030-76908-6_31

Describing Regular Languages

You've seen how to describe sets of strings—or languages—using different kinds of finite automata. An automaton describes a language by accepting the strings in the language and rejecting the rest. It decides whether or not to accept a string via a mechanical computation process based on the sequence of symbols in the string.

You're now going to learn a notation for describing sets of strings, called **regular expressions**, that is algebraic and descriptive rather than mechanical and computational. Regular expressions provide a convenient way of writing down a pattern that all strings in the language must fit. For example, the regular expression $(a|b)^*aaa(a|b)^*$ describes the set of strings over the alphabet $\{a, b\}$ that contain aaa.

Regular expressions look quite different from finite automata, but nevertheless, they have exactly the same expressive power: every regular expression describes a regular language, and every regular language can be described by a regular expression. The first of these facts will be shown by giving an algorithm for converting regular expressions to ε-NFAs. The second involves a method for converting NFAs to regular expressions. Putting this together with the previous chapters gives four ways of describing regular languages: DFAs, NFAs, ε-NFAs, and regular expressions.

Regular expressions are used to specify textual patterns in many applications. One use of regular expressions that you might have encountered already, but using a different notation, is specifying file names in operating system commands: in Linux, the command `rm *aaa*` will delete all files having a name that contains the substring `aaa`. The Linux `grep` command is used to find lines in a file that match a regular expression. Regular expression search is also commonly provided by text editors, and it is one of the main features of the Perl scripting language.

Examples

Any regular expression R over an alphabet Σ describes a language $L(R) \subseteq \Sigma^*$. A very simple example is a string of symbols like $abaab$, which describes the language $L(abaab) = \{abaab\}$ containing that string and nothing else. Similar are the regular expressions ε, describing the language $\{\varepsilon\}$ containing the empty string, and $\varnothing$ which describes the empty language.

The first example already demonstrates one of the operators for building more complicated regular expressions out of simpler ones, namely **concatenation**: if R and S are regular expressions then RS is a regular expression, with $L(RS) = \{xy \mid x \in L(R) \text{ and } y \in L(S)\}$. The regular expression $abaab$ is built from the regular expressions a and b with four uses of concatenation. The two regular expressions being concatenated are simply written next to each other, without an explicit operator. If $L(R) = L(S) = \{a, bb\}$ then $L(RS) = \{aa, bba, abb, bbbb\}$, consisting of strings formed by concatenating a string from $L(R)$ with a string from $L(S)$.

Another operator for building regular expressions is **union**, written with a vertical bar: if R and S are regular expressions then $R|S$ is a regular expression, with $L(R|S) = L(R) \cup L(S)$. For example, $abaab|bab$ describes the language $\{abaab, bab\}$. Parentheses can be used for grouping as in $(aba|b)ab$, which describes the same language. The regular expression $(a|\varepsilon)b(a|\varepsilon)ab$ describes the larger language $\{abaab, bab, abab, baab\}$ because the choices between a and ε in the two sub-expressions $a|\varepsilon$ are independent.

This difference is a little like the distinction between *procedural* and *declarative knowledge* in AI, see ▶ https://en.wikipedia.org/wiki/Procedural_knowledge#Artificial_intelligence.

Maybe you already guessed that from the fact that they are called *regular* expressions.

See Exercise 5 for a fifth way.

See ▶ https://en.wikipedia.org/wiki/Glob_(programming) for the use of regular expressions for specifying filenames. See ▶ https://en.wikipedia.org/wiki/Grep for `grep`, which is so well-known that it has become a verb in English, like "google", as in: "You can't grep dead trees."

$R|S$ is pronounced "R bar S". Some books write $R.S$ for concatenation and/or $R + S$ for union.

The final operator for building regular expressions is **closure**, which specifies zero or more repetitions: if R is a regular expression then its closure R^* is a regular expression, with $L(R^*) = L(R)^*$.

You've seen $*$ before: if Σ is an alphabet then Σ^* is the set of all strings over Σ. That is, Σ^* is the set of strings of length 0, plus the set of strings of length 1, plus the set of strings of length 2, etc. If $\Sigma = \{0, 1, 2, 3, 4, 5\}$ then $L(0|1|2|3|4|5) = \Sigma$ and $L((0|1|2|3|4|5)^*) = \Sigma^*$.

In general, if R is a regular expression then R^* describes the language containing strings formed by concatenating zero or more strings selected from R:

$$R^* = \varepsilon \mid R \mid RR \mid RRR \mid \ldots.$$

For example, $(a|bb)^*$ describes the infinite language

$$\{\varepsilon, a, bb, aa, abb, bba, bbbb, aaa, abba, \ldots\}.$$

Notice that again, each of the choices are independent. Choosing a string a or bb from R and then repeating it zero or more times would give the smaller language $\{\varepsilon, a, aa, aaa, \ldots, bb, bbbb, bbbbbb, \ldots\}$.

Here are some more examples:

- $a(a|b)^*b$ describes the language of strings over $\{a, b\}$ that begin with a and end with b.
- $(aa)^*|(aaa)^*$ describes the language of strings of as with length that is divisible by 2 or 3. On the other hand, $(aa|aaa)^*$ describes the language of strings of as with length ≥ 2 (i.e. with length $2m + 3n$ for some $m, n \geq 0$).
- $(a(b|c))^*$ describes strings of even length consisting of a in all even positions interspersed with b or c in all odd positions, $\{ax_1ax_2\ldots ax_n \mid x_1, \ldots, x_n \in \{b, c\}, n \geq 0\}$.
- $1^* \mid (1^*01^*01^*)^*$ describes strings of binary digits containing an even number of 0s.

Let's have a closer look at the final example. Any string of binary digits containing an even number of 0s either contains no 0s, or it contains an even but non-zero number of 0s. The regular expression 1^* covers the first case. A string in the second category is composed of substrings containing two 0s and zero or more 1s. The regular expression $1^*01^*01^*$ describes the set of such substrings—two 0s, separated and surrounded by any number of 1s—and so $(1^*01^*01^*)^*$ describes sequences of zero or more such substrings.

Simplifying Regular Expressions

There are many different regular expressions that specify any given language. Using laws for regular expressions—like the laws of Boolean algebra in Chap. 22—you can transform a regular expression to an equivalent one, for example to simplify it. The laws can also be used to show that two regular expressions are equivalent. They include the obvious associativity and identity properties of concatenation and union, commutativity of union, and distributivity of concatenation over union. More difficult to remember and use are the laws for closure, including the ones that relate closure to concatenation and union.

R^* is pronounced "R star". The closure operator $*$ is called **Kleene star** after the American mathematician Stephen Kleene (1909–1994), pronounced "cleanie", who invented regular expressions and the method for converting an NFA to a regular expression, see

▶ https://en.wikipedia.org/wiki/Stephen_Cole_Kleene.

In applications of regular expressions, more operators are often added, for example R^+ (one or more repetitions of R) as an abbreviation for RR^* and $R?$ (optional R) as an abbreviation for $R|\varepsilon$.

$$R \mid R = R = R \mid \varnothing \qquad R \mid S = S \mid R \qquad (R \mid S) \mid T = R \mid (S \mid T)$$
$$R\varepsilon = \varepsilon R = R \qquad R\varnothing = \varnothing R = \varnothing \qquad (RS)T = R(ST)$$
$$(R \mid S)T = RT \mid ST \qquad\qquad R(S \mid T) = RS \mid RT$$
$$\varepsilon^* = \varnothing^* = \varepsilon \qquad RR^* = R^*R \qquad (RS)^*R = R(SR)^*$$
$$R^*R^* = (R^*)^* = R^* = \varepsilon \mid RR^* \quad (R \mid S)^* = (R^*S^*)^* = (R^*S)^*R^* = (R^* \mid S^*)^*$$

Here's an example of using the laws to simplify the regular expression $(b \mid aa^*b) \mid (b \mid aa^*b)(a \mid aba)^*(a \mid aba)$. The part of the expression that changes in each step is underlined.

$$
\begin{array}{ll}
\underline{(b \mid aa^*b)} \mid (b \mid aa^*b)(a \mid aba)^*(a \mid aba) & \\
= \underline{(b \mid aa^*b)\varepsilon \mid (b \mid aa^*b)}(a \mid aba)^*(a \mid aba) & \text{since } R = R\varepsilon \\
= (b \mid aa^*b)\underline{(\varepsilon \mid (a \mid aba)^*(a \mid aba))} & \text{since } RS \mid RT = R(S \mid T) \\
= (b \mid aa^*b)\underline{(\varepsilon \mid (a \mid aba)(a \mid aba)^*)} & \text{since } R^*R = RR^* \\
= (\underline{b} \mid aa^*b)(a \mid aba)^* & \text{since } \varepsilon \mid RR^* = R^* \\
= (\underline{\varepsilon b} \mid aa^*b)(a \mid aba)^* & \text{since } R = \varepsilon R \\
= \underline{(\varepsilon \mid aa^*)}b(a \mid aba)^* & \text{since } RT \mid ST = (R \mid S)T \\
= \underline{a^*}b(a \mid aba)^* & \text{since } \varepsilon \mid RR^* = R^*
\end{array}
$$

Regular Expressions Describe Regular Languages

It turns out that all languages that can be described by regular expressions are regular. To see why, we need to establish a relationship between regular expressions and finite automata: that is, we need to show that any regular expression can be converted to a finite automaton. You've learned that all three versions of finite automata—DFAs, NFAs and ε-NFAs—have equivalent expressive power. Since there is a direct way of doing the conversion to ε-NFAs, that's what we'll do.

The conversion method can be seen as a recursive algorithm with six cases, one for each form of regular expression. The first three cases are the expressions a for $a \in \Sigma$, ε, and $\varnothing$. These are the base cases of the recursion since these regular expressions have no sub-expressions. It's easy to give the corresponding ε-NFAs M_a, M_ε, and $M_\varnothing$:

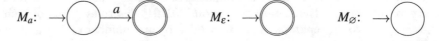

Then we need to consider the three operators for combining regular expressions. Each of these is a different recursive case. We'll start with concatenation since we've already seen the relevant construction on ε-NFAs in Chap. 29.

The regular expression RS is composed of two smaller regular expressions, R and S. Convert these to ε-NFAs M_R and M_S that accept the languages $L(R)$ and $L(S)$. The construction for concatenating ε-NFAs—in which all of the accepting states of M_R are connected via ε-transitions to all of the start states of M_S—can then be applied to M_R and M_S, giving an ε-NFA M_{RS} that accepts the language $L(RS)$:

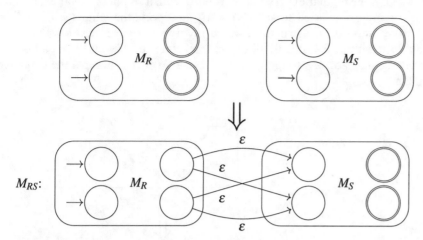

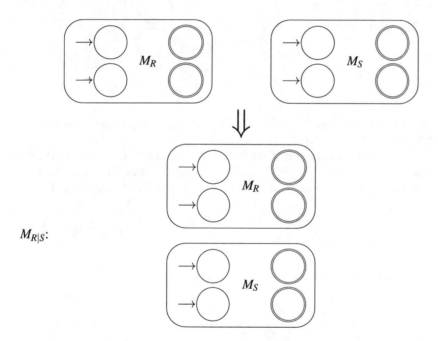

M_{RS}:

Combining ε-NFAs M_R and M_S to give the sum ε-NFA $M_{R|S}$ that accepts the language $L(R|S)$ is even easier: $M_{R|S}$ is just the union of M_R and M_S, combined into a single ε-NFA, with no change to either machine and no additional transitions:

$M_{R|S}$:

An alternative would be to convert M_R and M_S to DFAs using the powerset construction, and then apply the sum DFA construction to produce a DFA that accepts $L(R|S)$.

The algorithm for converting a regular expression to an NFA is due to 1983 Turing Award winner Ken Thompson (1943−), see ▶ https://en.wikipedia.org/wiki/Thompson's_construction. Thompson was also responsible (with Dennis Ritchie) for the Unix operating system, a precursor of Linux.

Finally, the regular expression R^* is built from a smaller regular expression R that can be converted to an ε-NFA M_R accepting the language $L(R)$. The following iteration construction, which involves adding a new start state q_0 with ε-transitions from q_0 to all of the start states of M_R and from all of the accepting states of M_R to q_0, gives an ε-NFA M_{R^*} that accepts the language $L(R^*)$:

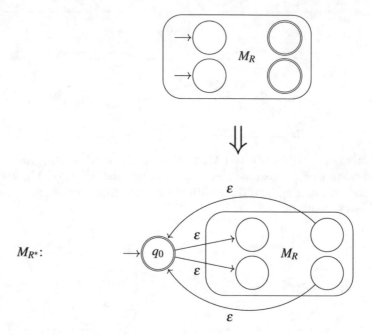

The iteration construction—of M_{R^*} from M_R—also shows that the set of regular languages is **closed under iteration**.

To show how the conversion method works, let's try converting the regular expression 1* | (1*01*01*)* (describing strings of binary digits containing an even number of 0s) to an ε-NFA. We'll start with ε-NFAs for 0 and 1:

We can apply the iteration construction above to construct M_{1^*} from M_1:

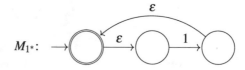

Applying the concatenation construction four times to combine M_{1^*}, M_0, M_{1^*}, M_0, and M_{1^*} gives $M_{1^*01^*01^*}$:

$M_{1*01*01*}$:

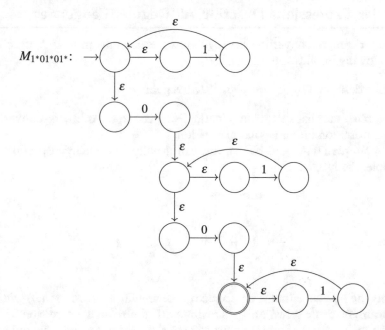

Applying the iteration construction to this gives $M_{(1*01*01*)*}$, and then applying the sum construction to M_{1*} and $M_{(1*01*01*)*}$ gives the final result, $M_{1*|(1*01*01*)*}$:

$M_{1*|(1*01*01*)*}$:

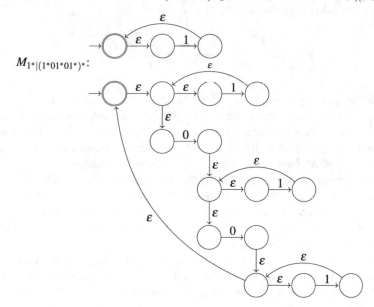

This algorithm tends to produce large ε-NFAs that are correct but far from minimal. A simple two-state DFA that accepts the same language is M' on page 299:

M':

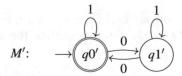

Regular Expressions Describe All Regular Languages

The algorithm for converting any regular expression into an ε-NFA establishes the following implication:

> L is described by R implies L is regular

Because DFAs, NFAs and ε-NFAs have the same expressive power, it's enough to do this for DFAs only, but the same method works for NFAs and ε-NFAs too.

To establish the "backward" implication, we need to show how to convert any finite automaton into a regular expression.

For simple DFAs and NFAs, this can usually be done by inspection. For example, the NFA

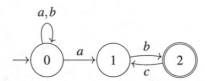

accepts the language that is described by the regular expression $(a|b)^*ab(cb)^*$. You can get this expression by starting with the regular expression ab that describes direct paths from the start state to the final state, and then add $(a|b)^*$ at the beginning to account for the first loop and $(cb)^*$ at the end to account for the second one, via the state 1.

Here's another simple example. The DFA

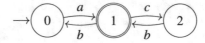

accepts the language that is described by the regular expression $a(ba \mid cb)^*$. Notice that two loops on the same state (in this case, the loop ba from state 1 through state 0, and the loop cb from state 1 through state 2) corresponds to the *closure of the union* of the two loops: the regular expressions $a(ba)^*(cb)^*$, $a(cb)^*(ba)^*$ and $a((ba)^* \mid (cb)^*)$ describe different languages!

When things get more complicated, you need to take a more systematic approach. The first step is to write down a set of simultaneous equations, one for each state n, describing the set of strings that would be accepted if n were an accepting state—we'll call that set L_n—in terms of the sets of strings corresponding to the other states.

There will be one case for each incoming transition, and we also need to record that the start state(s) can be reached without reading any input. For example, here are the equations for the DFA above:

$$L_0 = \varepsilon \mid L_1 b \qquad\qquad \text{state 0 is initial and } 1 \xrightarrow{b} 0$$
$$L_1 = L_0 a \mid L_2 b \qquad\quad 0 \xrightarrow{a} 1 \text{ and } 2 \xrightarrow{b} 1$$
$$L_2 = L_1 c \qquad\qquad\qquad 1 \xrightarrow{c} 2$$

You get the result by solving for L_1, because state 1 is the accepting state. Let's start by substituting for L_0 and L_2 in the equation for L_1:

$$L_1 = (\varepsilon \mid L_1 b)a \mid L_1 cb$$

and then apply distributivity and the identity law for concatenation:

$$L_1 = \varepsilon a \mid L_1 ba \mid L_1 cb$$
$$= a \mid L_1(ba \mid cb)$$

Now it looks like we're stuck, or maybe we've made a mistake: we have an equation for L_1 in terms of L_1. What do we do now?

Don't worry: this situation arises for any NFA that contains a loop. What the equation for L_1 is saying is that state 1 can be reached from the start state by reading input a, or from state 1 by reading either ba or cb. That looks right.

Having got this far, you won't be surprised to learn that there's a trick for solving equations of just this form:

Arden's Rule. If R and S are regular expressions then $X = R \mid XS$ has a solution $X = RS^*$. If $\varepsilon \notin L(S)$ then that's the only solution.

Applying Arden's Rule gives the same result as we originally got by inspection:

$$L_1 = a(ba \mid cb)^*$$

This NFA has just one accepting state. If there were more, you would need to solve for each of them and then take the union of those regular expressions.

That was easy! Let's try the same method on a much more challenging example:

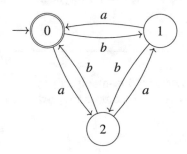

Here are the equations for this DFA:

$$L_0 = \varepsilon \mid L_1\, a \mid L_2\, b \qquad\qquad \text{state 0 is initial, } 1 \xrightarrow{a} 0 \text{ and } 2 \xrightarrow{b} 0$$
$$L_1 = L_2\, a \mid L_0\, b \qquad\qquad\qquad\ 2 \xrightarrow{a} 1 \text{ and } 0 \xrightarrow{b} 1$$
$$L_2 = L_0\, a \mid L_1\, b \qquad\qquad\qquad\ 0 \xrightarrow{a} 2 \text{ and } 1 \xrightarrow{b} 2$$

We'll start by substituting L_2 into the equation for L_1, collecting terms involving L_0, then applying Arden's Rule:

$$\begin{aligned}
L_1 &= (L_0\, a \mid L_1\, b)a \mid L_0\, b \\
&= L_0\, aa \mid L_1\, ba \mid L_0\, b \\
&= L_0(aa \mid b) \mid L_1\, ba \\
&= L_0(aa \mid b)(ba)^*
\end{aligned}$$

It's completely okay to apply Arden's Rule in a situation like this where the equation involves variables—in this case L_0—other than the one being solved for.

Arden's Rule is due to American electrical engineer and computer scientist Dean Arden (1925–2018), a member of the team that built and programmed Whirlwind, the first real-time digital computer. See ▶ https://en.wikipedia.org/wiki/Arden's_rule.

We didn't bother checking the requirement that $\varepsilon \notin L(ba \mid cb)$, but it holds, and it will always hold for the kind of examples you will encounter.

Now we substitute L_1 into the equation for L_2:

$$L_2 = L_0\, a \mid L_0(aa \mid b)(ba)^*b$$
$$= L_0(a \mid (aa \mid b)(ba)^*b)$$

Finally, we substitute L_1 and L_2 into the equation for L_0, collect terms involving L_0, and apply Arden's Rule to get the result:

$$L_0 = \varepsilon \mid L_0(aa \mid b)(ba)^*a \mid L_0(a \mid (aa \mid b)(ba)^*b)b$$
$$= \varepsilon \mid L_0(aa \mid b)(ba)^*a \mid L_0\, ab \mid L_0(aa \mid b)(ba)^*bb$$
$$= \varepsilon \mid L_0\, ab \mid L_0(aa \mid b)(ba)^*(a \mid bb)$$
$$= \varepsilon \mid L_0(ab \mid (aa \mid b)(ba)^*(a \mid bb))$$
$$= \varepsilon(ab \mid (aa \mid b)(ba)^*(a \mid bb))^*$$
$$= (ab \mid (aa \mid b)(ba)^*(a \mid bb))^*$$

A different but equivalent result can be produced by following a different sequence of reasoning steps. The bottom line is that this method can be used to convert any finite automaton into a regular expression describing the same language. That means that any regular language can be described using a regular expression, which establishes the reverse implication that was our goal.

Exercises

1. Prove that $RR^* = R^*R$.
2. For each of the following equations, either prove it using the laws of regular expressions or give a counterexample to show that it is false.

   (a) $R^* = \varepsilon \mid R \mid R^*RR$
   (b) $(R^*S)^* = (SR^*)^*$
   (c) $(R \mid S)^* = R^* \mid S^*$

3. Use the laws of regular expressions to prove that $a(ba)^*b \mid (ab)^* = (ab)^*$.
4. Define an algebraic data type `Regexp` for representing a simplified form of regular expressions which omits the closure operator. Then:

   (a) Write a function `language :: Regexp -> [String]` which, given a regular expression, returns its language in the form of a list of strings without duplicates.
   (b) Write a function `simplify :: Regexp -> Regexp` that converts a regular expression to an equivalent simpler regular expression by use of the laws $R \mid R = R$, $R \mid \varnothing = \varnothing \mid R = R$, $R\varepsilon = \varepsilon R = R$ and $R\varnothing = \varnothing R = \varnothing$.

5. A BNF grammar (see page 156) is *right-linear* if all of its rules are of the form $A ::= a$ or $A ::= aB$ or $A ::= \varepsilon$ where A, B are non-terminal symbols and a is a terminal symbol.
   Show that every right-linear grammar generates a regular language. **Hint:** Construct an NFA in which there is a state corresponding to every non-terminal symbol. You will need to designate one of the non-terminal symbols as the "start symbol".
   Show that every regular language is generated by a *right-linear* grammar. **Hint:** From a DFA, construct a grammar with a non-terminal symbol corresponding to each state.
6. Give a formal definition of the iteration construction that, given an ε-NFA M, produces a ε-NFA M^* such that $L(M^*) = L(M)^*$. **Hint:** See the definition of the concatenation construction on page 321.
7. Convert $a(a|b)^*b$ to an ε-NFA.

If you aren't familiar with the terminology "terminal"/"non-terminal symbol" and/or with the language generated by a grammar, see ► https://en.wikipedia.org/wiki/Formal_grammar.

8. Use Arden's Rule to convert the following DFAs to regular expressions that describe the same languages:

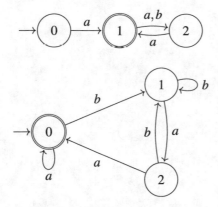

9. Produce a regular expression for the complement of the regular expression $a(a|b)^*b$ via the following steps:

   (a) convert $a(a|b)^*b$ to an ε-NFA (see Exercise 7);
   (b) convert that ε-NFA to a DFA;
   (c) take the complement of that DFA; then
   (d) using Arden's Rule, convert that DFA into a regular expression.

10. Extend the language of regular expressions with operators for complement and intersection, and show that the resulting language is still only able to describe regular languages. **Hint:** See the constructions which showed that the set of regular languages is closed under complement and intersection.

Non-Regular Languages

Contents

© The Author(s), under exclusive license to Springer Nature Switzerland AG 2021
D. Sannella et al., *Introduction to Computation*, Undergraduate Topics
in Computer Science, https://doi.org/10.1007/978-3-030-76908-6_32

Boundaries of Expressibility

Having learned a lot about regular languages, different ways of describing them, and a little about their applications, you will have rightly come to understand that they are important. As you've seen, the union of two regular languages is a regular language, and the same goes for intersection, complement, concatenation, and iteration. So, given just a few simple regular languages—for example, $\{a\}$ and $\{b\}$—you can build a large number of new regular languages.

But not all languages are regular. It's easy to give an example of a non-regular language, and not difficult to explain, at an intuitive level, why it can't be regular. Formalising this reasoning leads to a useful method for showing that other languages are not regular.

In science and Mathematics, useful insight is often achieved by exploring boundaries: limits on what can be expressed using given notations; limits on the kinds of problems that can be solved using given methods; and limits on the functions that can be computed using given amounts of time and/or space. Understanding the features that distinguish regular languages from non-regular languages sheds light on what can be done with regular languages and motivates the investigation of ways of describing and processing languages that are not regular. We'll not pursue that exploration here, but what you've learned up to this point provides a good foundation for going further.

> There are classes of languages that go beyond the regular languages but can be expressed using methods that are only a little more powerful, see ▶ https://en.wikipedia.org/wiki/Chomsky_hierarchy.

Accepting Infinite Languages Using a Finite Number of States

Many of the examples of finite automata you've seen—DFAs, NFAs, and ε-NFAs—accept infinite languages. An example is the DFA M_4 on page 286 which accepts strings of binary digits that contain an odd number of 1s:

> We'll concentrate on DFAs in this chapter because the explanation is clearest in that case.

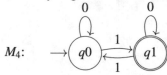

Any infinite language over a finite alphabet must include strings that are longer than n, for any $n \geq 0$. It's interesting to look at what happens for strings in such a language that are longer than the number of states in a DFA that accepts the language.

For M_4, 011010 is such a string, with the following computation: $q0 \xrightarrow{0} q0 \xrightarrow{1} q1 \xrightarrow{1} q0 \xrightarrow{0} q0 \xrightarrow{1} q1 \xrightarrow{0} q1$. Because the number of states visited by that computation is greater than the number of states in M_4, the computation must visit one or more of the states at least twice. Here, $q0$ is visited 4 times and $q1$ is visited 3 times.

> This is a consequence of the **pigeonhole principle**, see ▶ https://en.wikipedia.org/wiki/Pigeonhole_principle: if there are more pigeons (states visited during a computation) than holes (states in M_4), then some holes must contain more than one pigeon.

Let's look at the repeated visits of $q1$. The computation can be broken up into three phases:

1. from the start state $q0$ to the first visit of $q1$, reading 01;
2. from the first visit of $q1$ to the second visit of $q1$, reading 101;
3. after the second visit of $q1$, reading 0

> Recall that $q \xrightarrow{a} q'$ means that $(q, a, q') \in \delta$. Here we're using an extension of this notation: $q \xrightarrow{s} q'$ means that $\delta^*(q, s) = q'$, for $s \in \Sigma^*$.

We can write this as $q0 \xrightarrow{01} q1 \xrightarrow{101} q1 \xrightarrow{0} q1$.

The fact that the computation includes a repetition of $q1$, with $q1 \xrightarrow{101} q1$, means that M_4 must also accept an infinite number of similar strings that include any number of repetitions of the substring 101:

- 01 0, with the computation $q0 \xrightarrow{01} q1 \xrightarrow{0} q1$ (0 repetitions);
- 01 101 101 0, with the computation $q0 \xrightarrow{01} q1 \xrightarrow{101} q1 \xrightarrow{101} q1 \xrightarrow{0} q1$ (2 repetitions);
- 01 101 101 101 0, with the computation $q0 \xrightarrow{01} q1 \xrightarrow{101} q1 \xrightarrow{101} q1 \xrightarrow{101} q1 \xrightarrow{0} q1$ (3 repetitions);
- and so on.

That is, M_4 must accept $01(101)^i 0$, for every $i \geq 0$.

The same reasoning applies to the repeated visits of $q0$: because $q0 \xrightarrow{0} q0 \xrightarrow{11010} q1$, M_4 must accept $0^i 11010$, for every $i \geq 0$. (Seen in terms of three phases of computation, we have $q0 \xrightarrow{\varepsilon} q0 \xrightarrow{0} q0 \xrightarrow{11010} q1$ which means that M_4 must accept $\varepsilon 0^i 11010$, for every $i \geq 0$.) For reasons that will become clear later we'll concentrate just on the first two repeated visits of states, but the same also applies to any pair of repeated visits. And of course, the same reasoning applies to any DFA that accepts an infinite language, not just M_4.

A Non-Regular Language

A simple example of a non-regular language is $L_1 = \{a^n b^n \mid n \geq 0\}$: the strings over $\{a, b\}$ consisting of a sequence of as followed by a sequence of bs, where the number of as and bs match.

Before showing that L_1 cannot be regular, let's try to describe it using a DFA. We'll need states to count the number of as at the beginning, and then corresponding states to count the number of bs at the end, accepting the input if both numbers match. Something like this:

Observe that L_1 is related to the language of arithmetic expressions with balanced parentheses. Think of a and b as representing left and right parentheses, respectively, with the "content" of the arithmetic expression removed, and with the simplifying assumption that all of the left parentheses precede all of the right parentheses.

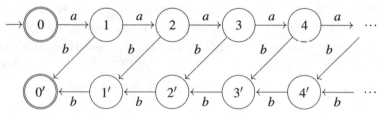

The use of ellipses ($\cdots$) mean that this isn't a DFA, but nevertheless the diagram makes the idea clear. Ignoring the part of the diagram involving ellipses, it's easy to check that this DFA accepts the language $\{a^n b^n \mid 4 \geq n \geq 0\}$. To accept the language $\{a^n b^n \mid m \geq n \geq 0\}$, the diagram just needs to be extended to the right to include states $0, \ldots, m$ and $0' \ldots, (m-1)'$.

Of course, the problem is that accepting L_1 without an upper limit on n would require extending the diagram infinitely far to the right, and that's impossible because of the requirement that a DFA has a finite set of states. The number of states, whatever it is, imposes a limit on the number of as that can be counted. If there are more as than that, there's no way to be sure that the number of as and bs match.

We can turn this idea into a proof by contradiction that L_1 is not regular, using the fact that computations that are longer than the number of states in a DFA must contain repeated states.

Proof. Assume that L_1 is regular; then there is a DFA M_1 such that $L(M_1) = L_1$. Let k be the number of states in M_1. Consider the string $x = a^k b^k \in L_1$. Since the number of states in the computation of a^k is greater than k, there must be at least one repeated state during the computation that accepts x while reading this first part of x. The computation must be $q_0 \xrightarrow{a^r} q \xrightarrow{a^s} q \xrightarrow{a^t b^k} q'$, where q_0 is the initial state, q is the first repeated state, $r+s+t = k$ and $s > 0$, and q' is an accepting state. Then M_1 must also accept $a^{r+t} b^k$ via the computation $q_0 \xrightarrow{a^r} q \xrightarrow{a^t b^k} q'$. But $a^{r+t} b^k \notin L_1$ because $r + t \neq k$, so $L(M_1) \neq L_1$. This is a contradiction, so the assumption that L_1 is regular was false.

This proof shows that M_1 must incorrectly accept $a^{r+t} b^k$ since the part of the computation between the first two visits of q can be removed without affecting acceptance. Repeating this part of the computation rather than removing it won't affect acceptance either, meaning that M_1 will also incorrectly accept $a^{r+2s+t} b^k$, $a^{r+3s+t} b^k$, etc.

The Pumping Lemma

The steps in the proof showing that $L_1 = \{a^n b^n \mid n \geq 0\}$ is not regular can be used to prove that many other languages are not regular. The proof method is captured by the following lemma:

$|x|$ is the length of the string x.	**Pumping Lemma.** Let M be a DFA with k states, and let $x \in L(M)$. If $|x| \geq k$ then there are strings u, v, w such that:

i. $uvw = x$;
ii. $|v| > 0$;
iii. $|uv| \leq k$; and
iv. $uv^i w \in L(M)$ for every $i \geq 0$.

Proof. Since $|x| \geq k$, the computation of M that accepts x must visit at least $k + 1$ states. It follows that some state q must be visited at least twice in the first $k + 1$ steps of that computation. That computation can therefore be written as $q_0 \xrightarrow{u} q \xrightarrow{v} q \xrightarrow{w} q'$, where q_0 is the start state, q' is an accepting state, $uvw = x$, $|v| > 0$ and $|uv| \leq k$. Since $q \xrightarrow{v} q$ ends in the same state as it starts, this part of the computation can be repeated—"pumped"—any number of times, including zero, giving $q_0 \xrightarrow{u} q \xrightarrow{v^i} q \xrightarrow{w} q'$ for every $i \geq 0$. This shows that M accepts $uv^i w$ for every $i \geq 0$.

We can apply the Pumping Lemma to give a simpler proof that $L_1 = \{a^n b^n \mid n \geq 0\}$ is not regular:

Proof. Assume that L_1 is regular; then there is a DFA M_1 such that $L(M_1) = L_1$. Let k be the number of states in M_1. Consider the string $x = a^k b^k \in L_1$. By the Pumping Lemma, there are strings u, v, w such that $uvw = x$, $|v| > 0$, $|uv| \leq k$, and $uv^i w \in L(M_1)$ for every $i \geq 0$. Therefore $u = a^r$, $v = a^s$ and $w = a^t b^k$ for $s > 0$, and $a^{r+is+t} b^k \in L(M_1)$ for every $i \geq 0$. But $a^{r+is+t} b^k \notin L_1$ unless $i = 1$, since $s > 0$, so $L(M_1) \neq L_1$. This is a contradiction, so the assumption that L_1 is regular was false.

Proving That a Language Is Not Regular

The Pumping Lemma is a useful tool for proving that languages are not regular. Let's use it to prove that another language, $L_2 = \{a^n \mid n = m^2 \text{ for some } m \geq 0\}$, is not regular.

Proof. Assume that L_2 is regular; then there is a DFA M_2 such that $L(M_2) = L_2$. Let k be the number of states in M_2. Consider the string $x = a^{(k^2)} \in L(M_2)$. Since $|x| \geq k$, by the Pumping Lemma there are strings u, v, w such that $uvw = x$, $|v| > 0$, $|uv| \leq k$, and $uv^i w \in L(M_2)$ for every $i \geq 0$. That is, $u = a^r$, $v = a^s$ and $w = a^t$ where $s > 0$, $r + s \leq k$, $r + s + t = k^2$, and $uv^i w = a^{r+is+t} \in L(M_2)$ for every $i \geq 0$. Let $i = 2$; then $r + is + t = k^2 + s$. We have $k^2 < k^2 + s$ (since $s > 0$) and $k^2 + s < k^2 + 2k + 1 = (k+1)^2$ (since $r + s \leq k$). This shows that $uv^i w$ is not in L_2, so $L(M_2) \neq L_2$. This is a contradiction, so the assumption that L_2 is regular was false.

The strategy for using the Pumping Lemma to prove that a language L is not regular is always the same:

1. Begin by assuming that L is regular. The goal is to use the Pumping Lemma to reach a contradiction from this assumption.
2. Because of the assumption that L is regular, there must be some DFA M such that $L(M) = L$. Let k be the number of states in M.
3. Choose some string $x \in L$ with $|x| \geq k$.
4. Apply the Pumping Lemma to break up x into u, v and w that satisfy properties i–iii.
5. Pick a value of $i \geq 0$ which allows you to prove that $uv^i w \notin L$.
6. Property iv of the Pumping Lemma guarantees that $uv^i w \in L(M)$ for every $i \geq 0$, so this contradicts the assumption that $L = L(M)$.
7. Having reached a contradiction, conclude that the initial assumption—that L is regular— was false.

Having this fixed strategy makes it easy to structure the proof. Nevertheless, getting the details right can be tricky. Here are some tips and potential pitfalls:

- Pitfall: in step 2, you don't know anything about M, other than $L(M) = L$, and you can't assume anything about the value of k.
- Tip: the choice of x in step 3 is crucial. You can pick any $x \in L$ you want. If possible, pick x whose first k symbols are the same, in order to simplify the reasoning in steps 4–5.
- Tip: in step 4, if you have chosen x appropriately then from properties i–iii in the Pumping Lemma you can conclude the form that u, v and w must have, which is useful in step 5.
- Pitfall: in step 4, the only information you have about u, v and w is properties i–iii.
- Tip: the choice of i in step 5 is sometimes crucial. Any choice other than $i = 1$ will work for some languages, while in other cases you need to select i carefully to make it possible to prove that $uv^i w \notin L$.
- Pitfall: in step 5, the only things you can use to prove that $uv^i w \notin L$ is your choice of i, what you can conclude about u, v and w from properties i–iii, and the definition of L.

Exercises

1. If $L \subseteq L'$ then L being regular says nothing about whether or not L' is regular, and vice versa. Give examples of all four regular/non-regular combinations.
2. Let Σ be an alphabet. Prove that every finite language $L \subseteq \Sigma^*$ is regular.
3. Use the Pumping Lemma to show that $L_3 = \{a^m b^n \mid m > n\}$ is not a regular language.
4. Use the Pumping Lemma to show that $L_4 = \{xx \mid x \in \{a, b\}^*\}$ is not a regular language.

A palindrome is a string that reads the same backward as forward, like the Finnish word "saippuakivikauppias".

5. Use the Pumping Lemma to show that $L_5 = \{x \in \{a, b\}^* \mid x \text{ is a palindrome}\}$ is not a regular language.

6. Use the Pumping Lemma to show that $L_6 = \{a^n b^{2n} \mid n \geq 0\}$ is not a regular language.

7. Use the Pumping Lemma to show that $L_7 = \{a^n \mid n \text{ is not prime}\}$ is not a regular language. **Hint:** It's easier to prove that the complement of L_7 is not regular.

8. Find the errors in the following attempt to prove that

$$L_8 = \{x \in \{a, b\}^* \mid \text{the number of } a\text{s and } b\text{s in } x \text{ are the same}\}$$

is not a regular language:

Proof. Assume that L_8 is regular; then there is a DFA M_8 such that $L(M_8) = L_8$. Suppose that M_8 has $k = 10$ states. Consider the string $x = a^5 b^5 \in L(M_8)$. Since $|x| \geq k$, by the Pumping Lemma there are strings u, v, w such that $uvw = x$, $|v| > 0$, $|uv| \leq k$, and $uv^i w \in L(M_8)$ for every $i \geq 0$. That is, $u = a^r$, $v = a^s$ and $w = a^t b^5$ where $s > 0$, $r + s \leq k$, $r + s + t = 5$, and $uv^i w = a^{r+is+t} \in L(M_8)$ for every $i \geq 0$. Let $i = 0$; then $r + is + t = 5 - s \neq 5$ because $s > 0$. This shows that $uv^i w$ is not in L_8, so $L(M_8) \neq L_8$. This is a contradiction, so the assumption that L_8 is regular was false.

Supplementary Information

Appendix: The Haskell Ecosystem

Haskell is the work of a substantial group of computer scientists and software developers, and it has a large and enthusiastic worldwide community of users from industry and academia. Their combined efforts have produced useful tools, documentation, learning materials, and other resources for Haskell programmers.

The Haskell Website

A wide range of Haskell-related material is available at ▶ https://www.haskell. org/. This includes most of the items listed below, as well as links to books and videos about Haskell and other learning material.

The Haskell Language

This book uses Haskell 2010, which is defined in the Haskell 2010 Language Report (▶ https://www.haskell.org/onlinereport/haskell2010/). Beginners who have no previous experience with language definitions will probably find it a little hard to read, but it's the definitive source for detailed and complete information about Haskell's syntax and semantics.

The Haskell Compiler GHC and Interactive Environment GHCi

See ▶ https://en.wikipedia.org/wiki/ Glasgow_Haskell_Compiler for information about the history and architecture of GHC.

The de facto standard implementation of Haskell is GHC, the Glasgow Haskell Compiler, which is available for all of the usual computing platforms. GHC can be downloaded from ▶ https://www.haskell.org/ghc/, but it's better to install the Haskell Platform from ▶ https://www.haskell.org/platform/ which also includes the main Haskell library modules and tools.

The GHC User's Guide, which provides pragmatic information about using GHC, is available at ▶ https://downloads.haskell.org/ghc/latest/docs/ html/users_guide/. It includes information about extensions to Haskell, some of which—for example, InstanceSigs, to allow type signatures in instance declarations—have been mentioned in this book. Extensions are available via command-line switches when invoking GHC from the command line, or by including a line like

```
{-# LANGUAGE InstanceSigs #-}
```

at the top of your file.

The easiest way to use Haskell is by running GHCi, the GHC interactive environment. This allows you to load and run Haskell programs, including programs that have been previously compiled using GHC, and evaluate Haskell expressions. The examples of interactive Haskell use in this book show what happens when you run programs in GHCi. The GHC User's Guide also includes information about commands like :load (abbreviated :l) and :reload (:r) for loading Haskell programs, :type (:t) for finding the type of an expression, and :set (:s) for setting options, including :set +s for displaying run-time and memory statistics after an expression is evaluated.

The Haskell Library, Hackage, and Hoogle

The Haskell Prelude, which contains many familiar types, type classes, and functions, is specified in the Haskell 2010 Language Report and is imported automatically into every module. Also specified there is a set of standard library modules that are regarded as part of Haskell and can be imported as required. Thousands of additional modules are available from Hackage (▶ https://hackage.haskell.org/), the Haskell community's repository of open source Haskell software. Most of these aren't included in Haskell installations by default, and need to be installed if you want to use them.

Hoogle (▶ https://hoogle.haskell.org/) makes it easy to search the Haskell libraries, including the Prelude and some of Hackage. You can search by name or by approximate type signature. Once you've found what you're looking for, you can follow a link to check the Haskell definition.

The Edinburgh Haskell Prelude (▶ https://github.com/wadler/edprelude) contains facilities for pretty-printing user-defined algebraic data types. It also provides simplified types for some of Haskell's built-in functions, and consistently uses arbitrary-precision **Integer** in place of **Int**, to encourage use of types that aren't subject to overflow.

For instance, run the command
```
$ cabal install teeth
```
to install **Anatomy.Teeth**, which contains things like **CentralIncisor :: ToothType**.

You may notice a similarity between the name "Hoogle" and the name of another search engine.

The Cabal Installation Tool

Cabal ▶ https://www.haskell.org/cabal/ is a tool for managing and installing Haskell software packages. Cabal keeps track of dependencies between packages, and installing one package will automatically install all of the packages that it depends on.

See ▶ https://blockchain-projects. readthedocs.io/overflow.html for information about an integer overflow attack against the Ethereum blockchain which shut down ERC-20 trading.

A **package** is a distributable unit of Haskell code, for example a set of modules that belong together.

The Haskell Profiler and Debugger

GHC includes a profiler which allows you to find out which parts of your program are consuming the most time and space while running. Instructions for profiling are included in the GHC User's Guide.

GHCi includes a simple interactive debugger in which you can stop a computation at user-defined "breakpoints" and check the values of variables before resuming. Instructions for using the debugger are also included in the GHC User's Guide.

Index

Index

Printed in the United States
by Baker & Taylor Publisher Services